FACING THE HEAT BARRIER

A History of Hypersonics

T. A. HEPPENHEIMER

DOVER PUBLICATIONS, INC.
Mineola, New York

Bibliographical Note

This Dover edition, first published in 2018, is an unabridged republication of the work originally published in 2007 by the National Aeronautics and Space Administration, Washington DC, as part of "The NASA History Series."

International Standard Book Number
ISBN-13: 978-0-486-82763-6
ISBN-10: 0-486-82763-1

Manufactured in the United States by LSC Communications
82763101   2018
www.doverpublications.com

Table of Contents

Acknowledgements. v
Introduction. vii
Abbreviations and Acronyms xiii
Chapter 1: First Steps in Hypersonic Research 1
   German Work with High-Speed Flows. 2
   Eugen Sänger 8
   NACA-Langley and John Becker 12
Chapter 2: Nose Cones and Re-entry 23
   The Move Toward Missiles. 23
   Approaching the Nose Cone. 29
   Ablation 36
   Flight Test 42
Chapter 3: The X-15. 55
   Origins of the X-15 56
   The Air Force and High-Speed Flight. 61
   X-15: The Technology 70
   X-15: Some Results 80
Chapter 4: First Thoughts of Hypersonic Propulsion 91
   Ramjets As Military Engines. 91
   Origins of the Scramjet 98
   Combined-Cycle Propulsion Systems. 107
   Aerospaceplane. 112
Chapter 5: Widening Prospects for Re-entry. 133
   Winged Spacecraft and Dyna-Soar. 133
   The Technology of Dyna-Soar 143
   Heat Shields for Mercury and Corona 151
   Gemini and Apollo. 154
Chapter 6: Hypersonics and the Space Shuttle. 165
   Preludes: Asset and Lifting Bodies 165
   ASSET Flight Tests. 167
   Reusable Surface Insulation 174
   Designing the Shuttle. 178
   The Loss of *Columbia* 191
Chapter 7: The Fading, the Comeback. 197
   Scramjets Pass Their Peak. 199
   Scramjets at NASA-Langley 202
   The Advent of NASP 211
   The Decline of NASP 218

Chapter 8: Why NASP Fell Short. 229
   Aerodynamics. 229
   Propulsion 238
   Materials 244
Chapter 9: Hypersonics After NASP 257
   The X-33 and X-34 258
   Scramjets Take Flight 267
   Recent Advances in Fluid Mechanics 277
   Hypersonics and the Aviation Frontier 282
Bibliography. 289
NASA History Series 317
Index 327

Acknowledgements

It is a pleasure to note the numerous people who helped me with this book. My personal involvement in hypersonics dates to 1982. I wrote a number of free-lance articles, along with three book-length reviews, before beginning work on the present book in 1999. During these past two decades, several dozen people kindly granted interviews in the course of these assignments. This book draws on discussions with the following:

J. Leland Atwood, Robert Barthelemy, George Baum, Fred Billig, Richard Booton, Peter Bradshaw, William Cabot, Robert Cooper, Scott Crossfield, Paul Czysz, William Dannevik, Anthony duPont, James Eastham, John Erdos, Maxime Faget, George Gleghorn, Edward Hall, Lawrence Huebner, Antony Jameson, Robert Jones, Arthur Kantrowitz, James Keller, George Keyworth, William "Pete" Knight, John Lumley, Luigi Martinelli, Robert Mercier, Parviz Moin, Gerhard Neumann, Louis Nucci, Philip Parrish, John Pike, Heinz Pitsch, Jerry Rising, Anatol Roshko, Paul Rubbert, Ron Samborsky, Robert Sanator, George Schairer, David Scott, Christian Stemmer, Arthur Thomas, Steven Weinberg, and Robert Williams.

In the NASA History Division, NASA Chief Historian Steven Dick served effectively as my editor. NASA-Langley has an excellent library, where I received particular help from Sue Miller and Garland Gouger. In addition, Dill Hunley, the historian at NASA-Dryden, hosted me for a week-long visit. The archivist, Archie DiFante, gave similar strong support during my visits to Maxwell Air Force Base. The Science and Technology Corporation, administered my work under subcontract, for which I give thanks to Andrea Carden, Carla Coombs, Sue Crotts, Marion Kidwell, and George Wood.

Dennis Jenkins provided me with documents and answered a number of questions. The artists Don Dixon and Chris Butler, who helped me on previous book projects, provided valuable assistance on this one as well. In addition, as for previous books, Phyllis LaVietes served as my secretary.

This book reflects my interest in the National Aerospace Plane effort, which I covered as a writer beginning in 1985. It is a pleasure to recall my ongoing friendships with Robert Williams, who gave me access to sources; Fred Billig, who helped me learn the trade of hypersonics; and Arthur Kantrowitz, who was present at the beginning. These three stand out among the dozens of people with whom it has been my privilege to conduct interviews and discussions.

Introduction

As an approach to the concept of hypersonic flight, one may begin by thinking of a sequence of high-performing aircraft that have flown at successively higher speeds. At Mach 2, twice the speed of sound, typical examples included the F-104 fighter and the Concorde commercial airliner. Though dramatically rakish in appearance, they were built of aluminum, the most familiar of materials, and used afterburning turbojets for propulsion.[1]

At Mach 3 and higher, there was the Lockheed SR-71 that cruised at 85,000 feet. The atmosphere at such altitudes, three times higher than Mount Everest, has a pressure only one-fiftieth of that at sea level. Even so, this airplane experienced aerodynamic heating that brought temperatures above 500°F over most of its surface. In turn, this heating brought requirements that dominated the problems of engineering design. Aluminum was out as a structural material; it lost strength at that high temperature. Titanium had to be used instead. Temperature-resistant fuels and lubricants also became necessary. Even so, this aircraft continued to rely on afterburning turbojets for propulsion.[2]

At Mach 4, the heating became still more severe and the difficulties of design were more daunting. No version of the turbojet has served at such speeds; it has been necessary to use a ramjet or rocket. The X-7, a ramjet testbed craft of the 1950s, was built of steel and had better temperature resistance than the SR-71. Still, when it flew past Mach 4.3 in 1958, the heating became so severe that it produced structural failure and a breakup of the vehicle in flight.[3]

Yet Mach 4 still counts as merely supersonic flight, not as hypersonic. For more than half a century analysts have defined hypersonic speeds as Mach 5 and higher.[4] Only rocket-powered craft have flown so fast—and Mach 5 defines only the lower bound of the hypersonic regime. An important range of hypersonic speeds extends from Mach 20 to 25 and includes the velocities of long-range ballistic missiles and of satellites re-entering from orbit. Moreover, flight above Mach 35 was a matter of national concern during the Apollo program, for its piloted Command Module entered the atmosphere at such speeds when returning from the Moon.

Specifically, the hypersonic regime is defined as the realm of speed wherein the physics of flows is dominated by aerodynamic heating. This heating is far more intense than at speeds that are merely supersonic, even though these lesser velocities have defined the performance of the SR-71 and X-7.

Hypersonics nevertheless was a matter of practical military application before the term entered use. Germany's wartime V-2 rocket flew above Mach 5,[5] but steel proved suitable for its construction and aerodynamic heating played only a limited role in its overall design.[6] The Germans used wind-tunnel tests to ensure that this missile would remain stable in flight, but they did not view its speed regime as meriting a name of its own. Hsue-shen Tsien, an aerodynamicist at the California Institute of Technology, coined the term in 1946.[7] Since then, it has involved three significant areas of application.

The first was the re-entry problem, which came to the forefront during the mid-1950s. The Air Force by then was committed to developing the Atlas ICBM, which was to carry a nuclear warhead to Moscow. Left to itself, this warhead would have heated up like a meteor when it fell back into the atmosphere. It would not have burned up—it was too massive—but it certainly would have been rendered useless. Hence, it was necessary to devise a heat shield to protect it against this intense aerodynamic heating.

The successful solution to this problem opened the door to a host of other initiatives. The return of film-carrying capsules from orbit became routine, and turned strategic reconnaissance of the Soviet Union into an important element of national defense. Piloted space flight also became feasible, for astronauts now could hope to come back safely. Then, as the engineering methods for thermal protection were further improved, thoughts of a space shuttle began to flourish. They took shape as a reusable launch vehicle, the first of its kind.

Hypersonic technologies also became important as policy makers looked ahead to an era in which the speed and performance of fighters and bombers might increase without limit. This expectation led to the X-15. Though designed during the 1950s, this rocket-powered research airplane set speed and altitude marks that were not surpassed until the advent of the shuttle. Aerodynamic heating again defined its design requirements, and it was built of the nickel alloy Inconel X. It routinely withstood temperatures of 1200°F as it flew to Mach 6,[8] and reached altitudes high enough for some of its pilots to qualify as astronauts.

Only rocket engines could propel a vehicle at such speeds, but hypersonic propulsion has represented a third important area of application. Here the hope has persisted that innovative airbreathing engines—scramjets—might cope with intense aerodynamic heating while offering fuel economy far surpassing that of a rocket. Other work has emphasized airbreathing rockets, which could give improved performance by eliminating the need to carry liquid oxygen in a tank. These concepts have held their own importance. They lay behind the National Aerospace Plane (NASP) program of 1985-1995, which sought to lay groundwork for single-stage vehicles that were to use both types of engine and were to fly from a runway to orbit.

The Air Force historian Richard Hallion has written of a "hypersonic revolution," as if to place the pertinent technologies on par with the turbojet and liquid-propellant rocket.[9] The present book takes a more measured view. Work in hypersonics had indeed brought full success in the area of re-entry. Consequences have included strategic missiles, the Soviet and American man-in-space programs, the

INTRODUCTION

Corona program in strategic reconnaissance, Apollo, and the space shuttle. These activities deterred nuclear war, gained accurate estimates of the Soviet threat, sent astronauts to the Moon and brought them home, and flew to and from space in a reusable launch vehicle. This list covers many of the main activities of the postwar missile and space industry, and supports Hallion's viewpoint.

But in pursuing technical revolution, engineers succeed in actually solving their problems, as when the Apollo program sent men to the Moon. These people do not merely display brilliant ingenuity while falling short of success. Unfortunately, the latter has been the case in the important area of hypersonic propulsion.

The focus has involved the scramjet as a new engine. It has taken form as a prime mover in its own right, capable of standing alongside such engines as the turboprop and ramjet. Still, far more so than the other engines, the scramjet has remained in the realm of experiment. Turboprops powered the Lockheed Electra airliner, P-3 antisubmarine aircraft, and C-130 transport. Ramjets provided propulsion for the successful Bomarc and Talos antiaircraft missiles. But the scramjet has powered only such small experimental airplanes as the X-43A.

Why? From the outset, the scramjet has faced overwhelming competition from a successful alternative: the rocket. This has strongly inhibited funding and has delayed its development to a point at which it could be considered seriously. On paper, scramjets offer superior performance. They therefore drew attention in the mid-1980s, during the heyday of NASP, at a time when Air Force officials had become disenchanted with the space shuttle but faced huge prospective demand for access to space in President Reagan's Strategic Defense Initiative. For once, then, scramjets gained funding that served to push their development—and their performance fell well short of people's hopes.

Within this book, Chapter 1 covers the immediate postwar years, when America still had much to learn from the Europeans. It focuses on two individuals: Eugen Sänger, who gave the first proposal for a hypersonic bomber, and John Becker, who built America's first hypersonic wind tunnel.

Chapter 2 covers the first important area of hypersonic research and development, which supported the advent of strategic missiles during the 1950s. The focus was on solving the re-entry problem, and this chapter follows the story through flight tests of complete nose cones.

Chapter 3 deals with the X-15, which took shape at a time when virtually the whole of America's capability in hypersonics research was contained within Becker's 11-inch instrument. Today it is hard to believe that so bold and so successful a step in aviation research could stand on so slender a foundation. This chapter shows how it happened.

Chapter 4 introduces hypersonic propulsion and emphasizes the work of Antonio Ferri, an Italian aerodynamicist who was the first to give a credible concept for a scramjet engine. This chapter also surveys Aerospaceplane, a little-known program of paper studies that investigated the feasibility of flight to orbit using such engines.

The next two chapters cover important developments in re-entry that followed the ICBM. Chapter 5, "Widening Prospects for Re-Entry," shows how work in this area supported the manned space program while failing to offer a rationale for a winged spacecraft, Dyna-Soar. Chapter 6, "Hypersonics and the Shuttle," begins by outlining developments during the mid-1960s that made it plausible that NASA's reusable space transporter would be designed as a lifting body and built using hot structures. In fact, the shuttle orbiter came forth as a conventional airplane with delta wings, and was built with aluminum structure covered with thermal-protecting tiles. This discussion indicates how those things happened.

Chapter 7, "The Fading, the Comeback," shows how work with scramjets did not share the priority afforded to the topic of re-entry. Instead it faded, and by the late 1960s only NASA-Langley was still pursuing studies in this area. This ongoing effort nevertheless gave important background to the National Aerospace Plane— but it was not technical success that won approval for NASP. As noted, it was the Strategic Defense Initiative. Within the Strategic Defense Initiative, the scramjet amounted to a rabbit being pulled from a hat, to satisfy Air Force needs. NASP was not well-founded at the outset; it was more of a leap of faith.

Chapter 8, "Why NASP Fell Short," explains what happened. In summary, the estimated performance of its scramjet engine fell well below initial hopes, while the drag was higher than expected. Computational aerodynamics failed to give accurate estimates in critical technical areas. The ejector ramjet, a key element of the propulsion system, proved to lack the desired performance. In the area of materials, metallurgists scored an impressive success with a new type of titanium called Beta-21S. It had only half the density of the superalloys that had been slated for Dyna-Soar, but even greater weight savings would have been needed for NASP.

Finally, Chapter 9 discusses "Hypersonics After NASP." Recent developments include the X-33 and X-34 launch vehicles, which represent continuing attempts to build the next launch vehicle. Scramjets have lately taken flight, not only as NASA's X-43A but also in Russia and in Australia. In addition, the new topic of Large Eddy Simulation, in computational fluid mechanics, raises the prospect that analysts indeed may learn, at least on paper, just how good a scramjet may be.

What, in the end, can we conclude? During the past half-century, the field of hypersonics has seen three major initiatives: missile nose cones, the X-15, and NASP. Of these, only one—the X-15—reflected ongoing progress in aeronautics. The other two stemmed from advances in nuclear weaponry: the hydrogen bomb, which gave rise to the ICBM, and the prospect of an x-ray laser, which lay behind the Strategic Defense Initiative and therefore behind NASP.

This suggests that if hypersonics is to flourish anew, it will do so because of developments in the apparently unrelated field of nuclear technology.

INTRODUCTION

1. F-104: Gunston, *Fighters*, pp. 120-126. Concorde: Heppenheimer, *Turbulent*, pp. 202-203, 208.
2. Crickmore, *SR-71*, pp. 89-91, 95-99, 194.
3. Ritchie, "Evaluation." Steel: Miller, *X-Planes*, p. 119.
4. See, for example, Anderson, *History*, pp. 438-439.
5. Top speed of the V-2 is given as 1,600 meters per second (Dornberger, *V-2*, p. xix) and as 1,700 meters per second (Naval Research Laboratory, *Upper*, cited in Ley, *Rockets*, pp. 596-597); the speed of sound at the pertinent altitudes is 295 meters per second (Kuethe and Chow, *Foundations*, p. 518).
6. Ley, *Rockets*, p. 243; Neufeld, *Rocket*, pp. 85-94.
7. Tsien, "Similarity."
8. 1200°F: NASA SP-2000-4518, diagram, p. 25.
9. Hallion, *Hypersonic*.

Abbreviations and Acronyms

AAF	Army Air Forces
AAS	American Astronautical Society
ACES	Air Collection and Enrichment System
AEC	Atomic Energy Commission
AEDC	Arnold Engineering Development Center
AEV	Aerothermodynamic Elastic Vehicle (ASSET)
AFB	Air Force Base
AFSC	Air Force Systems Command
AGARD	Advisory Group for Aeronautical Research and Development
AIAA	American Institute for Aeronautics and Astronautics
AIM	Aerothermodynamic Integration Model (HRE)
APL	Applied Physics Laboratory (Johns Hopkins University)
APU	auxiliary power unit
ARDC	Air Research and Development Command (USAF)
ARS	American Rocket Society
ASD	Aeronautical Systems Division (USAF)
ASSET	Aerothermodynamic/elastic Structural Systems Environmental Tests
ASV	Aerothermodynamic Structural Vehicle (ASSET)
BMW	Bayerische Motoren-Werke
BTU	British Thermal Unit
CASI	Center for Aerospace Information
C_D	coefficient of drag
CDE	Concept Demonstrator Engine (NASP)
CFD	computational fluid dynamics
CFHT	Continuous Flow Hypersonic Tunnel (NACA-Langley)
CIA	Central Intelligence Agency
CIAM	Central Institute of Aviation Motors (Moscow)
C_l	coefficient of lift
CO_2	carbon dioxide
DARPA	Defense Advanced Research Projects Agency
DM	Deutschmark
DNS	Direct Numerical Simulation (CFD)
DOD	Department of Defense
DSB	Defense Science Board
DTIC	Defense Technical Information Center
DVL	Deutsche Versuchsanstalt für Luftfahrtforschung
ELV	expendable launch vehicle
°F	degrees Fahrenheit
FAI	Federation Aeronautique Internationale
FDL	Flight Dynamics Laboratory (USAF)
FY	Fiscal Year
g	force of gravity

GAO	General Accounting Office (United States Congress)
GASL	General Applied Science Laboratories, Inc.
GE	General Electric
HGS	homogeneous gas sample (shock tubes)
HRE	Hypersonic Research Engine
HXEM	Hyper-X Engine Module
HXFE	Hyper-X Flight Engine
HYWARDS	Hypersonic Weapons Research and Development Supporting System
IBM	International Business Machines
ICBM	intercontinental ballistic missile
IFTV	Incremental Flight Test Vehicle
ILRV	Integrated Launch and Re-entry Vehicle
ISABE	International Society for Air Breathing Propulsion
I_{sp}	specific impulse
ISTAR	Integrated System Test of an Airbreathing Rocket
K	degrees Kelvin
LACE	Liquid Air Cycle Engine
L/D	lift-to-drag ratio
LES	Large-Eddy Simulation (CFD)
LH_2	liquid hydrogen
LSCIR	Low Speed Component Integration Rig (Pratt & Whitney)
LSS	Low Speed System (NASP)
MBB	Messerschmitt-Boelkow-Blohm
MIT	Massachusetts Institute of Technology
MOL	Manned Orbiting Laboratory
MOU	Memorandum of Understanding
MSC	Manned Spacecraft Center (now Johnson Space Center)
NACA	National Advisory Committee for Aeronautics
NASA	National Aeronautics and Space Administration
NASP	National Aerospace Plane
NIFTA	Non-Integral Fuselage Tank Article (NASP)
NMASAP	NASP Materials and Structures Augmentation Program
OAL	Ordnance Astrophysics Laboratory
OMSF	Office of Manned Space Flight (NASA)
P & W	Pratt & Whitney
POBATO	propellants on board at takeoff
PRIME	Precision Recovery Including Maneuvering Entry
PROFAC	propulsive fluid accumulator
PSi	pounds per square inch
RCC	Reusable Carbon-Carbon
RENE	Rocket Engine Nozzle Ejector
ROLS	Recoverable Orbital Launch System
RSI	reusable surface insulation
SAB	Scientific Advisory Board (USAF)
SAGE	Semi-Automatic Ground Environment
SAM	Structures Assembly Model (HRE)
SAMPE	Society for Advancement of Materials and Process Engineering
SCRAM	Supersonic Combustion Ramjet Missile (APL)

SDI	Strategic Defense Initiative
SSME	Space Shuttle Main Engine
SSTO	single stage to orbit
SXPE	Subscale Parametric Engine (NASP)
TAV	Trans-Atmospheric Vehicle
TPS	thermal protection system
TZM	titanium-zirconium-molybdenum alloy
USAF	United States Air Force
VKF	Von Karman Facility (AEDC)
WADC	Wright Air Development Center (USAF)

1

First Steps in Hypersonic Research

Today's world of high-speed flight is international, with important contributions having recently been made in Japan, Australia, and Russia as well as in the United States. This was even truer during World War II, when Adolf Hitler sponsored development programs that included early jet fighters and the V-2 missile. America had its own research center at NACA's Langley Memorial Aeronautical Laboratory, but in important respects America was little more than an apt pupil of the wartime Germans. After the Nazis surrendered, the U.S. Army brought Wernher von Braun and his rocket team to this country, and other leading researchers found themselves welcome as well.

Liftoff of a V-2 rocket. (U.S. Army)

Some of their best work had supported the V-2, using a pair of tunnels that operated at Mach 4.4. This was just short of hypersonic, but these facilities made a key contribution by introducing equipment and research methods that soon found use in studying true hypersonic flows. At Peenemunde, one set of experiments introduced a wind-tunnel nozzle of specialized design and reached Mach 8.8, becoming the first to achieve such a speed. Other German work included the design of a 76,000-horsepower installation that might have reached Mach 10.

The technical literature also contained an introductory discussion of a possible application. It appeared within a wartime report by Austria's Eugen Sänger, who had proposed to build a hypersonic bomber that would extend its range by repeatedly skipping off the top of the atmosphere like a stone skipping over water. This concept did not enter the mainstream of postwar weapons development, which gave pride of place to the long-range ballistic missile. Still, Sänger's report introduced skipping entry as a new mode of high-speed flight, and gave a novel suggestion as to how wings could increase the range of a rocket-powered vehicle.

Within Langley, ongoing research treated flows that were merely supersonic. However, the scientist John Becker wanted to go further and conduct studies of hypersonic flows. He already had spent several years at Langley, thereby learning his trade as an aerodynamicist. At the same time he still was relatively young, which meant that much of his career lay ahead of him. In 1947 he achieved a major advance in hypersonics by building its first important research instrument, an 11-inch wind tunnel that operated at Mach 6.9.

German Work with High-Speed Flows

At the Technische Hochschule in Hannover, early in the twentieth century, the physicist Ludwig Prandtl founded the science of aerodynamics. Extending earlier work by Italy's Tullio Levi-Civita, he introduced the concept of the boundary layer. He described it as a thin layer of air, adjacent to a wing or other surface, that clings to this surface and does not follow the free-stream flow. Drag, aerodynamic friction, and heat transfer all arise within this layer. Because the boundary layer is thin, the equations of fluid flow simplified considerably, and important aerodynamic complexities became mathematically tractable.[1]

As early as 1907, at a time when the Wright Brothers had not yet flown in public, Prandtl launched the study of supersonic flows by publishing investigations of a steam jet at Mach 1.5. He now was at Göttingen University, where he built a small supersonic wind tunnel. In 1911 the German government founded the Kaiser-Wilhelm-Gesellschaft, an umbrella organization that went on to sponsor a broad range of institutes in many areas of science and engineering. Prandtl proposed to set up a center at Göttingen for research in aerodynamics and hydrodynamics, but World War I intervened, and it was not until 1925 that this laboratory took shape.

After that, though, work in supersonics went forward with new emphasis. Jakob Ackeret, a colleague of Prandtl, took the lead in building supersonic wind tunnels. He was Swiss, and he built one at the famous Eidgenossische Technische Hochschule in Zurich. This attracted attention in nearby Italy, where the dictator Benito Mussolini was giving strong support to aviation. Ackeret became a consultant to the Italian Air Force and built a second wind tunnel in Guidonia, near Rome. It reached speeds approaching 2,500 miles per hour (mph), which far exceeded those that were available anywhere else in the world.[2]

These facilities were of the continuous-flow type. Like their subsonic counterparts, they ran at substantial power levels and could operate all day. At the Technische Hochschule in Aachen, the aerodynamicist Carl Wiesenberger took a different approach in 1934 by building an intermittent-flow facility that needed much less power. This "blowdown" installation relied on an evacuated sphere, which sucked outside air through a nozzle at speeds that reached Mach 3.3.

This wind tunnel was small, having a test-section diameter of only four inches. But it set the pace for the mainstream of Germany's wartime supersonic research. Wieselberger's assistant, Rudolf Hermann, went to Peenemunde, the center of that country's rocket development, where in 1937 he became head of its new Aerodynamics Institute. There he built a pair of large supersonic tunnels, with 16-inch test sections, that followed Aachen's blowdown principle. They reached Mach 4.4, but not immediately. A wind tunnel's performance depends on its nozzle, and it took time to develop proper designs. Early in 1941 the highest working speed was Mach 2.5; a nozzle for Mach 3.1 was still in development. The Mach 4.4 nozzles were not ready until 1942 or 1943.[3]

The Germans never developed a true capability in hypersonics, but they came close. The Mach 4.4 tunnels introduced equipment and methods of investigation that carried over to this higher-speed regime. The Peenemunde vacuum sphere was constructed of riveted steel and had a diameter of 40 feet. Its capacity of a thousand cubic meters gave run times of 20 seconds.[4] Humidity was a problem; at Aachen, Hermann had learned that moisture in the air could condense when the air cooled as it expanded through a supersonic nozzle, producing unwanted shock waves that altered the anticipated Mach number while introducing nonuniformities in the direction and velocity of flow. At Peenemunde he installed an air dryer that used silica gel to absorb the moisture in the air that was about to enter his supersonic tunnels.[5]

Configuration development was at the top of his agenda. To the modern mind the V-2 resembles a classic spaceship, complete with fins. It is more appropriate to say that spaceship designs resemble the V-2, for that missile was very much in the forefront during the postwar years, when science fiction was in its heyday.[6] The V-2 needed fins to compensate for the limited effectiveness of its guidance, and their design was trickier than it looked. They could not be too wide, or the V-2 would be unable to pass through railroad tunnels. Nor could they extend too far below the body of the missile, or the rocket exhaust, expanding at high altitude, would burn them off.

The historian Michael Neufeld notes that during the 1930s, "no one knew how to design fins for supersonic flight." The A-3, a test missile that preceded the V-2, had proven to be *too* stable; it tended merely to rise vertically, and its guidance system lacked the authority to make it tilt. Its fins had been studied in the Aachen supersonic tunnel, but this problem showed up only in flight test, and for a time it was unclear how to go further. Hermann Kurzweg, Rudolf Hermann's assistant, investigated low-speed stability building a model and throwing it off the roof of his home. When that proved unsatisfactory, he mounted it on a wire, attached it to his car, and drove down an autobahn at 60 mph.

The V-2 was to fly at Mach 5, but for a time there was concern that it might not top Mach 1. The sound barrier loomed as potentially a real barrier, difficult to pierce, and at that time people did not know how to build a transonic wind tunnel that would give reliable results. Investigators studied this problem by building heavy iron models of this missile and dropping them from a Heinkel He-111 bomber. Observers watched from the ground; in one experiment, Von Braun himself piloted a plane and dove after the model to observe it from the air. The design indeed proved to be marginally unstable in the transonic region, but the V-2 had the thrust to power past Mach 1 with ease.

A second test missile, the A-5, also contributed to work on fin design. It supported development of the guidance system, but it too needed fins, and it served as a testbed for further flight studies. Additional flight tests used models with length of five feet that were powered with rocket engines that flew with hydrogen peroxide as the propellant.

These tests showed that an initial fin design given by Kurzweg had the best subsonic stability characteristics. Subsequently, extensive wind-tunnel work both at Peenemunde and at a Zeppelin facility in Stuttgart covered the V-2's complete Mach range and refined the design. In this fashion, the V-2's fins were designed with only minimal support from Peenemunde's big supersonic wind tunnels.[7] But these tunnels came into their own later in the war, when investigators began to consider how to stretch this missile's range by adding wings and thereby turning it into a supersonic glider.

Once the Germans came up with a good configuration for the V-2, they stuck with it. They proposed to use it anew in a two-stage missile that again sported fins that look excessively large to the modern eye, and that was to cross the Atlantic to strike New York.[8] But there was no avoiding the need for a new round of wind-tunnel tests in studying the second stage of this intercontinental missile, the A-9, which was to fly with swept wings. As early as 1935 Adolf Busemann, another

First Steps in Hypersonic Research

colleague of Prandtl, had proposed the use of such wings in supersonic flight.[9] Walter Dornberger, director of V-2 development, describes witnessing a wind-tunnel test of a model's stability.

The model had "two knifelike, very thin, swept-back wings." Mounted at its center of gravity, it "rotated at the slightest touch." When the test began, a technician opened a valve to start the airflow. In Dornberger's words,

A-4b missile ready for launch. (U.S. Army)

"The model moved abruptly, turning its nose into the oncoming airstream. After a few quickly damping oscillations of slight amplitude, it lay quiet and stable in the air that hissed past it at 4.4 times the speed of sound. At the nose, and at the edges of the wing supports and guide mechanism, the shock waves could be clearly seen as they traveled diagonally backward at a sharp angle.

As the speed of the airflow fell off and the test ended, the model was no longer lying in a stable position. It made a few turns around its center of gravity, and then it came to a standstill with the nose pointing downward. The experiment Dr. Hermann had wished to show me had succeeded perfectly. This projectile, shaped like an airplane, had remained absolutely stable at a supersonic speed range of almost 3,500 mph."[10]

Work on the A-9 languished for much of the war, for the V-2 offered problems aplenty and had far higher priority. But in 1944, as the Allies pushed the Germans out of France and the Russians closed in from the east, Dornberger and Von Braun faced insistent demands that they pull a rabbit from a hat and increase the V-2's range. The rabbit was the A-9, with its wings promising a range of 465 miles, some three times that of the standard V-2.[11]

Peenemunde's Ludwig Roth proceeded to build two prototypes. The V-2 was known to its builders as the A-4, and Roth's A-9 now became the A-4b, a designation that allowed it to share in the high priority of that mainstream program. The A-4b took shape as a V-2 with swept wings and with a standard set of fins that included slightly enlarged air vanes for better control. Certainly the A-4b needed all the help it could get, for the addition of wings had made it highly sensitive to winds.

The first A-4b launch took place late in December 1944. It went out of control and crashed as the guidance system failed to cope with its demands. Roth's rocketeers tried again a month later, and General Dornberger describes how this flight went much better:

"The rocket, climbing vertically, reached a peak altitude of nearly 50 miles at a maximum speed of 2,700 mph. [It] broke the sound barrier without trouble. It flew with stability and steered automatically at both subsonic and supersonic speeds. On the descending part of the trajectory, soon after the rocket leveled out at the upper limit of the atmosphere and began to glide, a wing broke. This structural failure resulted from excessive aerodynamic loads."[12]

This shot indeed achieved its research goals, for it was to demonstrate successful launch and acceleration through the sound barrier, overcoming drag from the wings, and it did these things. Gliding flight was not on the agenda, for while wind-tunnel tests could demonstrate stability in a supersonic glide, they could not guard against atmosphere entry in an improper attitude, with the A-4b tumbling out of control.[13]

Yet while the Germans still had lessons to learn about loads on a supersonic aircraft in flight, they certainly had shown that they knew their high-speed aerodynamics. One places their achievement in perspective by recalling that all through the 1950s a far wealthier and more technically capable United States pursued a vigorous program in rocket-powered aviation without coming close to the A-4b's performance. The best American flight, of an X-2 in 1956, approached 2,100 mph—and essentially duplicated the German failure as it went out of control, killing the pilot and crashing. No American rocket plane topped the 2,700 mph of the A-4b until the X-15 in 1961.[14]

Hence, without operating in the hypersonic regime, the Peenemunde wind tunnels laid important groundwork as they complemented such alternative research techniques as dropping models from a bomber and flying scale models under rocket power. Moreover, the Peenemunde aerodynamicist Siegfried Erdmann used his center's facilities to conduct the world's first experiments with a hypersonic flow.

In standard operation, at speeds up to Mach 4.4, the Peenemunde tunnels had been fed with air from the outside world, at atmospheric pressure. Erdmann knew that a hypersonic flow needed more, so he arranged to feed his tunnel with compressed air. He also fabricated a specialized nozzle and aimed at Mach 8.8, twice the standard value. His colleague Peter Wegener describes what happened:

"Everything was set for the first-ever hypersonic flow experiment. The highest possible pressure ratio across the test section was achieved by evacuating the sphere to the limit the remaining pump could achieve. The supply of the nozzle—in contrast to that at lower Mach numbers—was now provided by air at a pressure of about 90 atmospheres…. The experiment was initiated by opening the fast-acting valve. The flow of brief duration looked perfect as viewed via the optical system.

Beautiful photographs of the flow about wedge-shaped models, cylinders, spheres, and other simple shapes were taken, photographs that looked just as one would expect from gas dynamics theory."[15]

These tests addressed the most fundamental of issues: How, concretely, does one operate a hypersonic wind tunnel? Supersonic tunnels had been bedeviled by condensation of water vapor, which had necessitated the use of silica gel to dry the air. A hypersonic facility demanded far greater expansion of the flow, with consequent temperatures that were lower still. Indeed, such flow speeds brought the prospect of condensation of the air itself.

Conventional handbooks give the liquefaction temperatures of nitrogen and oxygen, the main constituents of air, respectively as 77 K and 90 K. These refer to conditions at atmospheric pressure; at the greatly rarefied pressures of flow in a hypersonic wind tunnel, the pertinent temperatures are far lower.[16] In addition, Erdmann hoped that his air would "supersaturate," maintaining its gaseous state because of the rapidity of the expansion and hence of the cooling.

This did not happen. In Wegener's words, "Looking at the flow through the glass walls, one could see a dense fog. We know now that under the conditions of this particular experiment, the air had indeed partly condensed. The fog was made up of air droplets or solid air particles forming a cloud, much like the water clouds we see in the sky."[17] To prevent such condensation, it proved necessary not only to feed a hypersonic wind tunnel with compressed air, but to heat this air strongly.

One thus is entitled to wonder whether the Germans would have obtained useful results from their most ambitious wind-tunnel project, a continuous-flow system that was designed to achieve Mach 7, with a possible extension to Mach 10. Its power ratings pointed to the advantage of blowdown facilities, such as those of Peenemunde. The Mach 4.4 Peenemunde installations used a common vacuum sphere, evacuation of which relied on pumps with a total power of 1,100 horsepower. Similar power levels were required to dry the silica gel by heating it, after it became moist. But the big hypersonic facility was to have a one-meter test section and demanded 76,000 horsepower, or 57 megawatts.[18]

Such power requirements went beyond what could be provided in straightforward fashion, and plans for this wind tunnel called for it to use Germany's largest hydroelectric plant. Near Kochel in Bavaria, two lakes—the Kochelsee and Walchensee—are separated in elevation by 660 feet. They stand close together, providing an ideal site for generating hydropower, and a hydro plant at that location had gone into operation in 1925, generating 120 megawatts. Since the new wind tunnel would use half of this power entirely by itself, the power plant was to be enlarged, with additional water being provided to the upper lake by a tunnel through the mountains to connect to another lake.[19]

In formulating these plans, as with the A-4b, Germany's reach exceeded its grasp. Moreover, while the big hypersonic facility was to have generous provision for drying its air, there was nothing to prevent the air from condensing, which would have thrown the data wildly off.[20] Still, even though they might have had to learn their lessons in the hard school of experience, Germany was well on its way toward developing a true capability in hypersonics by the end of World War II. And among the more intriguing concepts that might have drawn on this capability was one by the Austrian rocket specialist Eugen Sänger.

EUGEN SÄNGER

Born in 1905, he was of the generation that came of age as ideas of space flight were beginning to germinate. Sänger's own thoughts began to take shape while he was still in grammar school. His physics teacher gave him, as a Christmas present, a copy of a science-fiction novel, *Auf Zwei Planeten* ("On Two Planets"). "I was about 16 years old," Sänger later recalled. "Naturally I read this novel avidly, and thereafter dreamed of doing something like this in my own lifetime." He soon broadened his readings with the classic work of Hermann Oberth. "I had to pass my examination in mechanics," he continued, "and had, therefore, made a particular study of this and related subjects. Then I also started to check and recalculate in detail everything in Oberth's book, and I became convinced that here was something that one could take seriously."

He then attended the Technische Hochschule in Vienna, where he tried to win a doctoral degree in 1928 by submitting a dissertation on the subject of rocket-powered aircraft. He did not get very far, later recalling that his professor told him, "If you try, today, to take your doctor degree in spaceflight, you will most probably be an old man with a long beard before you have succeeded in obtaining it." He turned his attention to a more conventional topic, the structural design of wings for aircraft, and won his degree a year later. But his initial attempt at a dissertation had introduced him to the line of study that he pursued during the next decade and then during the war.

In 1933 he turned this dissertation into a book, *Raketenflugtechnik*. It was the first text in this new field. He wrote of a rocket plane burning liquid oxygen and petrol, which was to reach Mach 10 along with altitudes of 60 to 70 miles. This concept was significant at the time, for the turbojet engine had not yet been invented, and futurists, such as Aldous Huxley who wrote *Brave New World*, envisioned rockets as the key to high-speed flight in centuries to come.[21]

Sänger's altitudes became those of the X-15, a generation later. The speed of his concept was markedly higher. He included a three-view drawing. Its wings were substantially larger than those of eventual high-performance aircraft, although these wings gave his plane plenty of lift at low speed, during takeoff and landing. Its tail surfaces also were far smaller than those of the X-15, for he did not know about the

First Steps in Hypersonic Research

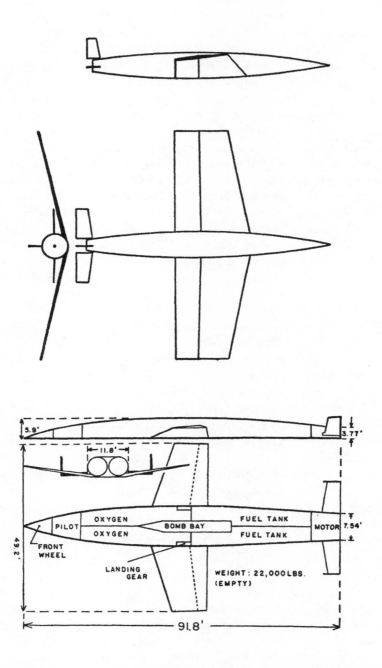

Rocket aircraft of Eugen Sänger. Top, the Silbervogel. Bottom, the Amerika-Bomber that was to use a skipping entry. Note that both were low-wing monoplanes. (Courtesy of Willy Ley)

stability problems that loomed in supersonic flight. Still, he clearly had a concept that he could modify through further study.

In 1934, writing in the magazine *Flug* ("Flight"), he used an exhaust velocity of 3,700 meters per second and gave a velocity at a cutoff of Mach 13. His *Silbervogel*, Silver Bird, now was a boost-glide vehicle, entering a steady glide at Mach 3.5 and covering 5,000 kilometers downrange while descending from 60 to 40 kilometers in altitude.

He stayed on at the Hochschule and conducted rocket research. Then in 1935, amid the Depression, he lost his job. He was in debt to the tune of DM 2,000, which he had incurred for the purpose of publishing his book, but he remained defiant as he wrote, "Nevertheless, my silver birds will fly!" Fortunately for him, at that time Hitler's Luftwaffe was taking shape, and was beginning to support a research establishment. Sänger joined the DVL, the German Experimental Institute for Aeronautics, where he worked as technical director of rocket research. He did not go to Peenemunde and did not deal with the V-2, which was in the hands of the Wehrmacht, not the Luftwaffe. But once again he was employed, and he soon was out of debt.

He also began collaborating with the mathematician Irene Bredt, whom he later married. His *Silbervogel* remained on his mind as he conducted performance studies with help from Bredt, hoping that this rocket plane might evolve into an *Amerika-Bomber*. He was aware that when transitioning from an initial ballistic trajectory into a glide, the craft was to re-enter the atmosphere at a shallow angle. He then wondered what would happen if the angle was too steep.

He and Bredt found that rather than enter a glide, the vehicle might develop so much lift that it would fly back to space on a new ballistic arc, as if bouncing off the atmosphere. Stones skipping over water typically make several such skips, and Sänger found that his winged craft would do this as well. With a peak speed of 3.73 miles per second, compared with 4.9 miles per second as the Earth's orbital velocity, it could fly halfway around the world and land in Japan, Germany's wartime ally. At 4.4 miles per second, the craft could fly completely around the world and land in Germany.[22]

Sänger wrote up their findings in a document of several hundred pages, with the title (in English) of "On a Rocket Propulsion for Long Distance Bombers." In December 1941 he submitted it for publication—and won a flat rejection the following March. This launched him into a long struggle with the Nazi bureaucracy, as he sought to get his thoughts into print.

His rocket craft continued to show a clear resemblance to his *Silbervogel* of the previous decade, for he kept the basic twin-tailed layout even as he widened the fuselage and reduced the size of the wings. Its bottom was flat to produce more lift, and his colleagues called it the *Platteisen*, the Flatiron. But its design proved to be patentable, and in June 1942 he received a piece of bright news as the government awarded him a Reichspatent concerning "Gliding Bodies for Flight Velocities Above Mach 5." As he continued to seek publication, he won support from an influential professor, Walter Georgii. He cut the length of his manuscript in half. Finally, in September 1944 he learned that his document would be published as a Secret Command Report.

The print run came to fewer than a hundred copies, but they went to the people who counted. These included the atomic-energy specialist Werner Heisenberg, the planebuilder Willy Messerschmitt, the chief designer Kurt Tank at Focke-Wulf, Ernst Heinkel of Heinkel Aircraft, Ludwig Prandtl who still was active, as well as Wernher von Braun and his boss, General Dornberger. Some copies reached the Allies after the Nazi surrender, with three of them being taken to Moscow. There their content drew attention from the dictator Josef Stalin, who ordered a full translation. He subsequently decided that Sänger and Bredt were to be kidnapped and brought to Moscow.

At that time they were in Paris, working as consultants for the French air force. Stalin sent two agents after them, accompanied by his own son. They nevertheless remained safe; the Soviets never found them. French intelligence agents learned about the plot and protected them, and in any case, the Soviets may not have been looking very hard. One of them, Grigory Tokaty-Tokayev, was the chief rocket scientist in the Soviet air force. He defected to England, where he wrote his memoirs for the *Daily Express* and then added a book, *Stalin Means War*.

Sänger, for his part, remained actively involved with his rocket airplane. He succeeded in publishing some of the material from his initial report that he had had to delete. He also won professional recognition, being chosen in 1951 as the first president of the new International Astronautical Federation. He died in 1964, not yet 60. But by then the X-15 was flying, while showing more than a casual resemblance to his *Silbervogel* of 30 years earlier. His Silver Bird indeed had flown, even though the X-15 grew out of ongoing American work with rocket-powered aircraft and did not reflect his influence. Still, in January of that year—mere weeks before he died—the trade journal *Astronautics & Aeronautics* published a set of articles that presented new concepts for flight to orbit. These showed that the winged-rocket approach was alive and well.[23]

What did he contribute? He was not the first to write of rocket airplanes; that palm probably belongs to his fellow Austrian Max Valier, who in 1927 discussed how a trimotor monoplane of the day, the Junkers G-23, might evolve into a rocket ship. This was to happen by successively replacing the piston motors with rocket engines and reducing the wing area.[24] In addition, World War II saw several military rocket-plane programs, all of which were piloted. These included Germany's Me-163 and Natter antiaircraft weapons as well as Japan's Ohka suicide weapon, the

Cherry Blossom, which Americans called Baka, "Fool." The rocket-powered Bell X-1, with which Chuck Yeager first broke the sound barrier, also was under development well before war's end.[25]

Nor did Sänger's 1944 concept hold military value. It was to be boosted by a supersonic rocket sled, which would have been both difficult to build and vulnerable to attack. Even then, and with help from its skipping entry, it would have been a single-stage craft attaining near-orbital velocity. No one then, 60 years ago, knew how to build such a thing. Its rocket engine lay well beyond the state of the art. Sänger projected a mass-ratio, or ratio of fueled to empty weight, of 10—with the empty weight including that of the wings, crew compartment, landing gear, and bomb load. Structural specialists did not like that. They also did not like the severe loads that skipping entry would impose. And after all this *Sturm und drang*, the bomb load of 660 pounds would have been militarily useless.[26]

But Sänger gave a specific design concept for his rocket craft, presenting it in sufficient detail that other engineers could critique it. Most importantly, his skipping entry represented a new method by which wings might increase the effectiveness of a rocket engine. This contribution did not go away. The train of thought that led to the Air Force's Dyna-Soar program, around 1960, clearly reflected Sänger's influence. In addition, during the 1980s the German firm of Messerschmitt-Boelkow-Blohm conducted studies of a reusable wing craft that was to fly to orbit as a prospective replacement for America's space shuttle. The name of this two-stage vehicle was Sänger.[27]

NACA-LANGLEY AND JOHN BECKER

During the war the Germans failed to match the Allies in production of airplanes, but they were well ahead in technical design. This was particularly true in the important area of jet propulsion. They fielded an operational jet fighter, the Me-262, and while the Yankees were well along in developing the Lockheed P-80 as a riposte, the war ended before any of those jets could see combat. Nor was the Me-262 a last-minute work of desperation. It was a true air weapon that showed better speed and acceleration than the improved P-80A in flight test, while demonstrating an equal rate of climb.[28] Albert Speer, Hitler's minister of armaments, asserted in his autobiographical *Inside the Third Reich* (1970) that by emphasizing production of such fighters and by deploying the Wasserfall antiaircraft missile that was in development, the Nazis "would have beaten back the Western Allies' air offensive against our industry from the spring of 1944 on."[29] The Germans thus might have prolonged the war until the advent of nuclear weapons.

Wartime America never built anything resembling the big Mach 4.4 wind tunnels at Peenemunde, but its researchers at least constructed facilities that could compare with the one at Aachen. The American installations did not achieve speeds to match Aachen's Mach 3.3, but they had larger test sections. Arthur Kantrowitz, a young physicist from Columbia University who was working at Langley, built a nine-inch tunnel that reached Mach 2.5 when it entered operation in 1942. (Aachen's had been four inches.) Across the country, at NACA's Ames Aeronautical Laboratory, two other wind tunnels entered service during 1945. Their test sections measured one by three feet, and their flow speeds reached Mach 2.2.[30]

The Navy also was active. It provided $4.5 million for the nation's first really large supersonic tunnel, with a test section six feet square. Built at NACA-Ames, operating at Mach 1.3 to 1.8, this installation used 60,000 horsepower and entered service soon after the war.[31] The Navy also set up its Ordnance Aerophysics Laboratory in Daingerfield, Texas, adjacent to the Lone Star Steel Company, which had air compressors that this firm made available. The supersonic tunnel that resulted covered a range of Mach 1.25 to 2.75, with a test section of 19 by 27.5 inches. It became operational in June 1946, alongside a similar installation that served for high-speed engine tests.[32]

Theorists complemented the wind-tunnel builders. In April 1947 Theodore von Karman, a professor at Caltech who was widely viewed as the dean of American aerodynamicists, gave a review and survey of supersonic flow theory in an address to the Institute of Aeronautical Sciences. His lecture, published three months later in the *Journal of the Aeronautical Sciences*, emphasized that supersonic flow theory now was mature and ready for general use. Von Karman pointed to a plethora of available methods and solutions that not only gave means to attack a number of important design problems but also gave independent approaches that could permit cross-checks on proposed solutions.

John Stack, a leading Langley aerodynamicist, noted that Prandtl had given a similarly broad overview of subsonic aerodynamics a quarter-century earlier. Stack declared, "Just as Prandtl's famous paper outlined the direction for the engineer in the development of subsonic aircraft, Dr. von Karman's lecture outlines the direction for the engineer in the development of supersonic aircraft."[33]

Yet the United States had no facility, and certainly no large one, that could reach Mach 4.4. As a stopgap, the nation got what it wanted by seizing German wind tunnels. A Mach 4.4 tunnel was shipped to the Naval Ordnance Laboratory in White Oak, Maryland. Its investigators had fabricated a Mach 5.18 nozzle and had conducted initial tests in January 1945. In 1948, in Maryland, this capability became routine.[34] Still, if the U.S. was to advance beyond the Germans and develop the true hypersonic capability that Germany had failed to achieve, the nation would have to rely on independent research.

The man who pursued this research, and who built America's first hypersonic tunnel, was Langley's John Becker. He had been at that center since 1936; during the latter part of the war he was assistant chief of Stack's Compressibility Research Division. He specifically was in charge of Langley's 16-Foot High-Speed Tunnel, which had fought its war by investigating cooling problems in aircraft motors as well as the design of propellers. This facility contributed particularly to tests of the B-50 bomber and to the aerodynamic shapes of the first atomic bombs. It also assisted development of the Pratt & Whitney R-2800 Double Wasp, a widely used piston engine that powered several important wartime fighter planes, along with the DC-6 airliner and the C-69 transport, the military version of Lockheed's Constellation.[35]

It was quite a jump from piston-powered warbirds to hypersonics, but Becker willingly made the leap. The V-2, flying at Mach 5, gave him his justification. In a memo to Langley's chief of research, dated 3 August 1945, Becker noted that planned facilities were to reach no higher than Mach 3. He declared that this was inadequate: "When it is considered that all of these tunnels will be used, to a large extent, to develop supersonic missiles and projectiles of types which have already been operated at Mach numbers as high as 5.0, it appears that there is a definite need for equipment capable of higher test Mach numbers."

Within this memo, he outlined a design concept for "a supersonic tunnel having a test section four-foot square and a maximum test Mach number of 7.0." It was to achieve continuous flow, being operated by a commercially-available compressor of 2,400 horsepower. To start the flow, the facility was to hold air within a tank that was compressed to seven atmospheres. This air was to pass through the wind tunnel before exhausting into a vacuum tank. With pressure upstream pushing the flow and with the evacuated tank pulling it, airspeeds within the test section would be high indeed. Once the flow was started, the compressor would maintain it.

A preliminary estimate indicated that this facility would cost $350,000. This was no mean sum, and Becker's memo proposed to lay groundwork by first building a model of the big tunnel, with a test section only one foot square. He recommended that this subscale facility should "be constructed and tested before proceeding with a four-foot-square tunnel." He gave an itemized cost estimate that came to $39,550, including $10,000 for installation and $6,000 for contingency.

Becker's memo ended in formal fashion: "Approval is requested to proceed with the design and construction of a model supersonic tunnel having a one-foot-square test section at Mach number 7.0. If successful, this model tunnel would not only provide data for the design of economical high Mach number supersonic wind tunnels, but would itself be a very useful research tool."[36]

On 6 August, three days after Becker wrote this memo, the potential usefulness of this tool increased enormously. On that day, an atomic bomb destroyed Hiroshima. With this, it now took only modest imagination to envision nuclear-tipped V-2s as weapons of the future. The standard V-2 had carried only a one-ton conventional warhead and lacked both range and accuracy. It nevertheless had been technically impressive, particularly since there was no way to shoot it down. But an advanced version with an atomic warhead would be far more formidable.

John Stack strongly supported Becker's proposal, which soon reached the desk of George Lewis, NACA's Director of Aeronautical Research. Lewis worked at NACA's Washington Headquarters but made frequent visits to Langley. Stack discussed the proposal with Lewis in the course of such a visit, and Lewis said, "Let's do it."

Just then, though, there was little money for new projects. NACA faced a postwar budget cut, which took its total appropriation from $40.9 million in FY 1945 to $24 million in FY 1946. Lewis therefore said to Stack, "John, you know I'm a sucker for a new idea, but don't call it a wind tunnel because I'll be in trouble with having to raise money in a formal way. That will necessitate Congressional review and approval. Call it a research project." Lewis designated it as Project 506 and obtained approval from NACA's Washington office on 18 December.[37]

A month later, in January 1946, Becker raised new issues in a memo to Stack. He was quite concerned that the high Mach would lead to so low a temperature that air in the flow would liquefy. To prevent this, he called for heating the air, declaring that "a temperature of 600°F in the pressure tank is essential." He expected to achieve this by using "a small electrical heater."

The pressure in that tank was to be considerably higher than in his plans of August. The tank would hold a pressure of 100 atmospheres. Instead of merely starting the flow, with a powered compressor sustaining in continuous operation, this pressure tank now was to hold enough air for operating times of 40 seconds. This would resolve uncertainties in the technical requirements for continuous operation. Continuous flows were still on the agenda but not for the immediate future. Instead, this wind tunnel was to operate as a blowdown facility.

Here, in outline, was a description of the installation as finally built. Its test section was 11 inches square. Its pressure tank held 50 atmospheres. It never received a compressor system for continuous flow, operating throughout its life entirely as a blowdown wind tunnel. But by heating its air, it indeed operated routinely at speeds close to Mach 7.[38]

Taking the name of 11-Inch Hypersonic Tunnel, it operated successfully for the first time on 26 November 1947. It did not heat its compressed air directly within the pressure tank, relying instead on an electric resistance heater as a separate component. This heater raised the air to temperatures as high as 900°F, eliminating air liquefaction in the test section with enough margin for Mach 8. Specialized experiments showed clearly that condensation took place when the initial temperature was not high enough to prevent it. Small particles promoted condensation by serving as nuclei for the formation of droplets. Becker suggested that such particles could have formed through the freezing of CO_2, which is naturally present in air. Subsequent research confirmed this conjecture.[39]

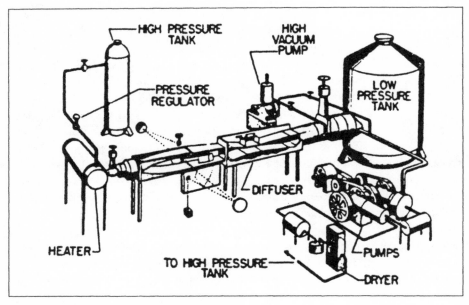

John Becker's 11-inch hypersonic wind tunnel. (NASA)

The facility showed initial early problems as well as a long-term problem. The early difficulties centered on the air heater, which showed poor internal heat conduction, requiring as much as five hours to reach a suitably uniform temperature distribution. In addition, copper tubes within the heater produced minute particles of copper oxide, due to oxidation of this metal at high temperature. These particles, blown within the hypersonic airstream, damaged test models and instruments. Becker attacked the problem of slow warmup by circulating hot air through the heater. To eliminate the problem of oxidation, he filled the heater with nitrogen while it was warming up.[40]

A more recalcitrant difficulty arose because the hot airflow, entering the nozzle, heated it and caused it to undergo thermal expansion. The change in its dimensions was not large, but the nozzle design was highly sensitive to small changes, with this expansion causing the dynamic pressure in the airflow to vary by up to 13 percent in the course of a run. Run times were as long as 90 seconds, and because of this, data taken at the beginning of a test did not agree with similar data recorded a minute later. Becker addressed this by fixing the angle of attack of each test model. He did not permit the angle to vary during a run, even though variation of this angle would have yielded more data. He also made measurements at a fixed time during each run.[41]

The wind tunnel itself represented an important object for research. No similar facility had ever been built in America, and it was necessary to learn how to use it most effectively. Nozzle design represented an early topic for experimental study. At Mach 7, according to standard tables, the nozzle had to expand by a ratio of 104.1 to 1. This nozzle resembled that of a rocket engine. With an axisymmetric design, a throat of one-inch diameter would have opened into a channel having a diameter slightly greater than 10 inches. However, nozzles for Becker's facility proved difficult to develop.

Conventional practice, carried over from supersonic wind tunnels, called for a two-dimensional nozzle. It featured a throat in the form of a narrow slit, having the full width of the main channel and opening onto that channel. However, for flow at Mach 7, this slit was to be only about 0.1 inch high. Hence, there was considerable interest in nozzles that might be less sensitive to small errors in fabrication.[42]

Initial work focused on a two-step nozzle. The first step was flat and constant in height, allowing the flow to expand to 10 inches wide in the horizontal plane and to reach Mach 4.36. The second step maintained this width while allowing the flow to expand to 10.5 inches in height, thus achieving Mach 7. But this nozzle performed poorly, with investigators describing its flow as "entirely unsatisfactory for use in a wind tunnel." The Mach number reached 6.5, but the flow in the test section was "not sufficiently uniform for quantitative wind-tunnel test purposes." This was due to "a thick boundary layer which developed in the first step" along the flat parallel walls set closely together at the top and bottom.[43]

A two-dimensional, single-step nozzle gave much better results. Its narrow slit-like throat indeed proved sensitive; this was the nozzle that gave the variation with time of the dynamic pressure. Still, except for this thermal-expansion effect, this nozzle proved "far superior in all respects" when compared with the two-step nozzle. In turn, the thermal expansion in time proved amenable to correction. This expansion occurred because the nozzle was made of steel. The commercially available alloy Invar had a far lower coefficient of thermal expansion. A new nozzle, fabricated from this material, entered service in 1954 and greatly reduced problems due to expansion of the nozzle throat.[44]

Another topic of research addressed the usefulness of the optical techniques used for flow visualization. The test gas, after all, was simply air. Even when it formed shock waves near a model under test, the shocks could not be seen with the unaided eye. Therefore, investigators were accustomed to using optical instruments when studying a flow. Three methods were in use: interferometry, schlieren, and shadowgraph. These respectively observed changes in air density, density gradient, and the rate of change of the gradient.

Such instruments had been in use for decades. Ernst Mach, of the eponymous Mach number, had used a shadowgraph as early as 1887 to photograph shock waves produced by a speeding bullet. Theodor Meyer, a student of Prandtl, used schlieren to visualize supersonic flow in a nozzle in 1908. Interferometry gave the most detailed photos and the most information, but an interferometer was costly and difficult to operate. Shadowgraphs gave the least information but were the least costly and easiest to use. Schlieren apparatus was intermediate in both respects and was employed often.[45]

Still, all these techniques depended on the flow having a minimum density. One could not visualize shock waves in a vacuum because they did not exist. Highly rarefied flows gave similar difficulties, and hypersonic flows indeed were rarefied. At Mach 7, a flow of air fell in pressure to less than one part in 4000 of its initial value, reducing an initial pressure of 40 atmospheres to less than one-hundredth of an atmosphere.[46] Higher test-section pressures would have required correspondingly higher pressures in the tank and upstream of the nozzle. But low test-section pressures were desirable because they were physically realistic. They corresponded to conditions in the upper atmosphere, where hypersonic missiles were to fly.

Becker reported in 1950 that the limit of usefulness of the schlieren method "is reached at a pressure of about 1 mm of mercury for slender test models at M = 7.0."[47] This corresponded to the pressure in the atmosphere at 150,000 feet, and there was interest in reaching the equivalent of higher altitudes still. A consultant, Joseph Kaplan, recommended using nitrogen as a test gas and making use of an afterglow that persists momentarily within this gas when it has been excited by an electrical discharge. With the nitrogen literally glowing in the dark, it became much easier to see shock waves and other features of the flow field at very low pressures.

"The nitrogen afterglow appears to be usable at static pressures as low as 100 microns and perhaps lower," Becker wrote.[48] This corresponded to pressures of barely a ten-thousandth of an atmosphere, which exist near 230,000 feet. It also corresponded to the pressure in the test section of a blowdown wind tunnel with air in the tank at 50 atmospheres and the flow at Mach 13.8.[49] Clearly, flow visualization would not be a problem.

Condensation, nozzle design, and flow visualization were important topics in their own right. Nor were they merely preliminaries. They addressed an important reason for building this tunnel: to learn how to design and use subsequent hypersonic facilities. In addition, although this 11-inch tunnel was small, there was much interest in using it for studies in hypersonic aerodynamics.

This early work had a somewhat elementary character, like the hypersonic experiments of Erdmann at Peenemunde. When university students take initial courses in aerodynamics, their textbooks and lab exercises deal with simple cases such as flow over a flat plate. The same was true of the first aerodynamic experiments with the 11-inch tunnel. The literature held a variety of theories for calculating lift, drag, and pressure distributions at hypersonic speeds. The experiments produced data that permitted comparison with theory—to check their accuracy and to determine circumstances under which they would fail to hold.

One set of tests dealt with cone-cylinder configurations at Mach 6.86. These amounted to small and simplified representations of a missile and its nose cone. The test models included cones, cylinders with flat ends, and cones with cylindrical afterbodies, studied at various angles of attack. For flow over a cone, the British researchers Geoffrey I. Taylor and J. W. Maccoll published a treatment in 1933. This quantitative discussion was a cornerstone of supersonic theory and showed its merits anew at this high Mach number. An investigation showed that it held "with a high degree of accuracy."

The method of characteristics, devised by Prandtl and Busemann in 1929, was a standard analytical method for designing surfaces for supersonic flow, including wings and nozzles. It was simple enough to lend itself to hand computation, and it gave useful results at lower supersonic speeds. Tests in the 11-inch facility showed that it continued to give good accuracy in hypersonic flow. For flow with angle of attack, a theory put forth by Antonio Ferri, a leading Italian aerodynamicist, produced "very good results." Still, not all preexisting theories proved to be accurate. One treatment gave good results for drag but overestimated some pressures and values of lift.[50]

Boundary-layer effects proved to be important, particularly in dealing with hypersonic wings. Tests examined a triangular delta wing and a square wing, the latter having several airfoil sections. Existing theories gave good results for lift and drag at modest angles of attack. However, predicted pressure distributions were often in error. This resulted from flow separation at high angles of attack—and from the presence of thick laminar boundary layers, even at zero angle of attack. These finds held high significance, for the very purpose of a hypersonic wing was to generate a pressure distribution that would produce lift, without making the vehicle unstable and prone to go out of control while in flight.

The aerodynamicist Charles McLellan, who had worked with Becker in designing the 11-inch tunnel and who had become its director, summarized the work within the *Journal of the Aeronautical Sciences*. He concluded that near Mach 7, the aerodynamic characteristics of wings and bodies "can be predicted by available theoretical methods with the same order of accuracy usually obtainable at lower speeds, at least for cases in which the boundary layer is laminar."[51]

At hypersonic speeds, boundary layers become thick because they sustain large temperature changes between the wall and the free stream. Mitchel Bertram, a colleague of McLellan, gave an approximate theory for the laminar hypersonic boundary layer on a flat plate. Using the 11-inch tunnel, he showed good agreement between his theory and experiment in several significant cases. He noted that boundary-layer effects could increase drag coefficients at least threefold, when compared with values using theories that include only free-stream flow and ignore the boundary layer. This emphasized anew the importance of the boundary layer in producing hypersonic skin friction.[52]

These results were fundamental, both for aerodynamics and for wind-tunnel design. With them, the 11-inch tunnel entered into a brilliant career. It had been built as a pilot facility, to lay groundwork for a much larger hypersonic tunnel that could sustain continuous flows. This installation, the Continuous Flow Hypersonic Tunnel (CFHT), indeed was built. Entering service in 1962, it had a 31-inch test section and produced flows at Mach 10.[53]

Still, it took a long time for this big tunnel to come on line, and all through the 1950s the 11-inch facility continued to grow in importance. At its peak, in 1961, it conducted more than 2,500 test runs, for an average of 10 per working day. It remained in use until 1972.[54] It set the pace with its use of the blowdown principle, which eliminated the need for costly continuous-flow compressors. Its run times proved to be adequate, and the CFHT found itself hard-pressed to offer much that was new. It had been built for continuous operation but found itself used in a blowdown mode most of the time. Becker wrote that his 11-inch installation "far exceeded" the CFHT "in both the importance and quality of its research output." He described it as "the only 'pilot tunnel' in NACA history to become a major research facility in its own right."[55]

Yet while the work of this wind tunnel was fundamental to the development of hypersonics, in 1950 the field of hypersonics was not fundamental to anything in particular. Plenty of people expected that America in time would build missiles and aircraft for flight at such speeds, but in that year no one was doing so. This soon changed, and the key year was 1954. In that year the Air Force embraced the X-15, a hypersonic airplane for which studies in the 11-inch tunnel proved to be essential. Also in that year, advances in the apparently unrelated field of nuclear weaponry brought swift and emphatic approval for the development of the ICBM. With this, hypersonics vaulted to the forefront of national priority.

1. Anderson, *History*, pp. 251-255.
2. Wegener, *Peenemunde*, pp. 23-24, 167; Von Karman and Edson, *Wind*, p. 221.
3. Wegener, *Peenemunde*, pp. 22-23; Neufeld, *Rocket*, pp. 87-88.
4. Neufeld, *Rocket*, p. 87.
5. Wegener, *Peenemunde*, pp. 24-25; Shapiro, *Compressible*, pp. 203-04.
6. See, for example, Miller and Durant, *Worlds*, pp. 9, 17, 23.
7. Neufeld, *Rocket*, pp. 86, 88-91 (quote, p. 89). Zeppelin: Wattendorf, *German*, p. 19.
8. Wegener, *Peenemunde*, photos following p. 84.
9. Von Karman and Edson, *Wind*, pp. 218-19.
10. Dornberger, *V-2*, pp. 122-23, 127-28.
11. Hallion, *Hypersonic*, pp. xvi, xviii; Neufeld, *Rocket*, pp. 248-50, 283.
12. Neufeld, *Rocket*, pp. 250-51; Dornberger, *V-2*, p. 268.
13. Neufeld, *Rocket*, p. 250.
14. NASA SP-4303, pp. 77, 316, 330.
15. Wegener, *Peenemunde*, p. 70.
16. Lukasiewicz, *Experimental*, pp. 71-76.
17. Wegener, *Peenemunde*, pp. 70-71.
18. Neufeld, *Rocket*, p. 87; Wegener, *Peenemunde*, p. 24; Wattendorf, *German*, p. 4.
19. Wegener, *Peenemunde*, pp. 32, 75, photos following p. 84.
20. Wattendorf, *German*, p. 4.
21. *Spaceflight*, May 1973, pp. 166-71 (quotes, pp. 168, 170); Huxley, *Brave*, pp. 58, 59, 61.
22. *Spaceflight*, May 1973, pp. 166, 171-72 (quote, p. 166); Ley, *Rockets*, pp. 533-537.
23. *Spaceflight*, May 1973, pp. 171-72, 175-76; Ley, *Rockets*, pp. 533-34, 535; Ordway and Sharpe, *Rocket Team*, pp. 327-28.
24. *Spaceflight*, May 1973, pp. 168-69.
25. Ley, *Rockets*, pp. 514-19, 524; Allen and Polmar, *Downfall*, pp. 103, 226.
26. Ordway and Sharpe, *Rocket Team*, p. 329; Jenkins, *Space Shuttle*, p. 2.
27. "Sänger." MBB brochure, August 1986.
28. Boyne, *Arrow*, p. 139.
29. Speer, *Inside*, pp. 364-66.
30. Anderson, *History*, p. 435; NASA: SP-440, pp. 51-52; SP-4305, p. 467.
31. NASA: SP-440, p. 52; SP-4302, pp. 63-64.
32. AIAA Paper 79-0219, pp. 3-4.
33. *Journal of the Aeronautical Sciences*, July 1947. pp. 373-409 (Stack quote, p. 406).
34. Hermann, "Supersonic," p. 439; Anspacher et al., *Legacy*, pp. 209-10.
35. Becker: Professional resume; NASA SP-4305, p. 54. R-2800 engine: "Dependable Engines" (Pratt & Whitney).

36 Memo, Becker to Chief of Research, 3 August 1945 (includes quotes); see also NASA SP-4305, pp. 344-346.
37 John Becker interview by Walter Bonney, March 1973 (quotes, p. 4). NACA budget: NASA SP-4305, p. 428. Project approval noted in memo, Becker to Stack, 16 January 1946.
38 Memo, Becker to Stack, 16 January 1946 (includes quotes).
39 *Journal of Applied Physics*, July 1950, pp. 619-21; NACA TN 3302. Air heater: NASA SP-4305, p. 471.
40 Becker, memo for record, 23 January 1948.
41 AIAA Paper 88-0230, p. 6.
42 NACA TN 2171, p. 3.
43 Ibid., pp. 6, 21 (quotes, pp. 1, 19).
44 NACA TN 2223 (quote, p. 11); AIAA Paper 88-0230, pp. 6-7.
45 Shapiro, *Compressible*, pp. 59-68. Photos by Mach and Meyer: Anderson, *History*, pp. 376, 382.
46 Shapiro, *Compressible*, table, p. 620.
47 *Journal of Applied Physics*, July 1950, pp. 619-28 (quote, p. 621).
48 Ibid. (quote, p. 622).
49 230,000 feet: Shapiro, *Compressible*, table, pp. 612-13. Mach 13.8: calculated from Shapiro, *Compressible*, eq. 4.14b, p. 83.
50 NACA RM L51J09 (quotes, p. 19).
51 NACA RM L51D17; *Journal of the Aeronautical Sciences*, October 1951, pp. 641-48 (quote, p. 648).
52 NACA TN 2773.
53 AIAA Paper 88-0230, pp. 8-9; NACA: SP-440, pp. 94-95; TM X-1130, p. 27.
54 AIAA Paper 88-0230, p. 7.
55 Becker, handwritten notes, January 1989 (includes quotes); NASA: SP-4305, p. 471; RP-1132, p. 256.

2

NOSE CONES AND RE-ENTRY

The ICBM concept of the early 1950s, called Atlas, was intended to carry an atomic bomb as a warhead, and there were two things wrong with this missile. It was unacceptably large and unwieldy, even with a warhead of reduced weight. In addition, to compensate for this limited yield, Atlas demanded unattainable accuracy in aim. But the advent of the hydrogen bomb solved both problems. The weight issue went away because projected H-bombs were much lighter, which meant that Atlas could be substantially smaller. The accuracy issue also disappeared. Atlas now could miss its target by several miles and still destroy it, by the simple method of blowing away everything that lay between the aim point and the impact point.

Studies by specialists, complemented by direct tests of early H-bombs, brought a dramatic turnaround during 1954 as Atlas vaulted to priority. At a stroke, its designers faced the re-entry problem. They needed a lightweight nose cone that could protect the warhead against the heat of atmosphere entry, and nothing suitable was in sight. The Army was well along in research on this problem, but its missiles did not face the severe re-entry environment of Atlas and its re-entry studies were not directly applicable.

The Air Force approached this problem systematically. It began by working with the aerodynamicist Arthur Kantrowitz, who introduced the shock tube as an instrument that could momentarily reproduce flow conditions that were pertinent. Tests with rockets, notably the pilotless X-17, complemented laboratory experiments. The solution to the problem of nose-cone design came from George Sutton, a young physicist who introduced the principle of ablation. Test nose cones soon were in flight, followed by prototypes of operational versions.

THE MOVE TOWARD MISSILES

In August 1945 it took little imagination to envision that the weapon of the future would be an advanced V-2, carrying an atomic bomb as the warhead and able to cross oceans. It took rather more imagination, along with technical knowledge, to see that this concept was so far beyond the state of the art as not to be worth pursuing. Thus, in December Vannevar Bush, wartime head of the Office of Scientific Research and Development, gave his views in congressional testimony:

> "There has been a great deal said about a 3,000 miles high-angle rocket. In my opinion, such a thing is impossible for many years. The people have been writing these things that annoy me, have been talking about a 3,000 mile high-angle rocket shot from one continent to another, carrying an atomic bomb and so directed as to be a precise weapon which would land exactly on a certain target, such as a city. I say, technically, I don't think anyone in the world knows how to do such a thing, and I feel confident that it will not be done for a very long period of time to come. I think we can leave that out of our thinking."[1]

Propulsion and re-entry were major problems, but guidance was worse. For intercontinental range, the Air Force set the permitted miss distance at 5,000 feet and then at 1,500 feet. The latter equaled the error of experienced bombardiers who were using radar bombsights to strike at night from 25,000 feet. The view at the Pentagon was that an ICBM would have to do as well when flying all the way to Moscow. This accuracy corresponded to hitting a golf ball a mile and having it make a hole in one. Moreover, each ICBM was to do this entirely through automatic control.[2]

The Air Force therefore emphasized bombers during the early postwar years, paying little attention to missiles. Its main program, such as it was, called for a missile that was neither ballistic nor intercontinental. It was a cruise missile, which was to solve its guidance problem by steering continually. The first thoughts dated to November 1945. At North American Aviation, chief engineer Raymond Rice and chief scientist William Bollay proposed to "essentially add wings to the V-2 and design a missile fundamentally the same as the A-9."

Like the supersonic wind tunnel at the Naval Ordnance Laboratory, here was another concept that was to carry a German project to completion. The initial design had a specified range of 500 miles,[3] which soon increased. Like the A-9, this missile—designated MX-770—was to follow a boost-glide trajectory and then extend its range with a supersonic glide. But by 1948 the U.S. Air Force had won its independence from the Army and had received authority over missile programs with ranges of 1,000 miles and more. Shorter-range missiles remained the concern of the Army. Accordingly, late in February, Air Force officials instructed North American to stretch the range of the MX-770 to a thousand miles.

A boost-glide trajectory was not well suited for a doubled range. At Wright Field, the Air Force development center, Colonel M. S. Roth proposed to increase the range by adding ramjets.[4] This drew on work at Wright, where the Power Plant

Nose Cones and Re-entry

Laboratory had a Nonrotating Engine Branch that was funding development of both ramjets and rocket engines. Its director, Weldon Worth, dealt specifically with ramjets.[5] A modification of the MX-770 design added two ramjet engines, mounting them singly at the tips of the vertical fins.[6] The missile also received a new name: Navaho. This reflected a penchant at North American for names beginning with "NA."[7]

Then, within a few months during 1949 and 1950, the prospect of world war emerged. In 1949 the Soviets exploded their first atomic bomb. At nearly the same time, China's Mao Zedong defeated the Nationalists of Chiang Kai-shek and proclaimed the People's Republic of China. The Soviets had already shown aggressiveness by subverting the democratic government of Czechoslovakia and by blockading Berlin. These new developments raised the prospect of a unified communist empire armed with the industry that had defeated the Nazis, wielding atomic weapons, and deploying the limitless manpower of China.

President Truman responded both publicly and with actions that were classified. In January 1950 he announced a stepped-up nuclear program, directing "the Atomic Energy Commission to continue its work on all forms of atomic weapons, including the so-called hydrogen or super bomb." In April he gave his approval to a secret policy document, NSC-68. It stated that the United States would resist communist expansion anywhere in the world and would devote up to twenty percent of the gross national product to national defense.[8] Then in June, in China's back yard, North Korea invaded the South, and America again was at war.

These events had consequences for the missile program, as the design and mission of Navaho changed dramatically during 1950. Bollay's specialists, working with Air Force counterparts, showed that they could anticipate increases in its range to as much as 5,500 nautical miles. Conferences among Air Force officials, held at the Pentagon in August, set this intercontinental range as a long-term goal. A letter from Major General Donald Putt, Director of Research and Development within the Air Materiel Command, became the directive instructing North American to pursue this objective. An interim version, Navaho II, with range of 2,500 nautical miles, appeared technically feasible. The full-range Navaho III represented a long-term project that was slated to go forward as a parallel effort.

The thousand-mile Navaho of 1948 had taken approaches based on the V-2 to their limit. Navaho II, the initial focus of effort, took shape as a two-stage missile with a rocket-powered booster. The booster was to use two such engines, each with thrust of 120,000 pounds. A ramjet-powered second stage was to ride it during initial ascent, accelerating to the supersonic speed at which the ramjet engines could produce their rated thrust. This second stage was then to fly onward as a cruise missile, at a planned flight speed of Mach 2.75.[9]

A rival to Navaho soon emerged. At Convair, structural analyst Karel Bossart held a strong interest in building an ICBM. As a prelude, he had built three rockets in the shape of a subscale V-2 and had demonstrated his ideas for lightweight structure in flight test. The Rand Corporation, an influential Air Force think tank, had been keeping an eye on this work and on the burgeoning technology of missiles. In December 1950 it issued a report stating that long-range ballistic missiles now were in reach. A month later the Air Force responded by giving Bossart, and Convair, a new study contract. In August 1951 he christened this missile Atlas, after Convair's parent company, the Atlas Corporation.

The initial concept was a behemoth. Carrying an 8,000-pound warhead, it was to weigh 670,000 pounds, stand 160 feet tall by 12 feet in diameter, and use seven of Bollay's new 120,000-pound engines. It was thoroughly unwieldy and represented a basis for further studies rather than a concept for a practical weapon. Still, it stood as a milestone. For the first time, the Air Force had a concept for an ICBM that it could pursue using engines that were already in development.[10]

For the ICBM to compete with Navaho, it had to shrink considerably. Within the Air Force's Air Research and Development Command, Brigadier General John Sessums, a strong advocate of long-range missiles, proposed that this could be done by shrinking the warhead. The size and weight of Atlas were to scale in proportion with the weight of its atomic weapon, and Sessums asserted that new developments in warhead design indeed would give high yield while cutting the weight.

He carried his argument to the Air Staff, which amounted to the Air Force's board of directors. This brought further studies, which indeed led to a welcome reduction in the size of Atlas. The concept of 1953 called for a length of 110 feet and a loaded weight of 440,000 pounds, with the warhead tipping the scale at only 3,000 pounds. The number of engines went down from seven to five.[11]

There also was encouraging news in the area of guidance. Radio guidance was out of the question for an operational missile; it might be jammed or the ground-based guidance center might be destroyed in an attack. Instead, missile guidance was to be entirely self-contained. All concepts called for the use of sensitive accelerometers along with an onboard computer, to determine velocity and location. Navaho was to add star trackers, which were to null out errors by tracking stars even in daylight. In addition, Charles Stark Draper of MIT was pursuing inertial guidance, which was to use no external references of any sort. His 1949 system was not truly inertial, for it included a magnetic compass and a Sun-seeker. But when flight-tested aboard a B-29, over distances as great at 1,737 nautical miles, it showed a mean error of only 5 nautical miles.[12]

For Atlas, though, the permitted miss distance remained at 1,500 feet, with the range being 5500 nautical miles. The program plan of October 1953 called for a leisurely advance over the ensuing decade, with research and development being completed only "sometime after 1964," and operational readiness being achieved in 1965. The program was to emphasize work on the major components: propulsion, guidance, nose cone, lightweight structure. In addition, it was to conduct extensive ground tests before proceeding toward flight.[13]

This concept continued to call for an atomic bomb as the warhead, but by then the hydrogen bomb was in the picture. The first test version, named Mike, detonated at Eniwetok Atoll in the Pacific on 1 November 1952. Its fireball spread so far and fast as to terrify distant observers, expanding until it was more than three miles across. "The thing was enormous," one man said. "It looked as if it blotted out the whole horizon, and I was standing 30 miles away." The weapons designer Theodore Taylor described it as "so huge, so brutal—as if things had gone too far. When the heat reached the observers, it stayed and stayed and stayed, not for seconds but for minutes." Mike yielded 10.4 megatons, nearly a thousand times greater than the 13 kilotons of the Hiroshima bomb of 1945.

Mike weighed 82 tons.[14] It was not a weapon; it was a physics experiment. Still, its success raised the prospect that warheads of the future might be smaller and yet might increase sharply in explosive power. Theodore von Karman, chairman of the Air Force Scientific Advisory Board, sought estimates from the Atomic Energy Commission of the size and weight of future bombs. The AEC refused to release this information. Lieutenant General James Doolittle, Special Assistant to the Air Force Chief of Staff, recommended creating a special panel on nuclear weapons within the SAB. This took form in March 1953, with the mathematician John von Neumann as its chairman. Its specialists included Hans Bethe, who later won the Nobel Prize, and Norris Bradbury who headed the nation's nuclear laboratory at Los Alamos, New Mexico.

In June this group reported that a thermonuclear warhead with the 3,000-pound Atlas weight could have a yield of half a megaton. This was substantially higher than that of the pure-fission weapons considered previously. It gave renewed strength to the prospect of a less stringent aim requirement, for Atlas now might miss by far more than 1,500 feet and still destroy its target.

Three months later the Air Force Special Weapons Center issued its own estimate, anticipating that a hydrogen bomb of half-megaton yield could weigh as little as 1,500 pounds. This immediately opened the prospect of a further reduction in the size of Atlas, which might fall in weight from 440,000 pounds to as little as 240,000. Such a missile also would need fewer engines.[15]

Also during September, Bruno Augenstein of the Rand Corporation launched a study that sought ways to accelerate the development of an ICBM. In Washington, Trevor Gardner was Special Assistant for Research and Development, reporting to the Air Force Secretary. In October he set up his own review committee. He recruited von Neumann to serve anew as its chairman and then added a dazzling array of talent from Caltech, Bell Labs, MIT, and Hughes Aircraft. In Gardner's words, "The aim was to create a document so hot and of such eminence that no one could pooh-pooh it."[16]

He called his group the Teapot Committee. He wanted particularly to see it call for less stringent aim, for he believed that a 1,500-foot miss distance was preposterous. The Teapot Committee drew on findings by Augenstein's group at Rand, which endorsed a 1,500-pound warhead and a three-mile miss distance. The formal Teapot report, issued in February 1954, declared "the military requirement" on miss distance "should be relaxed from the present 1,500 feet to at least two, and probably three, nautical miles." Moreover, "the warhead weight might be reduced as far as 1,500 pounds, the precise figure to be determined after the Castle tests and by missile systems optimization."[17]

The latter recommendation invoked Operation Castle, a series of H-bomb tests that began a few weeks later. The Mike shot of 1952 had used liquid deuterium, a form of liquid hydrogen. It existed at temperatures close to absolute zero and demanded much care in handling. But the Castle series was to test devices that used lithium deuteride, a dry powder that resembled salt. The Mike approach had been chosen because it simplified the weapons physics, but a dry bomb using lithium promised to be far more practical.

The first such bomb was detonated on 1 March as Castle Bravo. It produced 15 megatons, as its fireball expanded to almost four miles in diameter. Other Castle H-bombs performed similarly, as Castle Romeo went to 11 megatons and Castle Yankee, a variant of Romeo, reached 13.5 megatons. "I was on a ship that was 30 miles away," the physicist Marshall Rosenbluth recalls about Bravo, "and we had this horrible white stuff raining out on us." It was radioactive fallout that had condensed from vaporized coral. "It was pretty frightening. There was a huge fireball with these turbulent rolls going in and out. The thing was glowing. It looked to me like a diseased brain." Clearly, though, bombs of the lithium type could be as powerful as anyone wished—and these test bombs were readily weaponizable.[18]

The Castle results, strongly complementing the Rand and Teapot reports, cleared the way for action. Within the Pentagon, Gardner took the lead in pushing for Atlas. On 11 March he met with Air Force Secretary Harold Talbott and with the Chief of Staff, General Nathan Twining. He proposed a sped-up program that would nearly double the Fiscal Year (FY) 1955 Atlas budget and would have the first missiles ready to launch as early as 1958. General Thomas White, the Vice Chief of Staff, weighed in with his own endorsement later that week, and Talbott responded by directing Twining to accelerate Atlas immediately.

White carried the ball to the Air Staff, which held responsibility for recommending approval of new programs. He told its members that "ballistic missiles were here to stay, and the Air Staff had better realize this fact and get on with it." Then on 14 May, having secured concurrence from the Secretary of Defense, White gave Atlas the highest Air Force development priority and directed its acceleration "to the maximum extent that technology would allow." Gardner declared that White's order meant "the maximum effort possible with no limitation as to funding."[19]

This was a remarkable turnaround for a program that at the moment lacked even a proper design. Many weapon concepts have gone as far as the prototype

NOSE CONES AND RE-ENTRY

stage without winning approval, but Atlas gained its priority at a time when the accepted configuration still was the 440,000-pound, five-engine concept of 1953. Air Force officials still had to establish a formal liaison with the AEC to win access to information on projected warhead designs. Within the AEC, lightweight bombs still were well in the future. A specialized device, tested in the recent series as Castle Nectar, delivered 1.69 megatons but weighed 6,520 pounds. This was four times the warhead weight proposed for Atlas.

But in October the AEC agreed that it could develop warheads weighing 1,500 to 1,700 pounds, with a yield of one megaton. This opened the door to a new Atlas design having only three engines. It measured 75 feet long and 10 feet in diameter, with a weight of 240,000 pounds—and its miss distance could be as great as five miles. This took note of the increased yield of the warhead and further eased the problem of guidance. The new configuration won Air Force approval in December.[20]

APPROACHING THE NOSE CONE

An important attribute of a nose cone was its shape, and engineers were reducing drag to a minimum by crafting high-speed airplanes that displayed the ultimate in needle-nose streamlining. The X-3 research aircraft, designed for Mach 2, had a long and slender nose that resembled a church steeple. Atlas went even further, with an early concept having a front that resembled a flagpole. This faired into a long and slender cone that could accommodate the warhead.[21]

This intuitive approach fell by the wayside in 1953, as the NACA-Ames aerodynamicists H. Julian Allen and Alfred Eggers carried through an elegant analysis of the motion and heating of a re-entering nose cone. This work showed that they were masters of the simplifying assumption. To make such assumptions successfully represents a high art, for the resulting solutions must capture the most essential aspects of the pertinent physics while preserving mathematical tractability. Their paper stands to this day as a landmark. Quite probably, it is the single most important paper ever written in the field of hypersonics.

They calculated total heat input to a re-entry vehicle, seeking shapes that would minimize this. That part of the analysis enabled them to critique the assertion that a slender and sharply-pointed shape was best. For a lightweight nose cone, which would slow significantly in the atmosphere due to drag, they found a surprising result: the best shape, minimizing the total heat input, was blunt rather than sharp.

The next issue involved the maximum rate of heat transfer when averaged over an entire vehicle. To reduce this peak heating rate to a minimum, a nose cone of realistic weight might be either very sharp or very blunt. Missiles of intermediate slenderness gave considerably higher peak heating rates and "were definitely to be avoided."

This result applied to the entire vehicle, but heat-transfer rates were highest at the nose-cone tip. It was particularly important to minimize the heating at the tip, and again their analysis showed that a blunt nose cone would be best. As Allen and Eggers put it, "not only should pointed bodies be avoided, but the rounded nose should have as large a radius as possible."[22]

How could this be? The blunt body set up a very strong shock wave, which produced intense heating of the airflow. However, most of this heat was carried away in the flow. The boundary layer served to insulate the vehicle, and relatively little of this heat reached its surface. By contrast, a sharp and slender nose cone produced a shock that stood very close to this surface. At the tip, the boundary layer was too thin to offer protection. In addition, skin friction produced still more heating, for the boundary layer now received energy from shock-heated air flowing close to the vehicle surface.[23]

This paper was published initially as a classified document, but it took time to achieve its full effect. The Air Force did not adopt its principle for nose-cone design until 1956.[24] Still, this analysis outlined the shape of things to come. Blunt heat shields became standard on the Mercury, Gemini, and Apollo capsules. The space shuttle used its entire undersurface as a heat shield that was particularly blunt, raising its nose during re-entry to present this undersurface to the flow.

Yet while analysis could indicate the general shape for a nose cone, only experiment could demonstrate the validity of a design. At a stroke, Becker's Mach 7 facility, which had been far in the forefront only recently, suddenly became inadequate. An ICBM nose cone was to re-enter the atmosphere at speeds above Mach 20. Its kinetic energy would vaporize five times its weight of iron. Temperatures behind the bow shock would reach 9000 K, hotter than the surface of the Sun. Research scientist Peter Rose wrote that this velocity would be "large enough to dissociate all the oxygen molecules into atoms, dissociate about half of the nitrogen, and thermally ionize a considerable fraction of the air."[25]

Though hot, the 9000 K air actually would be cool, considering its situation, because its energy would go into dissociating molecules of gas. However, the ions and dissociated atoms were only too likely to recombine at the surface of the nose cone, thereby delivering additional heat. Such chemical effects also might trip the boundary layer from laminar to turbulent flow, with the rate of heat transfer increasing substantially as a result. In the words of Rose:

> "The presence of free-atoms, electrons, and molecules in excited states can be expected to complicate heat transfer through the boundary layer by additional modes of energy transport, such as atom diffusion, carrying the energy of dissociation. Radiation by transition from excited energy states may contribute materially to radiative heat transfer. There is also a

possibility of heat transfer by electrons and ions. The existence of large amounts of energy in any of these forms will undoubtedly influence the familiar flow phenomena."[26]

Within the Air Force, the Aircraft Panel of the Scientific Advisory Board (SAB) issued a report in October 1954 that looked ahead to the coming decade:

"In the aerodynamics field, it seems to us pretty clear that over the next 10 years the most important and vital subject for research and development is the field of hypersonic flows; and in particular, hypersonic flows with [temperatures at a nose-cone tip] which may run up to the order of thousands of degrees. This is one of the fields in which an ingenious and clever application of the existing laws of mechanics is probably not adequate. It is one in which much of the necessary physical knowledge still remains unknown at present and must be developed before we arrive at a true understanding and competence. The reason for this is that the temperatures which are associated with these velocities are higher than temperatures which have been produced on the globe, except in connection with the nuclear developments of the past 10 or 15 years and that there are problems of dissociation, relaxation times, etc., about which the basic physics is still unknown."[27]

The Atlas program needed a new experimental technique, one that could overcome the fact that conventional wind tunnels produced low temperatures due to their use of expanding gases, and hence the pertinent physics and chemistry associated with the heat of re-entry were not replicated. Its officials found what they wanted at a cocktail party.

This social gathering took place at Cornell University around Thanksgiving of 1954. The guests included university trustees along with a number of deans and senior professors. One trustee, Victor Emanuel, was chairman of Avco Corporation, which already was closely involved in work on the ICBM. He had been in Washington and had met with Air Force Secretary Harold Talbott, who told him of his concern about problems of re-entry. Emanuel raised this topic at the party while talking with the dean of engineering, who said, "I believe we have someone right here who can help you."[28]

That man was Arthur Kantrowitz, a former researcher at NACA-Langley who had taken a faculty position at Cornell following the war. While at Langley during the late 1930s, he had used a $5,000 budget to try to invent controlled thermonuclear fusion. He did not get very far. Indeed, he failed to gain results that were sufficient even to enable him to write a paper, leaving subsequent pioneers in controlled fusion to start again from scratch. Still, as he recalls, "I continued my interest in high temperatures with the hope that someday I could find something that I could use to do fusion."[29]

In 1947 this led him to the shock tube. This instrument produced very strong shocks in a laboratory, overcoming the limits of wind tunnels. It used a driving gas at high pressure in a separate chamber. This gas burst through a thin diaphragm to generate the shock, which traveled down a long tube that was filled with a test gas. High-speed instruments could observe this shock. They also could study a small model immersed within the hot flow at high Mach that streamed immediately behind the shock.[30]

When Kantrowitz came to the shock tube, it already was half a century old. The French chemist Paul Vieille built the first such devices prior to 1900, using them to demonstrate that a shock wave travels faster than the speed of sound. He proposed that his apparatus could prove useful in studying mine explosions, which took place in shafts that resembled his long tubes.[31]

The next important shock-tube researcher, Britain's William Payman, worked prior to World War II. He used diaphragm-bursting pressures as high as 1100 pounds per square inch and introduced high-speed photography to observe the shocked flows. He and his colleagues used the shock tube for experimental verification of equations in gasdynamics that govern the motion of shock waves.[32]

At Princeton University during that war, the physicist Walter Bleakney went further. He used shock tubes as precision instruments, writing, "It has been found that successive 'shots' in the tube taken with the same initial conditions reproduce one another to a surprising degree. The velocity of the incident shock can be reproduced to 0.1 percent." He praised the versatility of the device, noting its usefulness "for studying a great variety of problems in fluid dynamics." In addition to observations of shocks themselves, the instrument could address "problems of detonation and allied phenomena. The tube may be used as a wind tunnel with a Mach number variable over an enormous range." This was the role it took during the ICBM program.[33]

At Cornell, Kantrowitz initiated a reach for high temperatures. This demanded particularly high pressure in the upstream chamber. Payman had simply used compressed air from a thick-walled tank, but Kantrowitz filled his upstream chamber with a highly combustible mix of hydrogen and oxygen. Seeking the highest temperatures, he avoided choosing air as a test gas, for its diatomic molecules absorbed energy when they dissociated or broke apart, which limited the temperature rise. He turned instead to argon, a monatomic gas that could not dissociate, and reached 18,000 K.

He was a professor at Cornell, with graduate students. One of them, Edwin Resler, wrote a dissertation in 1951, "High Temperature Gases Produced by Strong

Shock Waves." In Kantrowitz's hands, the versatility of this instrument appeared anew. With argon as the test gas, it served for studies of thermal ionization, a physical effect separate from dissociation in which hot atoms lost electrons and became electrically charged. Using nitrogen or air, the shock tube examined dissociation as well, which increased with the higher temperatures of stronger shocks. Higher Mach values also lay within reach. As early as 1952, Kantrowitz wrote that "it is possible to obtain shock Mach numbers in the neighborhood of 25 with reasonable pressures and shock tube sizes."[34]

Other investigators also worked with these devices. Raymond Seeger, chief of aerodynamics at the Naval Ordnance Laboratory, built one. R. N. Hollyer conducted experiments at the University of Michigan. At NACA-Langley, the first shock tube entered service in 1951. The Air Force also was interested. The 1954 report of the SAB pointed to "shock tubes and other devices for producing extremely strong shocks" as an "experimental technique" that could give new insights into fundamental problems of hypersonics.[35]

Thus, when Emanuel met Kantrowitz at that cocktail party, this academic physicist indeed was in a position to help the Atlas effort. He had already gained hands-on experience by conducting shock-tube experiments at temperatures and shock velocities that were pertinent to re-entry of an ICBM. Emanuel then staked him to a new shock-tube center, Avco Research Laboratory, which opened for business early in 1955.

Kantrowitz wanted the highest shock velocities, which he obtained by using lightweight helium as the driver gas. He heated the helium strongly by adding a mixture of gaseous hydrogen and oxygen. Too little helium led to violent burning with unpredictable detonations, but use of 70 percent helium by weight gave a controlled burn that was free of detonations. The sudden heating of this driver gas also ruptured the diaphragm.

Standard optical instruments, commonly used in wind-tunnel work, were available for use with shock tubes as well. These included the shadowgraph, schlieren apparatus, and Mach-Zehnder interferometer. To measure the speed of the shock, it proved useful to install ionization-sensitive pickups that responded to changes in electrical resistance as shock waves passed. Several such pickups, spaced along the length of the tube, gave good results at speeds up to Mach 16.

Within the tube, the shock raced ahead of the turbulent mix of driver gases. Between the shock and the driver gases lay a "homogeneous gas sample" (HGS), a cylindrical slug of test gas moving nearly with the speed of the shock. The measured speed of the shock, together with standard laws of gasdynamics, permitted a complete calculation of the pressure, temperature, and internal energy of the HGS. Even when the HGS experienced energy absorption due to dissociation of its constituent molecules, it was possible to account for this through a separate calculation.[36]

The HGS swept over a small model of a nose cone placed within the stress. The time for passage was of the order of 100 microseconds, with the shock tube thus operating as a "wind tunnel" having this duration for a test. This nevertheless was long enough for photography. In addition, specialized instruments permitted study of heat transfer. These included thin-gauge resistance thermometers for temperature measurements and thicker-gauge calorimeters to determine heat transfer rates.

Metals increase their electrical resistance in response to a temperature rise. Both the thermometers and the calorimeters relied on this effect. To follow the sudden temperature increase behind the shock, the thermometer needed a metal film that was thin indeed, and Avco researchers achieved a thickness of 0.3 microns. They did this by using a commercial product, Liquid Bright Platinum No. 05, from Hanovia Chemical and Manufacturing Company. This was a mix of organic compounds of platinum and gold, dissolved in oils. Used as a paint, it was applied with a brush and dried in an oven.

The calorimeters used bulk platinum foil that was a hundred times thicker, at 0.03 millimeters. This thickness diminished their temperature rise and allowed the observed temperature increase to be interpreted as a rate of heat transfer. Both the thermometers and calorimeters were mounted to the surface of nose-cone models, which typically had the shape of a hemisphere that faired smoothly into a cylinder at the rear. The models were made of Pyrex, a commercial glass that did not readily crack. In addition, it was a good insulator.[37]

The investigator Shao-Chi Lin also used a shock tube to study thermal ionization, which made the HGS electrically conductive. To measure this conductivity, Shao used a nonconducting shock tube made of glass and produced a magnetic field within its interior. The flow of the conducting HGS displaced the magnetic lines of force, which he observed. He calibrated the system by shooting a slug of metal having known conductivity through the field at a known speed. Measured HGS conductivities showed good agreement with values calculated from theory, over a range from Mach 10 to Mach 17.5. At this highest flow speed, the conductivity of air was an order of magnitude greater than that of seawater.[38]

With shock tubes generating new data, there was a clear need to complement the data with new solutions in aerodynamics and heat transfer. The original Allen-Eggers paper had given a fine set of estimates, but they left out such realistic effects as dissociation, recombination, ionization, and changes in the ratio of specific heats. Again, it was necessary to make simplifying assumptions. Still, the first computers were at hand, which meant that solutions did not have to be in closed form. They might be equations that were solvable electronically.

Recombination of ions and of dissociated diatomic molecules—oxygen and nitrogen—was particularly important at high Mach, for this chemical process could deliver additional heat within the boundary layer. Two simplified cases stood out. In "equilibrium flow," the recombination took place instantly, responding immediately to the changing temperature and pressure within the boundary layer. The extent of ionization and dissociation then were simple point functions of the temperature and pressure at any location, and they could be calculated directly.

The other limiting case was "frozen flow." One hesitates to describe a 9000 K airstream as "frozen," but here it meant that the chemical state of the boundary layer retained its condition within the free stream behind the bow shock. Essentially this means that recombination proceeded so slowly that the changing conditions within the boundary layer had no effect on the degrees of dissociation and ionization. These again could be calculated directly, although this time as a consequence of conditions behind the shock rather than in the boundary layer. Frozen flow occurred when the air was rarefied.

These approximations avoided the need to deal with the chemistry of finite reaction rates, wherein recombination would not instantly respond to the rapidly varying flow conditions across the thickness of a boundary layer but would lag behind the changes. In 1956 the aerodynamicist Lester Lees proposed a heat-transfer theory that specifically covered those two limiting cases.[39] Then in 1957, Kantrowitz's colleagues at Avco Research Laboratory went considerably further.

The Avco lab had access to the talent of nearby MIT. James Fay, a professor of mechanical engineering, joined with Avco's Frederick Riddell to treat anew the problem of heat transfer in dissociated air. Finite reaction-rate chemistry was at the heart of their agenda, and again they needed a simplifying assumption: that the airflow velocity was zero. However, this condition was nearly true at the forward tip of a nose cone, where the heating was most severe.

Starting with a set of partial differential equations, they showed that these equations reduced to a set of nonlinear ordinary differential equations. Using an IBM 650 computer, they found that a numerical solution of these nonlinear equations was reasonably straightforward. In dealing with finite-rate chemistry, they introduced a "reaction rate parameter" that attempted to capture the resulting effects. They showed that a re-entering nose cone could fall through 100,000 feet while transitioning from the frozen to the equilibrium regime. Within this transition region, the boundary layer could be expected to be partly frozen, near the free stream, and partly in equilibrium, near the wall.

The Fay-Riddell theory appeared in the February 1958 *Journal of the Aeronautical Sciences*. That same issue presented experimental results, also from Avco, that tested the merits of this treatment. The researchers obtained shock-tube data with shock Mach numbers as high as 17.5. At this Mach, the corresponding speed of 17,500 feet per second approached the velocity of a satellite in orbit. Pressures within the shock-tube test gas simulated altitudes of 20,000, 70,000, and 120,000 feet, with equilibrium flow occurring in the models' boundary layers even at the highest equivalent height above the ground.

Most data were taken with calorimeters, although data points from thin-gauge thermometers gave good agreement. The measurements showed scatter but fit neatly on curves calculated from the Fay-Riddell theory. The Lees theory underpredicted heat-transfer rates at the nose-cone tip, calling for rates up to 30 percent lower than those observed. Here, within a single issue of that journal, two papers from Avco gave good reason to believe that theoretical and experimental tools were at hand to learn the conditions that a re-entering ICBM nose cone would face during its moments of crisis.[40]

Still, this was not the same as actually building a nose cone that could survive this crisis. This problem called for a separate set of insights. These came from the U.S. Army and were also developed independently by an individual: George Sutton of General Electric.

Ablation

In 1953, on the eve of the Atlas go-ahead, investigators were prepared to consider several methods for thermal protection of its nose cone. The simplest was the heat sink, with a heat shield of thick copper absorbing the heat of re-entry. An alternative approach, the hot structure, called for an outer covering of heat-resistant shingles that were to radiate away the heat. A layer of insulation, inside the shingles, was to protect the primary structure. The shingles, in turn, overlapped and could expand freely.

A third approach, transpiration cooling, sought to take advantage of the light weight and high heat capacity of boiling water. The nose cone was to be filled with this liquid; strong g-forces during deceleration in the atmosphere were to press the water against the hot inner skin. The skin was to be porous, with internal steam pressure forcing the fluid through the pores and into the boundary layer. Once injected, steam was to carry away heat. It would also thicken the boundary layer, reducing its temperature gradient and hence its rate of heat transfer. In effect, the nose cone was to stay cool by sweating.[41]

Still, each of these approaches held difficulties. Though potentially valuable, transpiration cooling was poorly understood as a topic for design. The hot-structure concept raised questions of suitably refractory metals along with the prospect of losing the entire nose cone if a shingle came off. The heat-sink approach was likely to lead to high weight. Even so, it seemed to be the most feasible way to proceed, and early Atlas designs specified use of a heat-sink nose cone.[42]

The Army had its own activities. Its missile program was separate from that of the Air Force and was centered in Huntsville, Alabama, with the redoubtable Wernher von Braun as its chief. He and his colleagues came to Huntsville in 1950 and developed the Redstone missile as an uprated V-2. It did not need thermal protection, but the next missile would have longer range and would certainly need it.[43]

Von Braun was an engineer. He did not set up a counterpart of Avco Research Laboratory, but his colleagues nevertheless proceeded to invent their way toward a nose cone. Their concern lay at the tip of a rocket, but their point of departure came at the other end. They were accustomed to steering their missiles by using jet vanes, large tabs of heat-resistant material that dipped into the exhaust. These vanes then deflected the exhaust, changing the direction of flight. Von Braun's associates thus had long experience in testing materials by placing them within the blast of a rocket engine. This practice carried over to their early nose-cone work.[44]

The V-2 had used vanes of graphite. In November 1952, these experimenters began testing new materials, including ceramics. They began working with nose-cone models late in 1953. In July 1954 they tested their first material of a new type: a reinforced plastic, initially a hard melamine resin strengthened with glass fiber. New test facilities entered service in June 1955, including a rocket engine with thrust of 20,000 pounds and a jet diameter of 14.5 inches.[45]

The pace accelerated after November of that year, as Von Braun won approval from Defense Secretary Charles Wilson to proceed with development of his next missile. This was Jupiter, with a range of 1,500 nautical miles.[46] It thus was markedly less demanding than Atlas in its thermal-protection requirements, for it was to re-enter the atmosphere at Mach 15 rather than Mach 20 and higher. Even so, the Huntsville group stepped up its work by introducing new facilities. These included a rocket engine of 135,000 pounds of thrust for use in nose-cone studies.

The effort covered a full range of thermal-protection possibilities. Transpiration cooling, for one, raised unpleasant new issues. Convair fabricated test nose cones with water tanks that had porous front walls. The pressure in a tank could be adjusted to deliver the largest flow of steam when the heat flux was greatest. But this technique led to hot spots, where inadequate flow brought excessive temperatures. Transpiration thus fell by the wayside.

Heat sink drew attention, with graphite holding promise for a time. It was light in weight and could withstand high temperatures. But it also was a good heat conductor, which raised problems in attaching it to a substructure. Blocks of graphite also contained voids and other defects, which made them unusable.

By contrast, hot structures held promise. Researchers crafted lightweight shingles of tungsten and molybdenum backed by layers of polished corrugated steel and aluminum, to provide thermal insulation along with structural support. When the shingles topped 3,250°F, the innermost layer stayed cool and remained below 200°F. Clearly, hot structures had a future.

The initial work with a reinforced plastic, in 1954, led to many more tests of similar materials. Engineers tested such resins as silicones, phenolics, melamines, Teflon, epoxies, polyesters, and synthetic rubbers. Filler materials included soft glass, fibers of silicon dioxide and aluminum silicate, mica, quartz, asbestos, nylon, graphite, beryllium, beryllium oxide, and cotton.

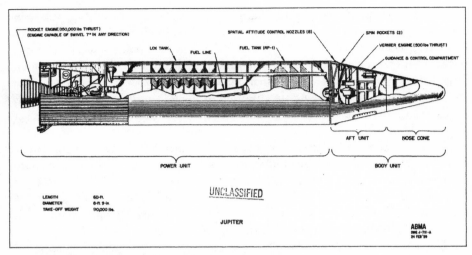

Jupiter missile with ablative nose cone. (U.S. Army)

Fiber-reinforced polymers proved to hold particular merit. The studies focused on plastics reinforced with glass fiber, with a commercially-available material, Micarta 259-2, demonstrating noteworthy promise. The Army stayed with this choice as it moved toward flight test of subscale nose cones in 1957. The first one used Micarta 259-2 for the plastic, with a glass cloth as the filler.[47]

In this fashion the Army ran well ahead of the Air Force. Yet the Huntsville work did not influence the Atlas effort, and the reasons ran deeper than interservice rivalry. The relevance of that work was open to question because Atlas faced a far more demanding re-entry environment. In addition, Jupiter faced competition from Thor, an Air Force missile of similar range. It was highly likely that only one would enter production, so Air Force designers could not merely become apt pupils of the Army. They had to do their own work, seeking independent approaches and trying to do better than Von Braun.

Amid this independence, George Sutton came to the re-entry problem. He had received his Ph.D. at Caltech in 1955 at age 27, jointly in mechanical engineering and physics. His only experience within the aerospace industry had been a summer job at the Jet Propulsion Laboratory, but he jumped into re-entry with both feet after taking his degree. He joined Lockheed and became closely involved in studying materials suitable for thermal protection. Then he was recruited by General Electric, leaving sunny California and arriving in snowy Schenectady, New York, early in 1956.

Heat sinks for Atlas were ascendant at that time, with Lester Lees's heat-transfer theory appearing to give an adequate account of the thermal environment. Sutton was aware of the issues and wrote a paper on heat-sink nose cones, but his work soon led him in a different direction. There was interest in storing data within a small capsule that would ride with a test nose cone and that might survive re-entry if the main cone were to be lost. This capsule needed its own thermal protection, and it was important to achieve light weight. Hence it could not use a heat sink. Sutton's management gave him a budget of $75,000 to try to find something more suitable.[48]

This led him to re-examine the candidate materials that he had studied at Lockheed. He also learned that other GE engineers were working on a related problem. They had built liquid propellant rocket engines for the Army's Hermes program, with these missiles being steered by jet vanes in the fashion of the V-2 and Redstone. The vanes were made from alternating layers of glass cloth and thermosetting resins. They had become standard equipment on the Hermes A-3, but some of them failed due to delamination. Sutton considered how to avoid this:

"I theorized that heating would char the resin into a carbonaceous mass of relatively low strength. The role of the fibers should be to hold the carbonaceous char to virgin, unheated substrate. Here, low thermal conductivity was essential to minimize the distance from the hot, exposed surface to the cool substrate, to minimize the mass of material that had to be held by the fibers as well as the degradation of the fibers. The char itself would eventually either be vaporized or be oxidized either by boundary layer oxygen or by CO_2 in the boundary layer. The fibers would either melt or also vaporize. The question was how to fabricate the material so that the fibers interlocked the resin, which was the opposite design philosophy to existing laminates in which the resin interlocks the fibers. I believed that a solution might be the use of short fibers, randomly oriented in a soup of resin, which was then molded into the desired shape. I then began to plan the experiments to test this hypothesis."[49]

Sutton had no pipeline to Huntsville, but his plan echoed that of Von Braun. He proceeded to fabricate small model nose cones from candidate fiber-reinforced plastics, planning to test them by immersion in the exhaust of a rocket engine. GE was developing an engine for the first stage of the Vanguard program; prototypes were at hand, along with test stands. Sutton arranged for an engine to produce an exhaust that contained free oxygen to achieve oxidation of the carbon-rich char.

He used two resins along with five types of fiber reinforcement. The best performance came with the use of Refrasil reinforcement, a silicon-dioxide fiber. Both resins yielded composites with a heat capacity of 6,300 BTU per pound or greater. This was astonishing. The materials had a density of 1.6 times that of water. Yet they absorbed more than six times as much heat, pound for pound, as boiling water![50]

Here was a new form of thermal protection: ablation. An ablative heat shield could absorb energy through latent heat, when melting or evaporating, and through sensible heat, with its temperature rise. In addition, an outward flow of ablating volatiles thickened the boundary layer, which diminished the heat flux. Ablation promised all the advantages of transpiration cooling, within a system that could be considerably lighter and yet more capable.[51]

Sutton presented his experimental results in June 1957 at a technical conference held at the firm of Ramo-Wooldridge in Los Angeles. This company was providing technical support to the Air Force's Atlas program management. Following this talk, George Solomon, one of that firm's leading scientists, rose to his feet and stated that ablation was the solution to the problem of thermal protection.

The Army thought so too. It had invented ablation on its own, considerably earlier and amid far deeper investigation. Indeed, at the moment when Sutton gave his talk, Von Braun was only two months away from a successful flight test of a subscale nose cone. People might argue whether the Soviets were ahead of the United States in missiles, but there was no doubt that the Army was ahead of the Air Force in nose cones. Jupiter was already slated for an ablative cone, but Thor was to use heat sink, as was the intercontinental Atlas.

Already, though, new information was available concerning transition from laminar to turbulent flow over a nose cone. Turbulent heating would be far more severe, and these findings showed that copper, the best heat-sink material, was inadequate for an ICBM. Materials testing now came to the forefront, and this work needed new facilities. A rocket-engine exhaust could reproduce the rate of heat transfer, but in Kantrowitz's words, "a rocket is not hot enough."[52] It could not duplicate the temperatures of re-entry.

A shock tube indeed gave a suitably hot flow, but its duration of less than a millisecond was hopelessly inadequate for testing ablative materials. Investigators needed a new type of wind tunnel that could produce a continuous flow, but at temperatures far greater than were available. Fortunately, such an installation did not have to reproduce the hypersonic Mach numbers of re-entry; it sufficed to duplicate the necessary temperatures within the flow. The instrument that did this was the arc tunnel.

It heated the air with an electric arc, which amounted to a man-made stroke of lightning. Such arcs were in routine use in welding; Avco's Thomas Brogan noted that they reached 6500 K, "a temperature which would exist at the [tip] of a blunt body flying at 22,000 feet per second." In seeking to develop an arc-heated wind tunnel, a point of departure lay in West Germany, where researchers had built a "plasma jet."[53]

This device swirled water around a long carbon rod that served as the cathode. The motion of the water helped to keep the arc focused on the anode, which was also of carbon and which held a small nozzle. The arc produced its plasma as a mix of very hot steam and carbon vapor, which was ejected through the nozzle. This invention achieved pressures of 50 atmospheres, with the plasma temperature at the nozzle exit being measured at 8000 K. The carbon cathode eroded relatively slowly, while the water supply was easily refilled. The plasma jet therefore could operate for fairly long times.[54]

At NACA-Langley, an experimental arc tunnel went into operation in May 1957. It differed from the German plasma jet by using an electric arc to heat a flow of air, nitrogen, or helium. With a test section measuring only seven millimeters square, it was a proof-of-principle instrument rather than a working facility. Still, its plasma temperatures ranged from 5800 to 7000 K, which was well beyond the reach of a conventional hypersonic wind tunnel.[55]

At Avco, Kantrowitz paid attention when he heard the word "plasma." He had been studying such ionized gases ever since he had tried to invent controlled fusion. His first arc tunnel was rated only at 130 kilowatts, a limited power level that restricted the simulated altitude to between 165,000 and 210,000 feet. Its hot plasma flowed from its nozzle at Mach 3.4, but when this flow came to a stop when impinging on samples of quartz, the temperature corresponded to flight velocities as high as 21,000 feet per second. Tests showed good agreement between theory and experiment, with measured surface temperatures of 2700 K falling within three percent of calculated values. The investigators concluded that opaque quartz "will effectively absorb about 4000 BTU per pound for ICBM and [intermediate-range] trajectories."[56]

In Huntsville, Von Braun's colleagues found their way as well to the arc tunnel. They also learned of the initial work in Germany. In addition, the small California firm of Plasmadyne acquired such a device and then performed experiments under contract to the Army. In 1958 Rolf Buhler, a company scientist, discovered that when he placed a blunt rod of graphite in the flow, the rod became pointed. Other investigators attributed this result to the presence of a cool core in the arc-heated jet, but Sutton succeeded in deriving this observed shape from theory.

This immediately raised the prospect of nose cones that after all might be sharply pointed rather than blunt. Such re-entry bodies would not slow down in the upper atmosphere, perhaps making themselves tempting targets for antiballistic missiles, but would continue to fall rapidly. Graphite still had the inconvenient features noted previously, but a new material, pyrolytic graphite, promised to ease the problem of its high thermal conductivity.

Pyrolytic graphite was made by chemical vapor deposition. One placed a temperature-resistant form in an atmosphere of gaseous hydrocarbons. The hot surface broke up the gas molecules, a process known as pyrolysis, and left carbon on the surface. The thermal conductivity then was considerably lower in a direction normal to the surface than when parallel to it. The low value of this conductivity, in the normal direction, made such graphite attractive.[57]

Having whetted their appetites with the 130-kilowatt facility, Avco went on to build one that was two orders of magnitude more powerful. It used a 15-megawatt power supply and obtained this from a bank of 2,000 twelve-volt truck batteries, with motor-generators to charge them. They provided direct current for run times of up to a minute and could be recharged in an hour.[58]

With this, Avco added the high-power arc tunnel to the existing array of hypersonic flow facilities. These included aerodynamic wind tunnels such as Becker's, along with plasma jets and shock tubes. And while the array of ground installations proliferated, the ICBM program was moving toward a different kind of test: full-scale flight.

Flight Test

The first important step in this direction came in January 1955, when the Air Force issued a letter contract to Lockheed that authorized them to proceed with the X-17. It took shape as a three-stage missile, with all three stages using solid-propellant rocket motors from Thiokol. It was to reach Mach 15, and it used a new flight mode called "over the top."

The X-17 was not to fire all three stages to achieve a very high ballistic trajectory. Instead it started with only its first stage, climbing to an altitude of 65 to 100 miles. Descent from such an altitude imposed no serious thermal problems. As it re-entered the atmosphere, large fins took hold and pointed it downward. Below 100,000 feet, the two upper stages fired, again while pointing downward. These stages accelerated a test nose cone to maximum speed, deep within the atmosphere. This technique prevented the nose cone from decelerating at high altitude, which would have happened with a very high ballistic flight path. Over-the-top also gave good control of both the peak Mach and of its altitude of attainment.

The accompanying table summarizes the results. Following a succession of subscale and developmental flights that ran from 1955 into 1956, the program conducted two dozen test firings in only eight months. The start was somewhat shaky as no more than two of the first six X-17s gained full success, but the program soon settled down to routine achievement. The simplicity of solid-propellant rocketry enabled the flights to proceed with turnaround times of as little as four days. Launches required no more than 40 active personnel, with as many as five such flights taking place within the single month of October 1956. All of them flew from a single facility: Pad 3 at Cape Canaveral.[59]

NOSE CONES AND RE-ENTRY

X-17 FLIGHT TESTS

Date	Nose-Cone Shape	Results
17 Jul 1956	Hemisphere	Mach 12.4 at 40,000 feet.
27 Jul 1956	Cubic Paraboloid	Third stage failed to ignite.
18 Aug 1956	Hemisphere	Missile exploded 18 sec. after launch.
23 Aug 1956	Blunt	Mach 12.4 at 38,500 feet.
28 Aug 1956	Blunt	Telemetry lost prior to apogee.
8 Sep 1956	Cubic Paraboloid	Upper stages ignited while ascending.
1 Oct 1956	Hemisphere	Mach 12.1 at 36,500 feet.
5 Oct 1956	Hemisphere	Mach 13.7 at 54,000 feet.
13 Oct 1956	Cubic Paraboloid	Mach 13.8 at 58,500 feet.
18 Oct 1956	Hemisphere	Mach 12.6 at 37,000 feet.
25 Oct 1956	Blunt	Mach 14.2 at 59,000 feet.
5 Nov 1956	Blunt (Avco)	Mach 12.6 at 41,100 feet.
16 Nov 1956	Blunt (Avco)	Mach 13.8 at 57,000 feet.
23 Nov 1956	Blunt (Avco)	Mach 11.3 at 34,100 feet.
3 Dec 1956	Blunt (Avco)	Mach 13.8 at 47,700 feet.
11 Dec 1956	Blunt Cone (GE)	Mach 11.4 at 34,000 feet.
8 Jan 1957	Blunt Cone (GE)	Mach 11.5 at 34,600 feet.
15 Jan 1957	Blunt Cone (GE)	Upper stages failed to ignite.
29 Jan 1957	Blunt Cone (GE)	Missile destroyed by Range Safety.
7 Feb 1957	Blunt Cone (GE)	Mach 14.4 at 57,000 feet.
14 Feb 1957	Hemisphere	Mach 12.1 at 35,000 feet.
1 Mar 1957	Blunt Cone (GE)	Mach 11.4 at 35,600 feet.
11 Mar 1957	Blunt (Avco)	Mach 11.3 at 35,500 feet.
21 Mar 1957	Blunt (Avco)	Mach 13.2 at 54,500 feet.

Source: "Re-Entry Test Vehicle X-17," pp. 30, 32.

Many nose cones approached or topped Mach 12 at altitudes below 40,000 feet. This was half the speed of a satellite, at altitudes where airliners fly today. One places this in perspective by noting that the SR-71 cruised above Mach 3, one-fourth this speed, and at 85,000 feet, which was more than twice as high. Thermal problems dominated its design, with this spy plane being built as a titanium hot structure. The X-15 reached Mach 6.7 in 1967, half the speed of an X-17 nose cone, and at

102,000 feet. Its structure was Inconel X heat sink, and it had further protection from a spray-on ablative. Yet it sustained significant physical damage due to high temperatures and never again approached that mark.[60]

Another noteworthy flight involved a five-stage NACA rocket that was to accomplish its own over-the-top mission. It was climbing gently at 96,000 feet when the third stage ignited. Telemetry continued for an additional 8.2 seconds and then suddenly cut off, with the fifth stage still having half a second to burn. The speed was Mach 15.5 at 98,500 feet. The temperature on the inner surface of the skin was 2,500°F, close to the melting point, with this temperature rising at nearly 5,300°F per second.[61]

How then did X-17 nose cones survive flight at nearly this speed, but at little more than one-third the altitude? They did not. They burned up in the atmosphere. They lacked thermal protection, whether heat sink or ablative (which the Air Force, the X-17's sponsor, had not invented yet), and no attempt was made to recover them. The second and third stages ignited and burned to depletion in only 3.7 seconds, with the thrust of these stages being 102,000 and 36,000 pounds, respectively.[62] Acceleration therefore was extremely rapid; exposure to conditions of very high Mach was correspondingly brief. The X-17 thus amounted to a flying shock tube. Its nose cones lived only long enough to return data; then they vanished into thin air.

Yet these data were priceless. They included measurements of boundary-layer transition, heat transfer, and pressure distributions, covering a broad range of peak Mach values, altitudes, and nose-cone shapes. The information from this program complemented the data from Avco Research Laboratory, contributing materially to Air Force decisions that selected ablation for Atlas (and for Titan, a second ICBM), while retaining heat sink for Thor.[63]

As the X-17 went forward during 1956 and 1957, the Army weighed in with its own flight-test effort. Here were no over-the-top heroics, no ultrashort moments at high Mach with nose cones built to do their duty and die. The Army wanted nothing less than complete tests of true ablating nose cones, initially at subscale and later at full scale, along realistic ballistic trajectories. The nose cones were to survive re-entry. If possible, they were to be recovered from the sea.

The launch vehicle was the Jupiter-C, another product of Von Braun. It was based on the liquid-fueled Redstone missile, which was fitted with longer propellant tanks to extend the burning time. Atop that missile rode two additional stages, both of which were built as clusters of small solid-fuel rockets.

The first flight took place from Cape Canaveral in September 1956. It carried no nose cone; this launch had the purpose of verifying the three-stage design, particularly its methods for stage separation and ignition. A dummy solid rocket rode atop this stack as a payload. All three stages fired successfully, and the flight broke all

NOSE CONES AND RE-ENTRY

Thor missile with heat-sink nose cone. (U.S. Air Force)

performance records. The payload reached a peak altitude of 682 miles and attained an estimated range of 3,335 miles.[64]

Nose-cone tests followed during 1957. Each cone largely duplicated that of the Jupiter missile but was less than one-third the size, having a length of 29 inches and maximum diameter of 20 inches. The weight was 314 pounds, of which 83

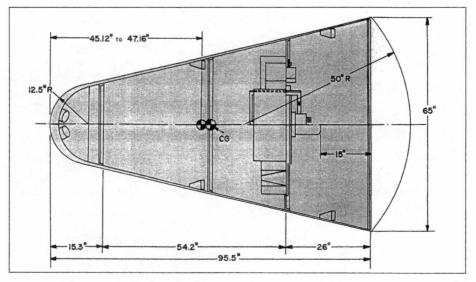

Jupiter nose cone. (U.S. Army)

pounds constituted the mix of glass cloth and Micarta plastic that formed the ablative material. To aid in recovery in the ocean, each nose cone came equipped with a balloon for flotation, two small bombs to indicate position for sonar, a dye marker, a beacon light, a radio transmitter—and shark repellant, to protect the balloon from attack.[65]

The first nose-cone flight took place in May. Telemetry showed that the re-entry vehicle came through the atmosphere successfully and that the ablative thermal protection indeed had worked. However, a faulty trajectory caused this nose cone to fall 480 miles short of the planned impact point, and this payload was not recovered.

Full success came with the next launch, in August. All three stages again fired, pushing the nose cone to a range of 1,343 statute miles. This was shorter than the planned range of Jupiter, 1,725 miles, but still this payload experienced 95 percent of the total heat transfer that it would have received at the tip for a full-range flight. The nose cone also was recovered, giving scientists their first close look at one that had actually survived.[66]

In November President Eisenhower personally displayed it to the nation. The Soviets had stirred considerable concern by placing two Sputnik satellites in orbit, thus showing that they already had an ICBM. Speaking on nationwide radio and television, Ike sought to reassure the public. He spoke of American long-range bombers and then presented his jewel: "One difficult obstacle on the way to producing a useful long-range weapon is that of bringing a missile back from outer space without its burning up like a meteor. This object here in my office is the nose cone of an experimental missile. It has been hundreds of miles into outer space and back. Here it is, completely intact."[67]

Jupiter then was in flight test and became the first missile to carry a full-size nose cone to full range.[68] But the range of Jupiter was far shorter than that of Atlas. The Army had taken an initial lead in nose-cone testing by taking advantage of its early start, but by the time of that flight—May 1958—all eyes were on the Air Force and on flight to intercontinental range.

Atlas also was in flight test during 1958, extending its range in small steps, but it still was far from ready to serve as a test vehicle for nose cones. To attain 5,000-mile range, Air Force officials added an upper stage to the Thor. The resulting rocket, the Thor-Able, indeed had the job of testing nose cones. An early model, from General Electric, weighed more than 600 pounds and carried 700 pounds of instruments.[69]

Two successful flights, both to full range, took place during July 1958. The first one reached a peak altitude of 1,400 miles and flew 5,500 miles to the South Atlantic. Telemetered data showed that its re-entry vehicle survived the fiery passage through the atmosphere, while withstanding four times the heat load of a Thor heat-sink nose cone. This flight carried a passenger, a mouse named Laska in honor of what soon became the 49th state. Little Laska lived through decelerations during re-entry that reached 60 g, due to the steepness of the trajectory, but the nose cone was not recovered and sank into the sea. Much the same happened two weeks later, with the mouse being named Wickie. Again the reentry vehicle came through the atmosphere successfully, but Wickie died for his country as well, for this nose cone also sank without being recovered.[70]

A new series of tests went forward during 1959, as General Electric introduced the RVX-1 vehicle. Weighing 645 pounds, 67 inches long with a diameter at the base of 28 inches, it was a cylinder with a very blunt nose and a conical afterbody for stability.[71] A flight in March used phenolic nylon as the ablator. This was a phenolic resin containing randomly oriented one-inch-square pieces of nylon cloth. Light weight was its strong suit; with a density as low as 72 pounds per cubic foot, it was only slightly denser than water. It also was highly effective as insulation. Following flight to full range, telemetered data showed that a layer only a quarter-inch thick could limit the temperature rise on the aft body, which was strongly heated, to less than 200ºF. This was well within the permissible range for aluminum, the most familiar of aerospace materials. For the nose cap, where the heating was strongest, GE installed a thick coating of molded phenolic nylon.[72]

Within this new series of flights, new guidance promised enhanced accuracy and a better chance of retrieval. Still, that March flight was not recovered, with another shot also flying successfully but again sinking beneath the waves. When the first recovery indeed took place, it resulted largely from luck.

Early in April an RVX-1 made a flawless flight, soaring to 764 miles in altitude and sailing downrange to 4,944 miles. Peak speed during re-entry was Mach 20, or 21,400 feet per second. Peak heating occurred at Mach 16, or 15,000 feet per second, and at 60,000 feet. The nose cone took this in stride, but searchers failed to detect its radio signals. An Avco man in one of the search planes saved the situation by spotting its dye marker. Aircraft then orbited the position for three hours until a recovery vessel arrived and picked it up.[73]

It was the first vehicle to fly to intercontinental range and return for inspection. Avco had specified its design, using an ablative heat shield of fused opaque quartz. Inspection of the ablated surface permitted comparison with theory, and the results were described as giving "excellent agreement." The observed value of maximum ablated thickness was 9 percent higher than the theoretical value. The weight loss of ablated material agreed within 20 percent, while the fraction of ablated material that vaporized during re-entry was only 3 percent higher than the theoretical value. Most of the differences could be explained by the effect of impurities on the viscosity of opaque quartz.[74]

A second complete success was achieved six weeks later, again with a range of 5,000 miles. Observers aboard a C-54 search aircraft witnessed the re-entry, acquired the radio beacon, and then guided a recovery ship to the site.[75] This time the nose-cone design came from GE. That company's project engineer, Walter Schafer, wanted to try several materials and to instrument them with breakwire sensors. These were wires, buried at various depths within the ablative material, that would break as it eroded away and thus disclose the rate of ablation. GE followed a suggestion from George Sutton and installed each material as a 60-degree segment around the cylinder and afterbody, with the same material being repeated every 180 degrees for symmetry.[76]

Within the fast-paced world of nose-cone studies, each year had brought at least one new flight vehicle. The X-17 had flown during 1956. For the Jupiter-C, success had come in 1957. The year 1958 brought both Jupiter and the Thor-Able. Now, in 1959, the nose-cone program was to gain final success by flying full-size re-entry vehicles to full range aboard Atlas.

The program had laid important groundwork in November 1958, when this missile first flew to intercontinental distance. The test conductor, with the hopeful name of Bob Shotwell, pushed the button and the rocket leaped into the night. It traced an arc above the Moon as it flew across the starry sky. It dropped its twin booster engines; then, continuing to accelerate, the brilliant light of its main engine faded. Now it seemed to hang in the darkness like a new star, just below Orion. Shotwell and his crew contained their enthusiasm for a full seven minutes; then they erupted in shouts. They had it; the missile was soaring at 16,000 miles per hour, bound for a spot near the island of St. Helena in the South Atlantic, a full 6,300 miles from the Cape. In Shotwell's words, "We knew we had done it. It was going like a bullet; nothing could stop it."[77]

NOSE CONES AND RE-ENTRY

Atlas could carry far heavier loads than Thor-Able, and its first nose cone reflected this. It was the RVX-2, again from General Electric, which had the shape of a long cone with a round tip. With a length of 147 inches and a width at the base of 64 inches, it weighed some 2,500 pounds. Once more, phenolic nylon was used for thermal protection. It flew to a range of 5,047 miles in July 1959 and was recovered. It thereby became the largest object to have been brought back following re-entry.[78]

Attention now turned to developmental tests of a nose cone for the operational Atlas. This was the Mark 3, also from GE. Its design returned to the basic RVX-1 configuration, again with a blunt nose at the front of a cylinder but with

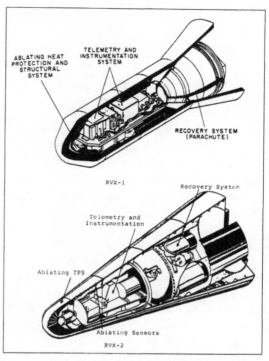

Nose cones used in flight test. Top, RVX-1; bottom, RVX-2. (U.S. Air Force)

a longer conical afterbody. It was slightly smaller than the RVX-2, with a length of 115 inches, diameter at the cylinder of 21 inches, and diameter at the base of 36 inches. Phenolic nylon was specified throughout for thermal protection, being molded under high pressure for the nose cap and tape-wound on the cylinder and afterbody. The Mark 3 weighed 2,140 pounds, making it somewhat lighter than the RVX-2. The low density of phenolic nylon showed itself anew, for of this total weight, only 308 pounds constituted ablative material.[79]

The Mark 3 launches began in October 1959 and ran for several months, with this nose cone entering operational service the following April.[80] The flights again were full-range, with one of them flying 5,000 miles to Ascension Island and another going 6,300 miles. Re-entry speeds went as high as 22,500 feet per second. Peak heat transfer occurred near Mach 14 and 40,000 feet in altitude, approximating the conditions of the X-17 tests. The air at that height was too thin to breathe, but the nose cone set up a shock wave that compressed the incoming flow, producing a wind resistance with dynamic pressure of more than 30 atmospheres. Temperatures at the nose reached 6,500°F.[81]

Each re-entry vehicle was extensively instrumented, mounting nearly two dozen breakwire ablation sensors along with pressure and temperature sensors. The latter were resistance thermometers employing 0.0003-inch tungsten wire, reporting temperatures to 2000ºF with an accuracy of 25 to 50°F. The phenolic nylon showed anew that it had the right stuff, for it absorbed heat at the rate of 3,000 BTU per pound, making it three times as effective as boiling water. A report from GE noted, "all temperature sensors located on the cylindrical section were at locations too far below the initial surface to register a temperature rise."[82]

With this, the main effort in re-entry reached completion, and its solution—ablation—had proved to be relatively simple. The process resembled the charring of wood. Indeed, Kantrowitz recalls Von Braun suggesting that it was possible to build a nose cone of lightweight balsa soaked in water and frozen. In Kantrowitz's words, "That might be a very reasonable ablator."[83]

Experience with ablation also contrasted in welcome fashion with a strong tendency of advanced technologies to rely on highly specialized materials. Nuclear energy used uranium-235, which called for the enormous difficulty of isotope separation, along with plutonium, which had to be produced in a nuclear reactor and then be extracted from highly radioactive spent fuel. Solid-state electronics depended on silicon or germanium, but while silicon was common, either element demanded refinement to exquisite levels of purity.

Ablation was different. Although wood proved inappropriate, once the basic concept was in hand the problem became one of choosing the best candidate from a surprisingly wide variety of possibilities. These generally were commercial plastics that served as binders, with the main heat resistance being provided by glass or silica. Quartz also worked well, particularly after being rendered opaque, while pyrolytic graphite exemplified a new material with novel properties.

The physicist Steven Weinberg, winner of a Nobel Prize, stated that a researcher never knows how difficult a problem is until the solution is in hand. In 1956 Theodore von Karman had described re-entry as "perhaps one of the most difficult problems one can imagine. It is certainly a problem that constitutes a challenge to the best brains working in these domains of modern aerophysics."[84] Yet in the end, amid all the ingenuity of shock tubes and arc tunnels, the fundamental insights derived from nothing deeper than testing an assortment of candidate materials in the blast of rocket engines.

1 Vannevar Bush, testimony before the Senate Special Committee on Atomic Energy, December 1945, quoted in Clarke, *Profiles*, p. 9 and *Time*, 22 May 1964, p. 25.
2 MacKenzie, *Inventing*, pp. 113-14. 1,500 feet: Neufeld, *Ballistic*, pp. 69-102 passim.
3 North American Aviation Report AL-1347, pp. 1-6 (quote, p. 4); Fahrney, *History*, p. 1291, footnote 3.
4 North American Aviation Report AL-1347, pp. 38-39; Colonel Roth: Letter, Colonel M. S. Roth to Power Plant Lab, Air Materiel Command, USAF, 11 February 1948.
5 Author interview, Colonel Edward Hall 29 August 1996, Folder 18649, NASA Historical Reference Collection, NASA History Division, Washington, D.C. 20546.
6 North American Aviation Report AL-1347, p. 39.
7 Other North American Aviation names of that era included the NAKA, NALAR, and NASTY battlefield missiles, the Navion private plane, and the NATIV research rocket; see "Thirty Years of Rocketdyne," photo, "Armament Rockets," 1962; Murray, *Atwood*, pp. 32, 44.
8 Rhodes, *Dark Sun*, p. 407 (includes quote); Manchester, *Caesar*, p. 642.
9 "Standard Missile Characteristics: XSM-64 Navaho"; Report AL-1347, op. cit., p. 88; *Development of the Navaho*, pp. 30-31; Augenstein, "Rand"; Fahrney, *History*, pp. 1296-98. General Putt: Letter, Major General Donald Putt to Commanding General, Air Materiel Command, USAF, 21 August 1950.
10 Neufeld, *Ballistic*, pp. 44-50, 68-70.
11 Ibid., pp. 70-79.
12 *Development of the Navaho*, pp. 40-46; *Journal of Guidance and Control*, September-October 1981, pp. 455-57.
13 Neufeld, *Ballistic*, pp. 73, 77-78 (quote, p. 78).
14 Rhodes, *Dark Sun*, pp. 495, 499, 505 (quotes, p. 509). Taylor quote: McPhee, *Curve*, p. 77. Hiroshima: Rhodes, *Atomic*, p. 711.
15 Von Karman and Edson, *Wind*, pp. 300-01; Neufeld, *Ballistic*, p. 98.
16 Neufeld, *Ballistic*, pp. 95, 98-99, 102; quote: Chapman, *Atlas*, p. 73.
17 AIAA Paper 67-838, pp. 12-13; Neufeld, *Ballistic*, p. 102 (quotes, p. 259).
18 Rhodes, *Dark Sun*, pp. 482-84, 541-42 (quote, p. 541).
19 Neufeld, *Ballistic*, pp. 104-06, 117 (quotes, pp. 105, 106); Rhodes, *Dark Sun*, p. 542.
20 Neufeld, *Ballistic*, p. 117; Emme, *History*, p. 151. Dimensions: Ley, *Rockets*, p. 400.
21 Miller, *X-Planes*, ch. 7; letter, Smith DeFrance to NACA Headquarters, 26 November 1952.
22 NACA Report 1381 (quotes, pp. 11, 13).
23 Anderson, *History*, pp. 440-41.
24 Neufeld, *Ballistic*, p. 79.
25 *Time*, 13 June 1960, p. 70. 9,000 K: *Journal of the Aeronautical Sciences*, February 1958, p. 88. Quote: Rose, "Physical," p. 1.
26 Rose, "Physical," p. 1.
27 Hallion, *Hypersonic*, pp. xxiii, xxvi.
28 *Time*, 13 June 1960 (includes quote). Arthur Kantrowitz: Author interview, 22 June 2001.

Folder 18649, NASA Historical Reference Collection, NASA History Division, Washington, D.C. 20546.
29   Heppenheimer, *Man-Made*, pp. 286-91. Quote: Author interview, Arthur Kantrowitz, 22 June 2001. Folder 18649, NASA Historical Reference Collection, NASA History Division, Washington, D.C. 20546.
30   Shapiro, *Compressible*, pp. 1007-11, 1027; Lukasiewicz, *Experimental*, ch. 13.
31   *Comptes Rendus*, Vol. 129 (1899), pp. 1128-30.
32   *Proceedings of the Royal Society*, Vol. A186 (1946), pp. 293-21.
33   *Review of Scientific Instruments*, November 1949, pp. 807-815 (quotes, pp. 813, 814).
34   *Journal of Applied Physics*, December 1952, pp. 1390-99 (quote, p. 1397).
35   Wegener, *Peenemunde*, p. 131; *Nature*, Vol. 171 (1953), pp. 395-96; NASA SP-4308, p. 134; quote: Hallion, *Hypersonic*, pp. xxvi-xxvii.
36   Rose, "Physical"; Kantrowitz, "Shock Tubes."
37   *Journal of the Aeronautical Sciences*, February 1958, pp. 86-97.
38   Rose, "Physical," pp. 11-12.
39   *Jet Propulsion*, April 1956, pp. 259-269.
40   *Journal of the Aeronautical Sciences*, February 1958: pp. 73-85, 86-97.
41   Brown et al., "Study," pp. 13-17.
42   Neufeld, *Ballistic*, pp. 78-79, 117.
43   Emme, *History*, pp. 7-110; Ordway and Sharpe, *Rocket Team*, pp. 366-73.
44   Ley, *Rockets*, pp. 237, 245.
45   "Re-Entry Studies," Vol. 1, pp. 21, 27, 29.
46   Grimwood and Strowd, *History*, pp. 10-13.
47   "Re-Entry Studies," Vol. 1, pp. 2, 24-25, 31, 37-45, 61.
48   *Journal of Spacecraft and Rockets*, January-February 1982, pp. 3-11.
49   Ibid. Extended quote, p. 5.
50   *Journal of the Aero/Space Sciences*, May 1960, pp. 377-85.
51   Kreith, *Heat Transfer*, pp. 538-45.
52   *Journal of Spacecraft and Rockets*, January-February 1982, p. 6. Quote: Author interview, Arthur Kantrowitz, 22 June 2001. Folder 18649, NASA Historical Reference Collection, NASA History Division, Washington, D.C. 20546.
53   NASA SP-440, p. 85; ARS Paper 724-58 (quote, p. 4).
54   ARS Paper 724-58.
55   NASA SP-4308, p. 133.
56   ARS Paper 724-58; *Journal of the Aero/Space Sciences*, July 1960, pp. 535-43 (quote, p. 535).
57   *Journal of Spacecraft and Rockets*, January-February 1982, pp. 8-9.
58   ARS Paper 838-59.
59   "Re-Entry Test Vehicle X-17." Over-the-top: NASA RP-1028, p. 443.
60   This flight is further discussed in ch. 6.

Nose Cones and Re-entry

61 NASA RP-1028, p. 488.
62 Miller, *X-Planes*, pp. 215, 217.
63 Ibid., p. 213; *Journal of Spacecraft and Rockets*, January-February 1982, p. 6.
64 "Re-Entry Studies," Vol. 1, pp. 69-71; Grimwood and Strowd, *History*, p. 156.
65 "Re-Entry Studies," Vol. 1, pp. 61, 64, 71, 73, 98.
66 Grimwood and Strowd, *History*, p. 156.
67 *Time*, 18 November 1957, pp. 19-20 (includes quote).
68 "Re-Entry Studies," Vol. 1, pp. 3, 5.
69 *Journal of the British Interplanetary Society*, May 1984, pp. 219-25; Hallion, *Hypersonic*, p. lxxii.
70 *Journal of the British Interplanetary Society*, May 1984, p. 219.
71 DTIC AD-320753, pp. 11-12.
72 DTIC AD-342828, pp. 16-19; AD-362539, pp. 11, 83.
73 *Journal of the British Interplanetary Society*, May 1984, p. 220. Trajectory data: DTIC AD-320753, Appendix A.
74 *AIAA Journal*, January 1963, pp. 41-45 (quote, p. 41).
75 *Journal of the British Interplanetary Society*, May 1984, p. 220.
76 *Journal of the British Interplanetary Society*, May 1984, p. 220; *Journal of Spacecraft and Rockets*, January-February 1982, p. 8.
77 *Time*, December 8, 1958, p. 15 (includes quote).
78 Hallion, *Hypersonic*, p. lxxii. Dimensions: DTIC AD-362539, p. 73. Weight: AD-832686, p. 20-23.
79 DTIC AD-362539, pp. 11, 39.
80 Hallion, *Hypersonic*, p. lxxii.
81 DTIC AD-362539, pp. 11-12. Peak heating occurred approximately at maximum dynamic pressure, at Mach 14 (see p. 12). Altitude, 40,000 feet, is from Figs. 2.6.
82 DTIC AD-362539, pp. 9, 12-13 (quote, p. 9).
83 Author interview, Arthur Kantrowitz, 22 June 2001. Folder 18649, NASA Historical Reference Collection, NASA History Division, Washington, D.C. 20546
84 Author interview, Steven Weinberg, 3 January 1990. Folder 18649, NASA Historical Reference Collection, NASA History Division, Washington, D.C. 20546. Quote: *Journal of Spacecraft and Rockets*, January-February 1982, p. 3.

3

THE X-15

Across almost half a century, the X-15 program stands out to this day not only for its achievements but for its audacity. At a time when the speed record stood right at Mach 2, the creators of the X-15 aimed for Mach 7—and nearly did it.* Moreover, the accomplishments of the X-15 contrast with the history of an X-planes program that saw the X-1A and X-2 fall out of the sky due to flight instabilities, and in which the X-3 fell short in speed because it was underpowered.[1]

The X-15 is all the more remarkable because its only significant source of aerodynamic data was Becker's 11-inch hypersonic wind tunnel. Based on that instrument alone, the Air Force and NACA set out to challenge the potential difficulties of hypersonic piloted flight. They succeeded, with this aircraft setting speed and altitude marks that were not surpassed until the advent of the space shuttle.

It is true that these agencies worked at a time of rapid advance, when performance was leaping forward at rates never approached either before or since. Yet there was more to this craft than a can-do spirit. Its designers faced specific technical issues and overcame them well before the first metal was cut.

The X-3 had failed because it proved infeasible to fit it with the powerful turbojet engines that it needed. The X-15 was conceived from the start as relying on rocket power, which gave it a very ample reserve.

Flight instability was already recognized as a serious concern. Using Becker's hypersonic tunnel, the aerodynamicist Charles McLellan showed that the effectiveness of tail surfaces could be greatly increased by designing them with wedge-shaped profiles.[2]

The X-15 was built particularly to study problems of heating in high-speed flight, and there was the question of whether it might overheat when re-entering the atmosphere following a climb to altitude. Calculations showed that the heating would remain within acceptable bounds if the airplane re-entered with its nose high. This would present its broad underbelly to the oncoming airflow. Here was a new application of the Allen-Eggers blunt-body principle, for an airplane with its nose up effectively became blunt.

---

*Official flight records are certified by the Federation Aeronautique Internationale. The cited accomplishments lacked this distinction, but they nevertheless represented genuine achievements.

The plane's designers also benefited from a stroke of serendipity. Like any airplane, the X-15 was to reduce its weight by using stressed-skin construction; its outer skin was to share structural loads with internal bracing. Knowing the stresses this craft would encounter, the designers produced straightforward calculations to give the requisite skin gauges. A separate set of calculations gave the skin thicknesses that were required for the craft to absorb its heat of re-entry without weakening. The two sets of skin gauges were nearly the same! This meant that the skin could do double duty, bearing stress while absorbing heat. It would not have to thicken excessively, thus adding weight, to cope with the heat.

Yet for all the ingenuity that went into this preliminary design, NACA was a very small tail on a very large dog in those days, and the dog was the Air Force. NACA alone lacked the clout to build anything, which is why one sees military insignia on photos of the X-planes of that era. Fortuitously, two new inventions—the twin-spool and the variable-stator turbojet—were bringing the Air Force face to face with a new era in flight speed. Ramjet engines also were in development, promising still higher speed. The X-15 thus stood to provide flight-test data of the highest importance—and the Air Force grabbed the concept and turned it into reality.

Origins of the X-15

Experimental aircraft flourished during the postwar years, but it was hard for them to keep pace with the best jet fighters. The X-1, for instance, was the first piloted aircraft to break the sound barrier. But only six months later, in April 1948, the test pilot George Welch did this in a fighter plane, the XP-86.[3] The layout of the XP-86 was more advanced, for it used a swept wing whereas the X-1 used a simple straight wing. Moreover, while the X-1 was a highly specialized research airplane, the XP-86 was a prototype of an operational fighter.

Much the same happened at Mach 2. The test pilot Scott Crossfield was the first to reach this mark, flying the experimental Douglas Skyrocket in November 1953.[4] Just then, Alexander Kartveli of Republic Aviation was well along in crafting the XF-105. The Air Force had ordered 37 of them in March 1953. It first flew in December 1955; in June 1956 an F-105 reached Mach 2.15. It too was an operational fighter, in contrast to the Skyrocket of two and a half years earlier.

Ramjet-powered craft were to do even better. Navaho was to fly near Mach 3. An even more far-reaching prospect was in view at that same Republic Aviation, where Kartveli was working on the XF-103. It was to fly at Mach 3.7 with its own ramjet, nearly 2,500 miles per hour (mph), with a sustained ceiling of 75,000 feet.[5]

Yet it was already clear that such aircraft were to go forward in their programs without benefit of research aircraft that could lay groundwork. The Bell X-2 was in development as a rocket plane designed to reach Mach 3, but although first thoughts of it dated to 1945, the program encountered serious delays. The airplane did not so much as fly past Mach 1 until 1956.[6]

Hence in 1951 and 1952, it already was too late to initiate a new program aimed at building an X-plane that could provide timely support for the Navaho and XF-103. The X-10 supported Navaho from 1954 to 1957, but it used turbojets rather than ramjets and flew at Mach 2. There was no quick and easy way to build aircraft capable of Mach 3, let alone Mach 4; the lagging X-2 was the only airplane that might do this, however belatedly. Yet it was already appropriate to look beyond the coming Mach 3 generation and to envision putative successors.

Maxwell Hunter, at Douglas Aircraft, argued that with fighter aircraft on their way to Mach 3, antiaircraft missiles would have to fly at Mach 5 to Mach 10.[7] In addition, Walter Dornberger, the wartime head of Germany's rocket program, now was at Bell Aircraft. He was directing studies of Bomi, Bomber Missile, a two-stage fully reusable rocket-powered bomber concept that was to reach 8,450 mph, or Mach 12.[8] At Convair, studies of intercontinental missiles included boost-glide concepts with much higher speeds.[9] William Dorrance, a company aerodynamicist, had not been free to disclose the classified Atlas concept to NACA but nevertheless declared that data at speeds up to Mach 20 were urgently needed.[10] In addition, the Rand Corporation had already published reports that envisioned spacecraft in orbit. The documents proposed that such satellites could serve for weather observation and for military reconnaissance.[11]

At Bell Aircraft, Robert Woods, a co-founder of the company, took a strong interest in Dornberger's ideas. Woods had designed the X-1, the X-1A that reached Mach 2.4, and the X-2. He also was a member of NACA's influential Committee on Aerodynamics. At a meeting of this committee in October 1951, he recommended a feasibility study of a "V-2 research airplane, the objective of which would be to obtain data at extreme altitudes and speeds and to explore the problems of re-entry into the atmosphere."[12] He reiterated this recommendation in a letter to the committee in January 1952. Later that month, he received a memo from Dornberger that outlined an "ionospheric research plane," capable of reaching altitudes of "more than 75 miles."[13]

NACA Headquarters sent copies of these documents to its field centers. This brought responses during May, as several investigators suggested means to enhance the performance of the X-2. The proposals included a rocket-powered carrier aircraft with which this research airplane was to attain "Mach numbers up to almost 10 and an altitude of about 1,000,000 feet,"[14] which the X-2 had certainly never been meant to attain. A slightly more practical concept called for flight to 300,000 feet.[15] These thoughts were out in the wild blue, but they showed that people at least were ready to think about hypersonic flight.

Accordingly, at a meeting in June 1952, the Committee on Aerodynamics adopted a resolution largely in a form written by another of its members, the Air Force science advisor Albert Lombard:

WHEREAS, The upper stratosphere is the important new flight region for military aircraft in the next decade and certain guided missiles are already under development to fly in the lower portions of this region, and

WHEREAS, Flight in the ionosphere and in satellite orbits in outer space has long-term attractiveness to military operations....

RESOLVED, That the NACA Committee on Aerodynamics recommends that (1) the NACA increase its program dealing with problems of unmanned and manned flight in the upper stratosphere at altitudes between 12 and 50 miles, and at Mach numbers between 4 and 10, and (2) the NACA devote a modest effort to problems associated with unmanned and manned flights at altitudes from 50 miles to infinity and at speeds from Mach number 10 to the velocity of escape from the Earth's gravity.

Three weeks later, in mid-July, the NACA Executive Committee adopted essentially the same resolution, thus giving it the force of policy.[16]

Floyd Thompson, associate director of NACA-Langley, responded by setting up a three-man study team. Their report came out a year later. It showed strong fascination with boost-glide flight, going so far as to propose a *commercial* aircraft based on a boost-glide Atlas concept that was to match the standard fares of current airliners. On the more immediate matter of a high-speed research airplane, this group took the concept of a boosted X-2 as a point of departure, suggesting that such a vehicle could reach Mach 3.7. Like the million-foot X-2 and the 300,000-foot X-2, this lay beyond its thermal limits. Still, this study pointed clearly toward an uprated X-2 as the next step.[17]

The Air Force weighed in with its views in October 1953. A report from the Aircraft Panel of its Scientific Advisory Board (SAB) discussed the need for a new research airplane of very high performance. The panelists stated that "the time was ripe" for such a venture and that its feasibility "should be looked into."[18] With this plus the report of the Langley group, the question of such a research plane went on the agenda of the next meeting of NACA's Interlaboratory Research Airplane Panel. It took place at NACA Headquarters in Washington in February 1954.

It lasted two days. Most discussions centered on current programs, but the issue of a new research plane indeed came up. The participants rejected the concept of an uprated X-2, declaring that it would be too small for use in high-speed studies. They concluded instead "that provision of an entirely new research airplane is desirable."[19]

This decision led quickly to a new round of feasibility studies at each of the four NACA centers: Langley, Ames, Lewis, and the High-Speed Flight Station. The study conducted at Langley was particularly detailed and furnished much of the basis for the eventual design of the X-15. Becker directed the work, taking responsibility for trajectories and aerodynamic heating. Maxime Faget addressed issues of propulsion. Three other specialists covered the topics of structures and materials, piloting, configuration, stability, and control.[20]

A performance analysis defined a loaded weight of 30,000 pounds. Heavier weights did not increase the peak speed by much, whereas smaller concepts showed a marked falloff in this speed. Trajectory studies then showed that this vehicle could reach a range of speeds, from Mach 5 when taking off from the ground to Mach 10 if launched atop a rocket-powered first stage. If dropped from a B-52 carrier, it would attain Mach 6.3.[21]

Concurrently with this work, prompted by a statement written by Langley's Robert Gilruth, the Air Force's Aircraft Panel recommended initiation of a research airplane that would reach Mach 5 to 7, along with altitudes of several hundred thousand feet. Becker's group selected a goal of Mach 7, noting that this would permit investigation of "extremely wide ranges of operating and heating conditions." By contrast, a Mach 10 vehicle "would require a much greater expenditure of time and effort" and yet "would add little in the fields of stability, control, piloting problems, and structural heating."[22]

A survey of temperature-resistant superalloys brought selection of Inconel X for the primary aircraft structure. This was a proprietary alloy from the firm of International Nickel, comprising 72.5 percent nickel, 15 percent chromium, 1 percent columbium, and iron as most of the balance. Its principal constituents all counted among the most critical materials used in aircraft construction, being employed in small quantities for turbine blades in jet engines. But Inconel X was unmatched in temperature resistance, holding most of its strength and stiffness at temperatures as high as 1200°F.[23]

Could a Mach 7 vehicle re-enter the atmosphere without exceeding this temperature limit? Becker's designers initially considered that during reentry, the airplane should point its nose in the direction of flight. This proved impossible; in Becker's words, "the dynamic pressures quickly exceeded by large margins the limit of 1,000 pounds per square foot set by structural considerations, and the heating loads became disastrous."

Becker tried to alleviate these problems by using lift during re-entry. According to his calculations, he obtained more lift by raising the nose—and the problem became far more manageable. He saw that the solution lay in having the plane enter the atmosphere with its nose high, presenting its flat undersurface to the air. It then would lose speed in the upper atmosphere, easing both the overheating and the aerodynamic pressure. The Allen-Eggers paper had been in print for nearly a year, and in Becker's words, "it became obvious to us that what we were seeing here was a new manifestation of H. J. Allen's 'blunt-body' principle. As we increased the angle of attack, our configuration in effect became more 'blunt.'" Allen and Eggers had

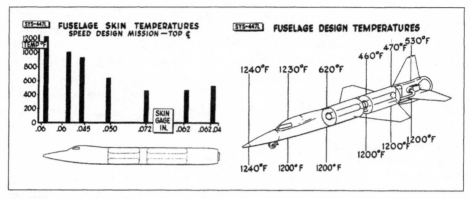

X-15 skin gauges and design temperatures. Generally, the heaviest gauges were required to meet the most severe temperatures. (NASA)

developed their principle for missile nose cones, but it now proved equally useful when applied to a hypersonic airplane.[24]

The use of this principle now placed a structural design concept within reach. To address this topic, Norris Dow, the structural analyst, considered the use of a heat-sink structure. This was to use Inconel X skin of heavy gauge to absorb the heat and spread it through this metal so as to lower its temperature. In addition, the skin was to play a structural role. Like other all-metal aircraft, the nascent X-15 was to use stressed-skin construction. This gave the skin an optimized thickness so that it could carry part of the aerodynamic loads, thus reducing the structural weight.

Dow carried through a design exercise in which he initially ignored the issue of heating, laying out a stressed-skin concept built of Inconel X with skin gauges determined only by requirements of mechanical strength and stiffness. A second analysis then took note of the heating, calculating new gauges that would allow the skin to serve as a heat sink. It was clear that if those gauges were large, adding weight to the airplane, then it might be necessary to back off from the Mach 7 goal so as to reduce the input heat load, thereby reducing the required thicknesses.

When Dow made the calculations, he received a welcome surprise. He found that the weight and thickness of a heat-absorbing structure were nearly the same as those of a simple aerodynamic structure! This meant that a hypersonic airplane, designed largely from consideration of aerodynamic loads, could provide heat-sink thermal protection as a bonus. It could do this with little or no additional weight.[25]

This, more than anything, was the insight that made the X-15 possible. Designers such as Dow knew all too well that ordinary aircraft aluminum lost strength beyond Mach 2, due to aerodynamic heating. Yet if hypersonic flight was to mean anything, it meant choosing a goal such as Mach 7 and then reaching this goal through the clever use of available heat-resistant materials. In Becker's study, the Allen-Eggers blunt-body principle reduced the re-entry heating to a level that Inconel X could accommodate.

The putative airplane still faced difficult issues of stability and control. Early in 1954 these topics were in the forefront, for the test pilot Chuck Yeager had nearly crashed when his X-1A fell out of the sky due to a loss of control at Mach 2.44. This problem of high-speed instability reflected the natural instability, at all Mach numbers, of a simple wing-body vehicle that lacked tail surfaces. Such surfaces worked well at moderate speeds, like the feathers of an arrow, but lost effectiveness with increasing Mach. Yeager's near-disaster had occurred because he had pushed just beyond a speed limit set by such considerations of stability. These considerations would be far more severe at Mach 7.[26]

Another Langley aerodynamicist, Charles McLellan, took up this issue by closely examining the airflow around a tail surface at high Mach. He drew on recent experimental results from the Langley 11-inch hypersonic tunnel, involving an airfoil with a cross section in the shape of a thin diamond. Analysis had indicated that most of the control effectiveness of this airfoil was generated by its forward wedge-shaped portion. The aft portion contributed little to its overall effectiveness because the pressures on that part of the surface were lower. Experimental tests had confirmed this.

McLellan now proposed to respond to the problem of hypersonic stability by using tail surfaces having airfoils that would be wedge-shaped along their entire length. In effect, such a surface would consist of a forward portion extending all the way to the rear. Subsequent tests in the 11-inch tunnel confirmed that this solution worked. Using standard thin airfoils, the new research plane would have needed tail surfaces nearly as large as the wings. The wedge shape, which saw use in the operational X-15, reduced their sizes to those of conventional tails.[27]

The group's report, dated April 1954, contemplated flight to altitudes as great as 350,000 feet, or 66 miles. (The X-15 went to 354,200 feet in 1963.)[28] This was well above the sensible atmosphere, well into an altitude range where flight would be ballistic. This meant that at that early date, Becker's study was proposing to accomplish piloted flight into space.

THE AIR FORCE AND HIGH-SPEED FLIGHT

This report did not constitute a design. However, it gave good reason to believe that such a design indeed was feasible. It also gave a foundation for briefings at which supporters of hypersonic flight research could seek to parlay the pertinent calculations into a full-blown program that would actually build and fly the new research planes. To do this, NACA needed support from the Air Force, which had a budget 300 times greater than NACA's. For FY 1955 the Air Force budget was $16.6 billion; NACA's was $56 million.[29]

Fortunately, at that very moment the Air Force was face to face with two major technical innovations that were upsetting all conventional notions of military flight. They faced the immediate prospect that aircraft would soon be flying at temperatures at which aluminum would no longer suffice. The inventions that brought this issue to the forefront were the dual-spool turbojet and the variable-stator turbojet—which call for a digression into technical aspects of jet propulsion.

Jet engines have functioned at speeds as high as Mach 3.3. However, such an engine must accelerate to reach that speed and must remain operable to provide control when decelerating from that speed. Engine designers face the problem of "compressor stall," which arises because compressors have numerous stages or rows of blades and the forward stages take in more air than the rear stages can accommodate. Gerhard Neumann of General Electric, who solved this problem, states that when a compressor stalls, the airflow pushes forward "with a big bang and the pilot loses all his thrust. It's violent; we often had blades break off during a stall."

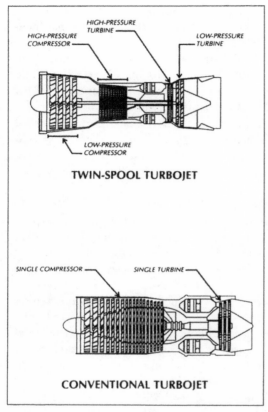

Twin-spool turbojet, amounting to two engines in one. It avoided compressor stall because its low-pressure compressor rotated somewhat slowly during acceleration, and hence pulled in less air. (Art by Don Dixon and Chris Butler)

An interim solution came from Pratt & Whitney, as the "twin-spool" engine. It separated the front and rear compressor stages into two groups, each of which could be made to spin at a proper speed. To do this, each group had its own turbine to provide power. A twin-spool turbojet thus amounted to putting one such engine inside another one. It worked; it prevented compressor stall, and it also gave high internal pressure that promoted good fuel economy. It thus was selected for long-range aircraft, including jet bombers and early commercial jet airliners. It also powered a number of fighters.

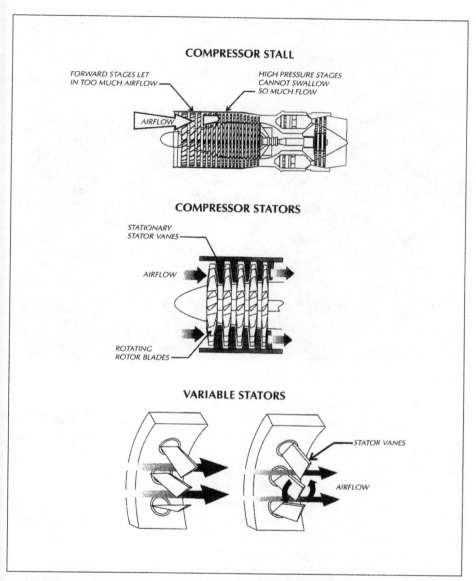

Gerhard Neumann's engine for supersonic flight. Top, high performance appeared unattainable because when accelerating, the forward compressor stages pulled in more airflow than the rear ones could swallow. Center, Neumann approached this problem by working with the stators, stationary vanes fitted between successive rows of rotating compressor blades. Bottom, he arranged for stators on the front stages to turn, varying their angles to the flow. When set crosswise to the flow, as on the right, these variable stators reduced the amount of airflow that their compressor stages would pull in. This solved the problem of compressor stall, permitting flight at Mach 2 and higher. (Art by Don Dixon and Chris Butler)

The F-104, which used variable stators. (U. S. Air Force)

But the twin-spool was relatively heavy, and there was much interest in avoiding compressor stall with a lighter solution. It came from Neumann in the form of the "variable-stator" engine. Within an engine's compressor, one finds rows of whirling blades. One also finds "stators," stationary vanes that receive airflow from those blades and direct the air onto the next set of blades. Neumann's insight was that the stators could themselves be adjusted, varied in orientation. At moderate speeds, when a compressor was prone to stall, the stators could be set crosswise to the flow, blocking it in part. At higher speeds, close to an engine's peak velocity, the stators could turn to present themselves edge-on to the flow. Very little of the airstream would be blocked, but the engine could still work as designed.[30]

The twin-spool approach had demanded nothing less than a complete redesign of the entire turbojet. The variable-stator approach was much neater because it merely called for modification of the forward stages of the compressor. It first flew as part of the Lockheed F-104, which was in development during 1953 and which then flew in March 1954. Early versions used engines that did not have variable stators, but the F-104A had them by 1958. In May of that year this aircraft reached 1,404 mph, setting a new world speed record, and set a similar altitude mark at 91,249 feet.[31]

The X-15

To place this in perspective, one must note the highly nonuniform manner in which the Air Force increased the speed of its best fighters after the war. The advent of jet propulsion itself brought a dramatic improvement. The author Tom Wolfe notes that "a British jet, the Gloster Meteor, jumped the official world speed record from 469 to 606 in a single day."[32] That was an increase of nearly thirty percent, but after that, things calmed down. The Korean War-era F-86 could break the sound barrier in a dive, but although it was the best fighter in service during that war, it definitely counted as subsonic. When the next-generation F-100A flew supersonic in level flight in May 1953, the event was worthy of note.[33]

By then, though, both the F-104 and F-105 were on order and in development. A twin-spool engine was already powering the F-100A, while the F-104 was to fly with variable stators. At a stroke, then, the Air Force found itself in another great leap upward, with speeds that were not to increase by a mere thirty percent but were to double.

There was more. There had been much to learn about aerodynamics in crafting earlier jets; the swept wing was an important example of the requisite innovations. But the new aircraft had continued to use aluminum structures. Still, the F-104 and F-105 were among the last aircraft that were to be designed using this metal alone. At higher speeds, it would be necessary to use other materials as well.

Other materials were already part of mainstream aviation, even in 1954. The Bell X-2 had probably been the first airplane to be built with heat-resistant metals, mounting wings of stainless steel on a fuselage of the nickel alloy K Monel. This gave it a capability of Mach 3.5. Navaho and the XF-103 were both to be built of steel and titanium, while the X-7, a ramjet testbed, was also of steel.[34] But all these craft were to fly near Mach 3, whereas the X-15 was to reach Mach 7. This meant that in an era of accelerating change, the X-15 was plausibly a full generation ahead of the most advanced designs that were under development.

The Air Force already had shown its commitment to support flight at high speed by building the Arnold Engineering Development Center (AEDC). Its background dated to the closing days of World War II, when leaders in what was then the Army Air Forces became aware that Germany had been well ahead of the United States in the fields of aerodynamics and jet propulsion. In March 1946, Brigadier General H. I. Hodes authorized planning an engineering center that would be the Air Force's own.

This facility was to use plenty of electrical power to run its wind tunnels, and a committee selected three possible locations. One was Grand Coulee near Spokane, Washington, but was ruled out as being too vulnerable to air attack. The second was Arizona's Colorado River, near Hoover Dam. The third was the hills north of Alabama, where the Tennessee Valley Authority had its own hydro dams. Senator Kenneth McKellar, the president pro tempore of the Senate and chairman of its

Armed Services Committee, won the new AEDC for his home state of Tennessee by offering to give the Air Force an existing military base, the 40,000-acre Camp Forrest. It was located near Tullahoma, far from cities and universities, but the Air Force was accustomed to operating in remote areas. It accepted this offer in April 1948, with the firm of ARO, Inc. providing maintenance and operation.[35]

There was no interest in reproducing the research facilities of NACA, for the AEDC was to conduct its own activities. Engine testing was to be a specialty, and the first facility at this center was an engine test installation that had been "liberated" from the German firm of BMW. But the Air Force soon was installing its own equipment, achieving its first supersonic flow within its Transonic Model Tunnel early in 1953. Then, during 1954, events showed that AEDC was ready to conduct engineering development on a scale well beyond anything that NACA could envision.[36]

That year saw the advent of the 16-Foot Propulsion Wind Tunnel, with a test section 16 feet square. NACA had larger tunnels, but this one approached Mach 3.5 and reached Mach 4.75 under special operating conditions. A Mach of 4.75 had conventionally been associated with the limited run times of blowdown tunnels, but this tunnel, known as 16S, was a continuous-flow facility. It was unparalleled for exercising full-scale engines for realistic durations over the entire supersonic range.[37]

In December 1956 it tested the complete propulsion package of the XF-103, which had a turbojet with an afterburner that functioned as a ramjet. This engine had a total length of 39 feet. But the test section within 16S had a length of 40 feet, which gave room to spare.[38] In addition, the similar Engine Test Facility accommodated the full-scale SRJ47 engine of Navaho, with a 51-inch diameter that made it the largest ramjet engine ever built.[39]

The AEDC also jumped into hypersonics with both feet. It already had an Engine Test Facility, a Gas Dynamics Facility (renamed the Von Karman Gas Dynamics Facility in 1959), and a Propulsion Wind Tunnel, the 16S. During 1955 it added a ramjet center to the Engine Test Facility, which many people regarded as a fourth major laboratory.[40] Hypersonic wind tunnels were also on the agenda. Two 50-inch installations were in store, to operate respectively at Mach 8 and Mach 10. Both were continuous-flow facilities that used a 92,500-horsepower compressor system. Tunnel B, the Mach 8 facility, became operational in October 1958. Tunnel C, the Mach 10 installation, prevented condensation by heating its air to 1,450°F using a combustion heater and a 12-megawatt resistance heater. It entered operation in May 1960.[41]

The AEDC also conducted basic research in hypersonics. It had not intended to do that initially; it had expected to leave such studies to NACA, with its name reflecting its mission of engineering development. But the fact that it was off in the wilds of Tullahoma did not prevent it from attracting outstanding scientists, some of whom went on to work in hypersonics.

The X-15

Facilities such as Tunnels B and C could indeed attain hypersonic speeds, but the temperatures of the flows were just above the condensation point of liquid air. There was much interest in achieving far greater temperatures, both to add realism at speeds below Mach 10 and to obtain Mach numbers well beyond 10. Beginning in 1953, the physicist Daniel Bloxsom used the exploding-wire technique, in which a powerful electric pulse vaporizes a thin wire, to produce initial temperatures as high as 5900 K.

This brought the advent of a new high-speed flow facility: the hotshot tunnel. It resembled the shock tube, for the hot gas was to burst a diaphragm and then reach high speeds by expanding through a nozzle. But its run times were considerably longer, reaching one-twentieth of a second compared to less than a millisecond for the shock tube. The first such instrument, Hotshot 1, had a 16-inch test section and entered service early in 1956. In March 1957, the 50-inch Hotshot 2 topped "escape velocity."[42]

Against this background, the X-15 drew great interest. It was to serve as a full-scale airplane at Mach 7, when the best realistic tests that AEDC could offer was full-scale engine test at Mach 4.75. Indeed, a speed of Mach 7 was close to the Mach 8 of Tunnel B. The X-15 also could anchor a program of hypersonic studies that soon would have hotshot tunnels and would deal with speeds up to orbital velocity and beyond. And while previous X-planes were seeing their records broken by jet fighters, it would be some time before any other plane flew at such speeds.

The thermal environment of the latest aircraft was driving designers to the use of titanium and steel. The X-15 was to use Inconel X, which had still better properties. This nickel alloy was to be heat-treated and welded, thereby developing valuable shop-floor experience in its use. In addition, materials problems would be pervasive in building a working X-15. The success of a flight could depend on the proper choice of lubricating oil.

The performance of the X-15 meant that it needed more than good aerodynamics. The X-2 was already slated to execute brief leaps out of the atmosphere. Thus, in September 1956 test pilot Iven Kincheloe took it to 126,200 feet, an altitude at which his ailerons and tail surfaces no longer functioned.[43] In the likely event that future interceptors were to make similar bold leaps, they would need reaction controls—which represented the first really new development in the field of flight control since the Wright Brothers.[44] But the X-15 was to use such controls and would show people how to do it.

The X-15 would also need new flight instruments, including an angle-of-attack indicator. Pilots had been flying with turn-and-bank indicators for some time, with these gyroscopic instruments enabling them to determine their attitude while flying blind. The X-15 was to fly where the skies were always clear, but still it needed to determine its angle with respect to the oncoming airflow so that the pilot could set up a proper nose-high attitude. This instrument would face the full heat load of reentry and had to work reliably.

It thus was not too much to call the X-15 a flying version of AEDC, and high-level Air Force representatives were watching developments closely. In May 1954 Hugh Dryden, Director of NACA, wrote a letter to Lieutenant General Donald Putt, who now was the Air Force's Deputy Chief of Staff, Development. Dryden cited recent work, including that of Becker's group, noting that these studies "will lead to specific preliminary proposals for a new research airplane." Putt responded with his own letter, stating that "the Scientific Advisory Board has done some thinking in this area and has formally recommended that the Air Force initiate action on such a program."[45]

The director of Wright Air Development Center (WADC), Colonel V. R. Haugen, found "unanimous" agreement among WADC reviews that the Langley concept was technically feasible. These specialists endorsed Langley's engineering solutions in such areas as choice of material, structure, thermal protection, and stability and control. Haugen sent his report to the Air Research and Development Command (ARDC), the parent of WADC, in mid-August. A month later Major General F. B. Wood, an ARDC deputy commander, sent a memo to Air Force Headquarters, endorsing the NACA position and noting its support at WADC. He specifically recommended that the Air Force "initiate a project to design, construct, and operate a new research aircraft similar to that suggested by NACA without delay."[46]

Further support came from the Aircraft Panel of the Scientific Advisory Board. In October it responded to a request from the Air Force Chief of Staff, General Nathan Twining, with its views:

> "[A] research airplane which we now feel is ready for a program is one involving manned aircraft to reach something of the order of Mach 5 and altitudes of the order of 200,000 to 500,000 feet. This is very analogous to the research aircraft program which was initiated 10 years ago as a joint venture of the Air Force, the Navy, and NACA. It is our belief that a similar co-operative arrangement would be desirable and appropriate now."[47]

The meetings contemplated in the Dryden-Putt correspondence were also under way. There had been one in July, at which a Navy representative had presented results of a Douglas Aircraft study of a follow-on to the Douglas Skyrocket. It was to reach Mach 8 and 700,000 feet.[48]

Then in October, at a meeting of NACA's Committee on Aerodynamics, Lockheed's Clarence "Kelly" Johnson challenged the entire postwar X-planes program. His XF-104 was already in flight, and he pulled no punches in his written statement:

"Our present research airplanes have developed startling performance only by the use of rocket engines and flying essentially in a vacuum. Testing airplanes designed for transonic flight speeds at Mach numbers between 2 and 3 has proven, mainly, the bravery of the test pilots and the fact that where there is no drag, the rocket engine can propel even mediocre aerodynamic forms at high Mach numbers.

I am not aware of any aerodynamic or power plant improvements to air-breathing engines that have resulted from our very expensive research airplane program. Our modern tactical airplanes have been designed almost entirely on NACA and other wind-tunnel data, plus certain rocket model tests...."[49]

Drawing on Lockheed experience with the X-7, an unpiloted high-speed missile, he called instead for a similar unmanned test aircraft as the way to achieve Mach 7. However, he was a minority of one. Everyone else voted to support the committee's resolution:

BE IT HEREBY RESOLVED, That the Committee on Aerodynamics endorses the proposal of the immediate initiation of a project to design and construct a research airplane capable of achieving speeds of the order of Mach number 7 and altitudes of several hundred thousand feet....[50]

The Air Force was also on board, and the next step called for negotiation of a Memorandum of Understanding, whereby the participants—which included the Navy—were to define their respective roles. Late in October representatives from the two military services visited Hugh Dryden at NACA Headquarters, bringing a draft of this document for discussion. It stated that NACA was to provide technical direction, the Air Force would administer design and construction, and the Air Force and Navy were to provide the funds. It concluded with the words, "Accomplishment of this project is a matter of national urgency."[51]

The draft became the final MOU, with little change, and the first to sign it was Trevor Gardner. He was a special assistant to the Air Force Secretary and had midwifed the advent of Atlas a year earlier. James Smith, Assistant Secretary of the Navy for Air, signed on behalf of that service, while Dryden signed as well. These signatures all were in place two days before Christmas of 1954. With this, the groundwork was in place for the Air Force's Air Materiel Command to issue a Request for Proposal and for interested aircraft companies to begin preparing their bids.[52]

As recently as February, all that anyone knew was that this new research aircraft, if it materialized, would be something other than an uprated X-2. The project had taken form with considerable dispatch, and the key was the feasibility study of Becker's group. An independent review at WADC confirmed its conclusions, whereupon Air Force leaders, both in uniform and in mufti, embraced the concept. Approval at the Pentagon then came swiftly.

In turn, this decisiveness demonstrated a willingness to take risks. It is hard today to accept that the Pentagon could endorse this program on the basis of just that one study. Moreover, the only hypersonic wind tunnel that was ready to provide supporting research was Becker's 11-inch instrument; the AEDC hypersonic tunnels were still several years away from completion. But the Air Force was in no mood to hold back or to demand further studies and analyses.

This service was pursuing a plethora of initiatives in jet bombers, advanced fighters, and long-range missiles. Inevitably, some would falter or find themselves superseded, which would lead to charges of waste. However, Pentagon officials knew that the most costly weapons were the ones that America might need and not have in time of war. Cost-benefit analysis had not yet raised its head; Robert McNamara was still in Detroit as a Ford Motor executive, and Washington was not yet a city where the White House would deliberate for well over a decade before ordering the B-1 bomber into limited production. Amid the can-do spirit of the 1950s, the X-15 won quick approval.

X-15: The Technology

Four companies competed for the main contract, covering design and construction of the X-15: Republic, Bell, Douglas, and North American. Each of them brought a substantial amount of hands-on experience with advanced aircraft. Republic, for example, had Alexander Kartveli as its chief designer. He was a highly imaginative and talented man whose XF-105 was nearly ready for first flight and whose XF-103 was in development. Republic had also built a rocket plane, the XF-91. This was a jet fighter that incorporated the rocket engine of the X-1 for an extra boost in combat. It did not go into production, but it flew in flight tests.

Still, Republic placed fourth in the competition. Its concept rated "unsatisfactory" as a craft for hypersonic research, for it had a thin outer fuselage skin that appeared likely to buckle when hot. The overall proposal rated no better than average in a number of important areas, while achieving low scores in Propulsion System and Tanks, Engine Installation, Pilot's Instruments, Auxiliary Power, and Landing Gear. In addition, the company itself was judged as no more than "marginal" in the key areas of Technical Qualifications, Management, and Resources. The latter included availability of in-house facilities and of an engineering staff not committed to other projects.[53]

Bell Aircraft, another contender, was the mother of research airplanes, having built the X-1 series as well as the X-2. This firm therefore had direct experience both with advanced heat-resistant metals and with the practical issues of powering piloted aircraft using liquid-fuel rocket engines. It even had an in-house group that was building such engines. Bell also was the home of the designers Robert Woods and Walter Dornberger, with the latter having presided over the V-2.

Dornberger's Bomi concept already was introducing the highly useful concept of hot structures. These used temperature-resistant alloys such as stainless steel. Wings might be covered with numerous small and very hot metal panels, resembling shingles, that would radiate heat away from the aircraft. Overheating would be particularly severe along the leading edges of wings; these could be water-cooled. Insulation could protect an internal structure that would withstand the stresses and forces of flight; active cooling could protect a pilot's cockpit and instrument compartment. Becker described these approaches as "the first hypersonic aircraft hot structures concepts to be developed in realistic meaningful detail."[54]

Even so, Bell ranked third. Historian Dennis Jenkins writes that within the proposal, "almost every innovation they proposed was hedged in such a manner as to make the reader doubt that it would work. The proposal itself seemed rather poorly organized and was internally inconsistent (i.e., weights and other figures frequently differed between sections)."[55] Yet the difficulties ran deeper and centered on the specifics of its proposed hot structure.

Bell adopted the insulated-structure approach, with the primary structure being of aluminum, the most familiar of aircraft materials and the best understood. Corrugated panels of Inconel X, mounted atop the aluminum, were to provide insulation. Freely-suspended panels of this alloy, contracting and expanding with ease, were to serve as the outer skin.

Yet this concept was quite unsuitable for the X-15, both on its technical merits and as a tool for research. A major goal of the program was to study aircraft structures at elevated temperatures, and this would not be possible with a primary structure of cool aluminum. There were also more specific deficiencies, as when Bell's thermal analysis assumed that the expanding panels of the outer shell would prevent leakage of hot air from the boundary layer. However, the evaluation made the flat statement, "leakage is highly probable." Aluminum might not withstand the resulting heating, with the loss of even one such panel leading perhaps to destructive heating. Indeed, the Bell insulated structure appeared so sensitive that it could be trusted to successfully complete only three of 13 reference flights.[56]

Another contender, Douglas Aircraft, had shared honors with Bell in building previous experimental aircraft. Its background included the X-3 and the Skyrocket, which meant that Douglas also had people who knew how to integrate a liquid rocket engine with an airplane. This company's concept came in second.

Its design avoided reliance on insulated structures, calling instead for use of a heat sink. The material was to be a lightweight magnesium alloy that had excellent

The North American X-15. (NASA)

heat capacity. Indeed, its properties were so favorable that it would reach temperatures of only 600°F, while an Inconel X heat-sink airplane would go to 1,200°F.

Again, though, this concept missed the point. Managers *wanted* a vehicle that could cope successfully with temperatures of 1,200°F, to lay groundwork for operational fighters that could fly well beyond Mach 3. In addition, the concept had virtually no margin for temperature overshoots. Its design limit of 600°F was right on the edge of a regime of which its alloy lost strength rapidly. At 680°F, its strength could fall off by 90 percent. With magnesium being flammable, there was danger of fire within the primary structure itself, with the evaluation noting that "only a small area raised to the ignition temperature would be sufficient to destroy the aircraft."[57]

Then there was North American, the home of Navaho. That missile had not flown, but its detailed design was largely complete and specified titanium in hot areas. This meant that that company knew something about using advanced metals. The firm also had a particularly strong rocket-engine group, which split off during 1955 to form a new corporate division called Rocketdyne. Indeed, engines built by that association had already been selected for Atlas.[58]

North American became the winner. It paralleled the thinking at Douglas by independently proposing its own heat-sink structure, with the material being Inconel X. This concept showed close similarities to that of Becker's feasibility study a year earlier. Still, this was not to say that the deck was stacked in favor of Becker's approach. He and his colleagues had pursued conceptual design in a highly impromptu fashion. The preliminary-design groups within industry were far more experienced, and it had appeared entirely possible that these experts, applying their seasoned judgment, might come up with better ideas. This did not happen. Indeed, the Bell and Douglas concepts failed even to meet an acceptable definition of the new research airplane. By contrast, the winning concept from North American amounted to a particularly searching affirmation of the work of Becker's group.[59]

How had Bell and Douglas missed the boat? The government had set forth performance requirements, which these companies both had met. In the words of the North American proposal, "the specification performance can be obtained with very moderate structural temperatures." However, "the airplane has been designed to tolerate much more severe heating in order to provide a practical temperature band within which exploration can be conducted."

In Jenkins's words, "the Bell proposal…was terrible—you walked away not entirely sure that Bell had committed themselves to the project. The exact opposite was true of the North American proposal. From the opening page you knew that North American understood what was trying to be accomplished with the X-15 program and had attempted to design an airplane that would help accomplish the task—not just meet the performance specifications (which did not fully describe the intent of the program)."[60] That intent was to build an aircraft that could accomplish research at 1,200°F and not merely meet speed and altitude goals.

The overall process of proposal evaluation cast the competing concepts in sharp relief, heightening deficiencies and emphasizing sources of potential difficulty. These proposals also received numerical scores, while another basis for comparison involved estimated program costs:

North American	81.5 percent	$56.1 million
Douglas Aircraft	80.1	36.4
Bell Aircraft	75.5	36.3
Republic Aviation	72.2	47.0

North American's concept thus was far from perfect, while Republic's represented a serious effort. In addition, it was clear that the Air Force—which was to foot most of the bill—was willing to pay for what it would get. The X-15 program thus showed budgetary integrity, with the pertinent agencies avoiding the temptation to do it on the cheap.[61]

On 30 September 1955, letters went out to North American as well as to the unsuccessful bidders, advising them of the outcome of the competition. With this, engineers now faced the challenge of building and flying the X-15 as a practical exercise in hypersonic technology. Accordingly, it broke new ground in such areas as metallurgy and fabrication, onboard instruments, reaction controls, pilot training, the pilot's pressure suit, and flight simulation.[62]

Inconel X, a nickel alloy, showed good ductility when fully annealed and had some formability. When severely formed or shaped, though, it showed work-hardening, which made the metal brittle and prone to crack. Workers in the shop addressed this problem by forming some parts in stages, annealing the workpieces by heating them between each stage. Inconel X also was viewed as a weldable alloy, but some welds tended to crack, and this problem resisted solution for some time. The solution lay in making welds that were thicker than the parent material. After being ground flat, their surfaces were peened—bombarded with spherical shot—and rolled flush with the parent metal. After annealing, the welds often showed better crack resistance than the surrounding Inconel X.

A titanium alloy was specified for the internal structure of the wings. It proved difficult to weld, for it became brittle by reacting with oxygen and nitrogen in the air. It therefore was necessary to enclose welding fixtures within enclosures that could be purged with an inert gas such as helium and to use an oxygen-detecting device to determine the presence of air. With these precautions, it indeed proved possible to weld titanium while avoiding embrittlement.[63]

Greases and lubricants posed their own problems. Within the X-15, journal and antifriction bearings received some protection from heat and faced operating temperatures no higher than 600°F. This nevertheless was considerably hotter than engineers were accustomed to accommodating. At North American, candidate lubricants underwent evaluation by direct tests in heated bearings. Good greases protected bearing shafts for 20,000 test cycles and more. Poor greases gave rise to severe wearing of shafts after as few as 350 cycles.[64]

In contrast to conventional aircraft, the X-15 was to fly out of the sensible atmosphere and then re-enter, with its nose high. It also was prone to yaw while in near-vacuum. Hence, it needed a specialized instrument to determine angles of attack and of sideslip. This took form as the "Q-ball," built by the Nortronics Division of Northrop Aircraft. It fitted into the tip of the X-15's nose, giving it the appearance of a greatly enlarged tip of a ballpoint pen.

The ball itself was cooled with liquid nitrogen to withstand air temperatures as high as 3,500°F. Orifices set within the ball, along yaw and pitch planes, measuring differential pressures. A servomechanism rotated the ball to equalize these pressures by pointing the ball's forward tip directly into the onrushing airflow. With the direction of this flow thus established, the pilot could null out any sideslip. He also could raise the nose to a desired angle of attack. "The Q-ball is a go-no go item," the test pilot Joseph Walker told *Time* magazine in 1961. "Only if she checks okay do we go."[65]

To steer the aircraft while in flight, the X-15 mounted aerodynamic controls. These retained effectiveness at altitudes well below 100,000 feet. However, they lost

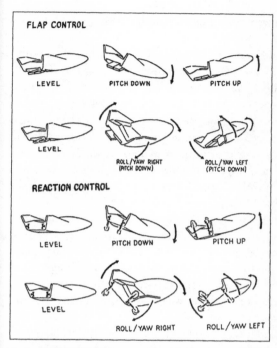

Attitude control of a hypersonic airplane using aerodynamic controls and reaction controls. (U.S. Air Force)

effectiveness between 90,000 and 100,000 feet. The X-15 therefore incorporated reaction controls, which were small thrusters fueled with hydrogen peroxide. Nose-mounted units controlled pitch and yaw. Other units, set near the wingtips, gave control of roll.

No other research airplane had ever flown with such thrusters, although the X-1B conducted early preliminary experiments and the X-2 came close to needing them in 1956. During a flight in September of that year, the test pilot Iven Kincheloe took it to 126,200 feet. At that altitude, its aerodynamic controls were useless. Kincheloe flew a ballistic arc, experiencing near-weightlessness for close to a minute. His airplane banked to the left, but he did not try to counter this movement, for he knew that his X-2 could easily go into a deadly tumble.[66]

In developing reaction controls, an important topic for study involved determining the airplane handling qualities that pilots preferred. Initial investigations used an analog computer as a flight simulator. The "airplane" was disturbed slightly; a man used a joystick to null out the disturbance, achieving zero roll, pitch, and yaw. These experiments showed that pilots wanted more control authority for roll than for pitch or yaw. For the latter, angular accelerations of 2.5 degrees per second squared were acceptable. For roll, the preferred control effectiveness was two to four times greater.

Flight test came next. The X-2 would have served splendidly for this purpose, but only two had been built, with both being lost in accidents. At NACA's High-Speed Flight Station, investigators fell back on the X-1B, which was less capable but still useful. In preparation for its flights with reaction controls, the engineers built a simulator called the Iron Cross, which matched the dimensions and inertial characteristics of this research plane. A pilot, sitting well forward along the central arm, used a side-mounted control stick to actuate thrusters that used compressed nitrogen. This simulator was mounted on a universal joint, which allowed it to move freely in yaw, pitch, and roll.

Reaction controls went into the X-1B late in 1957. The test pilot Neil Armstrong, who walked on the Moon 12 years later, made three flights in this research plane before it was grounded in mid-1958 due to cracks in its fuel tank. Its peak altitude during these three flights was 55,000 feet, where its aerodynamic controls readily provided backup. The reaction controls then went into an F-104, which reached 80,000 feet and went on to see much use in training X-15 pilots. When the X-15 was in flight, these pilots had to transition from aerodynamic controls to reaction controls and back again. The complete system therefore provided overlap. It began blending in the reaction controls at approximately 85,000 feet, with most pilots switching to reaction controls exclusively by 100,000 feet.[67]

Since the war, with aircraft increasing in both speed and size, it had become increasingly impractical for a pilot to exert the physical strength to operate a plane's ailerons and elevators merely by moving the control stick in the cockpit. Hydraulically-boosted controls thus were in the forefront, resembling power steering in a car. The X-15 used such hydraulics, which greatly eased the workload on a test pilot's muscles. These hydraulic systems also opened the way for stability augmentation systems of increasing sophistication.

Stability augmentation represented a new refinement of the autopilot. Conventional autopilots used gyroscopes to detect deviations from a plane's straight and level course. These instruments then moved an airplane's controls so as to null these deviations to zero. For high-performance jet fighters, the next step was stability augmentation. Such aircraft often were unstable in flight, tending to yaw or roll; indeed, designers sometimes enhanced this instability to make them more maneuverable. Still, it was quite wearying for a pilot to have to cope with this. A stability augmentation system made life in the cockpit much easier.

Such a system used rate gyros, which detected rates of movement in pitch, roll, and yaw at so many degrees per second. The instrument then responded to these rates, moving the controls somewhat like before to achieve a null. Each axis of this control had "gain," defining the proportion or ratio between a sensed rate of angular motion and an appropriate deflection of ailerons or other controls. Fixed-gain systems worked well; there also were variable-gain arrangements, with the pilot setting the value of gain within the cockpit. This addressed the fact that the airplane might need more gain in thin air at high altitude, to deflect these surfaces more strongly.[68]

The X-15 program built three of these aircraft. The first two used a stability augmentation system that incorporated variable gain, although in practice these aircraft flew well with constant values of gain, set in flight.[69] The third replaced it with a more advanced arrangement that incorporated something new: adaptive gain. This was a variable gain, which changed automatically in response to flight conditions. Within the Air Force, the Flight Control Laboratory at WADC had laid groundwork with a program dating to 1955. Adaptive-gain controls flew aboard F-94 and F-101 test aircraft. The X-15 system, the Minneapolis Honeywell MH-96, made its first flight in December 1961.[70]

How did it work? When a pilot moved the control stick, as when changing the pitch, the existing value of gain in the pitch channel caused the aircraft to respond at a certain rate, measured by a rate gyro. The system held a stored value of the optimum pitch rate, which reflected preferred handling qualities. The adaptive-gain control compared the measured and desired rates and used the difference to determine a new value for the gain. Responding rapidly, this system enabled the airplane to maintain nearly constant control characteristics over the entire flight envelope.[71]

The MH-96 made it possible to introduce the X-15's blended aerodynamic and reaction controls on the same control stick. This blending occurred automatically in response to the changing gains. When the gains in all three channels—roll, pitch, and yaw—reached 80 percent of maximum, thereby indicating an imminent loss of effectiveness in the aerodynamic controls, the system switched to reaction controls. During re-entry, with the airplane entering the sensible atmosphere, the system returned to aerodynamic control when all the gains dropped to 60 percent.[72]

The X-15 flight-control system thus stood three steps removed from the conventional stick-and-cable installations of World War II. It used hydraulically-boosted controls; it incorporated automatic stability augmentation; and with the MH-96, it introduced adaptive gain. Fly-by-wire systems lay ahead and represented the next steps, with such systems being built both in analog and digital versions.

Analog fly-by-wire systems exist within the F-16A and other aircraft. A digital system, as in the space shuttle, uses a computer that receives data both from the pilot and from the outside world. The pilot provides input by moving a stick or sidearm controller. These movements do not directly actuate the ailerons or rudder, as in days of old. Instead, they generate signals that tell a computer the nature of the desired maneuver. The computer then calculates a gain by applying control laws, which take account of the plane's speed and altitude, as measured by onboard instruments. The computer then sends commands down a wire to hydraulic actuators co-mounted with the controls to move or deflect these surfaces so as to comply with the pilot's wishes.[73]

The MH-96 fell short of such arrangements in two respects. It was analog, not digital, and it was a control system, not a computer. Like other systems executing automatic control, the MH-96 could measure an observed quantity such as pitch rate, compare it to a desired value, and drive the difference to zero. But the MH-96 was wholly incapable of implementing a control law, programmed as an algebraic expression that required values of airspeed and altitude. Hence, while the X-15 with

MH-96 stood three steps removed from the fighters of the recent war, it was two steps removed from the digital fly-by-wire control of the shuttle.

The X-15 also used flight simulators. These served both for pilot training and for development of onboard systems, including the reaction controls and the MH-96. The most important flight simulator was built by North American. It replicated the X-15 cockpit and included actual hydraulic and control-system hardware. Three analog computers implemented equations of motion that governed translation and rotation of the X-15 about all three axes, transforming pilot inputs into instrument displays.[74]

Flight simulators dated to the war. The famous Link Trainer introduced over half a million neophytes to their cockpits. The firm of Link Aviation added analog computers in 1949, within a trainer that simulated flight in a jet fighter.[75] In 1955, when the X-15 program began, it was not at all customary to use flight simulators to support aircraft design and development. But program managers turned to such simulators because they offered effective means to study new issues in cockpit displays, control systems, and aircraft handling qualities.

Flight simulation showed its value quite early. An initial X-15 design proved excessively unstable and difficult to control. The cure lay in stability augmentation. A 1956 paper stated that this had "heretofore been considered somewhat of a luxury for high-speed aircraft," but now "has been demonstrated as almost a necessity," in all three axes, to ensure "consistent and successful entries" into the atmosphere.[76]

The North American simulator, which was transferred to the NACA Flight Research Center, became critical in training X-15 pilots as they prepared to execute specific planned flights. A particular mission might take little more than 10 minutes, from ignition of the main engine to touchdown on the lakebed, but a test pilot could easily spend 10 hours making practice runs in this facility. Training began with repeated trials of the normal flight profile, with the pilot in the simulator cockpit and a ground controller close at hand. The pilot was welcome to recommend changes, which often went into the flight plan. Next came rehearsals of off-design missions: too much thrust from the main engine, too high a pitch angle when leaving the stratosphere.

Much time was spent practicing for emergencies. The X-15 had an inertial reference unit that used analog circuitry to display attitude, altitude, velocity, and rate of climb. Pilots dealt with simulated failures in this unit, attempting to complete the normal mission or, at least, execute a safe return. Similar exercises addressed failures in the stability augmentation system. When the flight plan raised issues of possible flight instability, tests in the simulator used highly pessimistic assumptions concerning stability of the vehicle. Other simulated missions introduced in-flight failures of the radio or Q-ball. Premature engine shutdowns imposed a requirement for safe landing on an alternate lakebed, which was available for emergency use.[77]

The simulations indeed were realistic in their cockpit displays, but they left out an essential feature: the g-loads, produced both by rocket thrust and by deceleration during re-entry. In addition, a failure of the stability augmentation system, during re-entry, could allow the airplane to oscillate in pitch or yaw. This would change its drag characteristics, imposing a substantial cyclical force.

To address such issues, investigators installed a flight simulator within the gondola of a centrifuge at the Naval Air Development Center in Johnsville, Pennsylvania. The gondola could rotate on two axes while the centrifuge as a whole was turning. It not only produced g-forces, but its g-forces increased during the simulated rocket burn. The centrifuge imposed such forces anew during reentry, while adding a cyclical component to give the effect of a yaw or pitch oscillation.[78]

Not all test pilots rode the centrifuge. William "Pete" Knight, who stood among the best, was one who did not. His training, coupled with his personal coolness and skill, enabled him to cope even with an extreme emergency. In 1967, during a planned flight to 250,000 feet, an X-15 experienced a complete electrical failure while climbing through 107,000 feet at Mach 4. This failure brought the shutdown of both auxiliary power units and hence of both hydraulic systems. Knight, the pilot, succeeded in restarting one of these units, which restored hydraulic power. He still had zero electrical power, but with his hydraulics, he now had both his aerodynamic and reaction controls. He rode his plane to a peak of 173,000 feet, re-entered the atmosphere, made a 180-degree turn, and glided to a safe landing on Mud Lake near Tonopah, Nevada.[79]

During such flights, as well as during some exercises in the centrifuge, pilots wore a pressure suit. Earlier models had already been good enough to allow the test pilot Marion Carl to reach 83,235 feet in the Douglas Skyrocket in 1953. Still, some of those versions left much to be desired. *Time* magazine, in 1952, discussed an Air Force model that allowed a pilot to breathe, but "with difficulty. His hands, not fully pressurized, swell up with blue venous blood. His throat is another trouble spot; the medicos have not yet learned how to pressurize a throat without strangling its owner."[80]

The David G. Clark Company, a leading supplier of pressure suits for Air Force flight crews, developed a greatly improved model for the X-15. Such suits tended to become rigid and hard to bend when inflated. This is also true of a child's long balloon, with an internal pressure that only slightly exceeds that of the atmosphere. The X-15 suit was to hold five pounds per square inch of pressure, or 720 pounds per square foot. The X-15 cockpit had its own counterbalancing pressure, but it could (and did) depressurize at high altitude. In such an event, the suit was to protect the test pilot rather than leave him immobile.

The solution used an innovative fabric that contracted in circumference while it stretched in length. With proper attention to the balance between these two effects, the suit maintained a constant volume when pressurized, enhancing a pilot's freedom of movement. Gloves and boots were detachable and zipped to this fabric. The helmet was joined to the suit with a freely-swiveling ring that gave full mobility to the head. Oxygen flowed into the helmet; exhalant passed through valves in a neck seal and pressurized the suit. Becker later described it as "the first practical full-pressure suit for pilot protection in space."[81]

Thus accoutered, protected for flight in near-vacuum, X-15 test pilots rode their rockets as they approached the edge of space and challenged the hypersonic frontier. They returned with results galore for project scientists—and for the nation.

X-15: SOME RESULTS

During the early 1960s, when the nation was agog over the Mercury astronauts, the X-15 pointed to a future in which piloted spaceplanes might fly routinely to orbit. The men of Mercury went water-skiing with Jackie Kennedy, but within their orbiting capsules, they did relatively little. Their flights were under automatic control, which left them as passengers along for the ride. Even a monkey could do it. Indeed, a chimpanzee named Ham rode a Redstone rocket on a suborbital flight in January 1961, three months before Alan Shepard repeated it before the gaze of an astonished world. Later that year another chimp, Enos, orbited the Earth and returned safely. The much-lionized John Glenn did this only later.[82]

In the X-15, by contrast, only people entered the cockpit. A pilot fired the rocket, controlled its thrust, and set the angle of climb. He left the atmosphere, soared high over the top of the trajectory, and then used reaction controls to set up his re-entry. All the while, if anything went wrong, he had to cope with it on the spot and work to save himself and the plane. He maneuvered through re-entry, pulled out of his dive, and began to glide. Then, while Mercury capsules were using parachutes to splash clumsily near an aircraft carrier, the X-15 pilot goosed his craft onto Rogers Dry Lake like a fighter.

All aircraft depend on propulsion for their performance, and the X-15's engine installations allow the analyst to divide its career into three eras. It had been designed from the start to use the so-called Big Engine, with 57,000 pounds of thrust, but delays in its development brought a decision to equip it with two XLR11 rocket engines, which had served earlier in the X-1 series and the Douglas Skyrocket. Together they gave 16,000 pounds of thrust.

Flights with the XLR11s ran from June 1959 to February 1961. The best speed and altitude marks were Mach 3.50 in February 1961 and 136,500 feet in August 1961. These closely matched the corresponding numbers for the X-2 during 1956: Mach 3.196, 126,200 feet.[83] The X-2 program had been ill-starred—it had had two operational aircraft, both of which were destroyed in accidents. Indeed, these research aircraft made only 20 flights before the program ended, prematurely, with the loss of the second flight vehicle. The X-15 with XLR11s thus amounted to X-2s that had been brought back from the dead, and that belatedly completed their intended flight program.

The Big Engine, the Reaction Motors XLR99, went into service in November 1960. It launched a program of carefully measured steps that brought the fall of one Mach number after another. A month after the last flight with XLR11s, in March 1961, the pilot Robert White took the X-15 past Mach 4. This was the first time a piloted aircraft had flown that fast, as White raised the speed mark by nearly a full Mach. Mach 5 fell, also to Robert White, four months later. In November 1961 White did it again, as he reached Mach 6.04. Once flights began with the Big Engine, it took only 15 of them to reach this mark and to double the maximum Mach that had been reached with the X-2.

Altitude flights were also on the agenda. The X-15 climbed to 246,700 feet in April 1962, matched this mark two months later, and then soared to 314,750 feet in July 1962. Again White was in the cockpit, and the Federation Aeronautique Internationale, which keeps the world's aviation records, certified this one as the absolute altitude record for its class. A year later, without benefit of the FAI, the pilot Joseph Walker reached 354,200 feet. He thus topped 100 kilometers, a nice round number that put him into space without question or cavil.[84]

The third era in the X-15's history took shape as an extension of the second one. In November 1962, with this airplane's capabilities largely demonstrated, a serious landing accident caused major damage and led to an extensive rebuild. The new aircraft, designated X-15A-2, retained the Big Engine but sported external tankage for a longer duration of engine burn. It also took on an ablative coating for enhanced thermal protection.

It showed anew the need for care in flight test. In mid-1962, and for that matter in 1966, the X-2's best speed stood at 4,104 miles per hour, or Mach 5.92. (Mach number depends on both vehicle speed and air temperature. The flight to Mach 6.04 reached 4,093 miles per hour.) Late in 1966, flying the X-15A-2 without the ablator, Pete Knight raised this to Mach 6.33. Engineers then applied the ablator and mounted a dummy engine to the lower fin, with Knight taking this craft to Mach 4.94 in August 1967. Then in October he tried for more.

But the X-15A-2, with both ablator and dummy engine, now was truly a new configuration. Further, it had only been certified with these additions in the flight to Mach 4.94 and could not be trusted at higher Mach. Knight took the craft to Mach 6.72, a jump of nearly two Mach numbers, and this proved to be too much. The ablator, when it came back, was charred and pitted so severely that it could not be restored for another flight. Worse, shock-impingement heating burned the engine off its pylon and seared a hole in the lower fin, disabling the propellant ejec-

X-15 with dummy Hypersonic Research Engine mounted to the lower fin. (NASA)

tion system and threatening the craft's vital hydraulics. No one ever tried to fly faster in the X-15.[85]

It soon retired with honor, for in close to 200 powered flights, it had operated as a true instrument of hypersonic research. Its flight log showed nearly nine hours above Mach 3, close to six hours above Mach 4, and 87 minutes above Mach 5.[86] It served as a flying wind tunnel and made an important contribution by yielding data that made it possible to critique the findings of experiments performed in ground-based tunnels. Tunnel test sections were small, which led to concern that their results might not be reliable when applied to full-size hypersonic aircraft. Such discrepancies appeared particularly plausible because wind tunnels could not reproduce the extreme temperatures of hypersonic flight.

The X-15 set many of these questions to rest. In Becker's words, "virtually all of the flight pressures and forces were found to be in excellent agreement with the low-temperature wind-tunnel predictions."[87] In addition to lift and drag, this good agreement extended as well to wind-tunnel values of "stability derivatives," which governed the aircraft's handling qualities and its response to the aerodynamic controls. Errors due to temperature became important only beyond Mach 10 and were negligible below such speeds.

The X-15

B-52 mother ship with X-15A-2. The latter mounted a dummy scramjet and carried external tanks as well as ablative thermal protection. (NASA)

But the X-15 brought surprises in boundary-layer flow and aerodynamic heating. There was reason to believe that this flow would remain laminar, being stabilized in this condition by heat flow out of the boundary layer. This offered hope, for laminar flow, as compared to turbulent, meant less skin-friction drag and less heating. Instead, the X-15 showed mostly turbulent boundary layers. These resulted from small roughnesses and irregularities in the aircraft skin surface, which tripped the boundary layers into turbulence. Such skin roughness commonly produced turbulent boundary layers on conventional aircraft. The same proved to be true at Mach 6.

The X-15 had a conservative thermal design, giving large safety margins to cope with the prevailing lack of knowledge. The turbulent boundary layers might have brought large increases in the heat-transfer rates, limiting the X-15's peak speed. But in another surprise, these rates proved to be markedly lower than expected. As a consequence, the measured skin temperatures often were substantially less than had been anticipated (based on existing theory as well as on wind-tunnel tests). These flight results, confirmed by repeated measurements, were also validated with further wind-tunnel work. They resisted explanation by theory, but a new empirical model used these findings to give a more accurate description of hypersonic heating. Because this model predicted less heating and lower temperatures, it permitted design of vehicles that were lighter in weight.[88]

An important research topic involved observation of how the X-15 itself would stand up to thermal stresses. The pilot Joseph Walker stated that when his craft was accelerating and heating rapidly, "the airplane crackled like a hot stove." This resulted from buckling of the skin. The consequences at times could be serious, as when hot air leaked into the nose wheel well and melted aluminum tubing while in flight. On other occasions, such leaks destroyed the nose tire.[89]

Fortunately, such problems proved manageable. For example, the skin behind the wing leading edge showed local buckling during the first flight to Mach 5.3. The leading edge was a solid bar of Inconel X that served as a heat sink, with thin slots or expansion joints along its length. The slots tripped the local airflow into turbulence, with an accompanying steep rise in heat transfer. This created hot spots, which led to the buckling. The cure lay in cutting additional expansion slots, covering them with thin Inconel tabs, and fastening the skin with additional rivets. The wing leading edge faced particularly severe heating, but these modifications prevented buckling as the X-15 went beyond Mach 6 in subsequent flights.

Buckling indeed was an ongoing problem, and an important way to deal with it lay in the cautious step-by-step program of advance toward higher speeds. This allowed problems of buckling to appear initially in mild form, whereas a sudden leap toward record-breaking performance might have brought such problems in forms so severe as to destroy the airplane. This caution showed its value anew as buckling problems proved to lie behind an ongoing difficulty in which the cockpit canopy windows repeatedly cracked.

An initial choice of soda-lime glass for these windows gave way to alumino-silicate glass, which had better heat resistance. The wisdom of this decision became clear in 1961, when a soda-lime panel cracked in the course of a flight to 217,000 feet. However, a subsequent flight to Mach 6.04 brought cracking of an alumino-silicate panel that was far more severe. The cause again was buckling, this time in the retainer or window frame. It was made of Inconel X; its buckle again produced a local hot spot, which gave rise to thermal stresses that even this heat-resistant glass could not withstand. The original retainers were replaced with new ones made of titanium, which had a significantly lower coefficient of thermal expansion. Again the problem disappeared.[90]

The step-by-step test program also showed its merits in dealing with panel flutter, wherein skin panels oscillated somewhat like a flag waving in the breeze. This brought a risk of cracking due to fatigue. Some surface areas showed flutter at conditions no worse than Mach 2.4 and dynamic pressure of 650 pounds per square foot, a rather low value. Wind-tunnel tests verified the flight results. Engineers reinforced the panels with skin doublers and longitudinal stiffeners to solve the problem. Flutter did not reappear, even at the much higher dynamic pressure of 2,000 pounds per square foot.[91]

THE X-15

Caution in flight test also proved beneficial in dealing with the auxiliary power units (APUs). The APU, built by General Electric, was a small steam turbine driven by hydrogen peroxide and rotating at 51,200 revolutions per minute. Each X-15 airplane mounted two of them for redundancy, with each unit using gears to drive an electric alternator and a pump for the hydraulic system. Either APU could carry the full electrical and hydraulic load, but failure of both was catastrophic. Lacking hydraulic power, a pilot would have been unable to operate his aerodynamic controls.

Midway through 1962 a sudden series of failures in a main gear began to show up. On two occasions, a pilot experienced complete gear failure and loss of one APU, forcing him to rely on the second unit as a backup. Following the second such flight, the other APU gear also proved to be badly worn. The X-15 aircraft then were grounded while investigators sought the source of the problem.

They traced it to a lubricating oil, one type of which had a tendency to foam when under reduced pressure. The gear failures coincided with an expansion of the altitude program, with most of the flights above 100,000 feet having taken place during 1962 and later. When the oil turned to foam, it lost its lubricating properties. A different type had much less tendency to foam; it now became standard. Designers also enclosed the APU gearbox within a pressurized enclosure. Subsequent flights again showed reliable APU operation, as the gear failures ceased.[92]

Within the X-15 flight-test program, the contributions of its research pilots were decisive. A review of the first 44 flights, through November 1961, showed that 13 of them would have brought loss of the aircraft in the absence of a pilot and of redundancies in onboard systems. The actual record showed that all but one of these missions had been successfully flown, with the lone exception ending in an emergency landing that also went well.[93]

Still there were risks. The dividing line between a proficient flight and a disastrous one, between life and death for the pilot, could be narrow indeed, and the man who fell afoul of this was Major Mike Adams. His career in the cockpit dated to the Korean War. He graduated from the Experimental Test Pilot School, ranking first in his class, and then was accepted for the Aerospace Research Pilot School. Yeager himself was its director; his faculty included Frank Borman, Tom Stafford, and Jim McDivitt, all of whom went on to win renown as astronauts. Yeager and his selection board picked only the top one percent of this school's applicants.[94]

Adams made his first X-15 flight in October 1966. The engine shut down prematurely, but although he had previously flown this craft only in a simulator, he successfully guided his plane to a safe landing on an emergency dry lakebed. A year later, in the fall of 1967, he trained for his seventh mission by spending 23 hours in the simulator. The flight itself took place on 15 November.

As he went over the top at 266,400 feet, his airplane made a slow turn to the right that left it yawing to one side by 15 degrees.[95] Soon after, Adams made his

85

mistake. His instrument panel included an attitude indicator with a vertical bar. He could select between two modes of display, whereby this bar could indicate either sideslip angle or roll angle. He was accustomed to reading it as a yaw or sideslip angle—but he had set it to display roll.

"It is most probable that the pilot misinterpreted the vertical bar and flew it as a sideslip indicator," the accident report later declared. Radio transmissions from the ground might have warned him of his faulty attitude, but the ground controllers had no data on yaw. Adams might have learned more by looking out the window, but he had been carefully trained to focus on his instruments. Three other cockpit indicators displayed the correct values of heading and sideslip angle, but he apparently kept his eyes on the vertical bar. He seems to have felt vertigo, which he had trained to overcome by concentrating on that single vertical needle.[96]

Mistaking roll for sideslip, he used his reaction controls to set up a re-entry with his airplane yawed at ninety degrees. This was very wrong; it should have been pointing straight ahead with its nose up. At Mach 5 and 230,000 feet, he went into a spin. He fought his way out of it, recovering from the spin at Mach 4.7 and 120,000 feet. However, some of his instruments had been knocked badly awry. His inertial reference unit was displaying an altitude that was more than 100,000 feet higher than his true altitude. In addition, the MH-96 flight-control system made a fatal error.

It set up a severe pitch oscillation by operating at full gain, as it moved the horizontal stabilizers up and down to full deflection, rapidly and repeatedly. This system should have reduced its gain as the aircraft entered increasingly dense atmosphere, but instead it kept the gain at its highest value. The wild pitching produced extreme nose-up and nose-down attitudes that brought very high drag, along with decelerations as great as 15 g. Adams found himself immobilized, pinned in his seat by forces far beyond what his plane could withstand. It broke up at 62,000 feet, still traveling at Mach 3.9. The wings and tail came off; the fuselage fractured into three pieces. Adams failed to eject and died when he struck the ground.[97]

"We set sail on this new sea," John Kennedy declared in 1962, "because there is new knowledge to be gained, and new rights to be won." Yet these achievements came at a price, which Adams paid in full.[98]

1. Miller, *X-Planes*, chs. 4, 6, 7.
2. NACA RM L54F21.
3. Gunston, *Fighters*, pp. 156-57; Anderson, *History*, pp. 430, 432.
4. NASA SP-4303, pp. 67-69.
5. Gunston, *Fighters*, pp. 193-94, 199.
6. Miller, *X-Planes*, ch. 6.
7. NASA RP-1028, p. 237.
8. *Spaceflight*: November 1979, p. 435; July-August 1980, pp. 270-72.
9. Neufeld, *Ballistic*, p. 70.
10. NASA RP-1028, p. 237.
11. Rand Corporation: Reports SM-11827, R-217.
12. Miller, *X-Planes*, p. 22; NASA SP-4305, p. 350. Quote: Minutes, Committee on Aerodynamics, 4 October 1951, p. 16.
13. Letter, Woods to Committee on Aerodynamics, 8 January 1952; memo, Dornberger to Woods, 18 January 1952 (includes quotes).
14. Drake and Carman, "Suggestion" (quote, p. 1).
15. NASA SP-4305, p. 354; memo, Stone to Chief of Research (Langley).
16. Drake and Carman, "Suggestion" (quote, p. 1).
17. Brown et al., "Study," p. 58.
18. *Astronautics & Aeronautics*, February 1964, p. 54 (includes quotes).
19. Memo, J. R. Crowley to NACA centers, 11 December 1953; minutes, Interlaboratory Research Airplane Projects Panel, 4-5 February 1954 (quote, p. 12).
20. Martin, "History," p. 4; *Astronautics & Aeronautics*, February, 1964, p. 54.
21. Becker et al., "Research," pp. 5-8.
22. Martin, "History," p. 3 with insert; Becker et al., "Research," p. 24.
23. Becker et al., "Research," p. 18. Alloy composition: memo, Rhode to Gilruth, 4 August 1954.
24. Hallion, *Hypersonic*, p. 386 (includes quotes).
25. *Astronautics & Aeronautics*, February 1964, p. 58; Brown, "Study," pp. 38-39.
26. *Astronautics & Aeronautics*, February 1964, p. 56.
27. Ibid., pp. 53, 54, 56-57; NACA RM L54F21.
28. Becker et al., "Research," pp. 8, 21.
29. Budgets: NACA: NASA SP-4305, p. 428. Air Force: Hansen, *Almanac*, p. 757.
30. Heppenheimer, *Turbulent*, pp. 149-51, 197-200 (quote, p. 197).
31. Gunston, *Fighters*, pp. 121-22.
32. Wolfe, *Right Stuff*, p. 39.
33. Gunston, *Fighters*, pp. 155-64, 170-76.
34. Miller, *X-Planes*, chs. 6, 11; *Acta Astronautica*, Volume 26 (1992), p. 743; Gunston, *Fighters*, pp. 193-94.

35 "History of the Arnold Engineering Development Center," 1 August 1944 to 1 January 1951; Von Karman and Edson, *Wind*, pp. 298-300.
36 "History of the Arnold Engineering Development Center," 1 January to 1 June 1953.
37 NASA RP-1132, pp. 199-199A.
38 Gunston, *Fighters*, pp. 193-94; *Air Enthusiast*, July-September 1978, pp. 206-07.
39 "History of the Arnold Engineering Development Center," 1 July to 1 December 1957; "Standard Missile Characteristics: XSM-64 Navaho," p. 3.
40 "History of the Arnold Engineering Development Center," 1 July to 1 December 1955; 1 July to 1 December 1959.
41 DTIC AD-299774, pp. 1-2.
42 DTIC AD-098980; *Quarterly Review:* Summer 1959, p. 81; Fall 1959, p. 57.
43 NASA SP-4303, p. 76.
44 Glenn Curtiss, rather than the Wrights, was the first to use ailerons. But in subsequent litigation, the Wrights established that his innovation in fact was covered under their patent, which had been issued in 1906. See Heppenheimer, *First Flight*, ch. 9.
45 Letters, Dryden to Putt, 4 May 1954; Putt to Dryden, 26 May 1954.
46 Hallion, *Hypersonic*, pp. 2-5 (includes quotes).
47 Ibid., p. xxiii   Extended quote, pp. xxvi, xxix.
48 Ibid., pp. 2-3; Martin, "History," pp. 5-6; "Agenda: Meeting to Consider Need for New Research Aircraft," 9 July 1954.
49 Minutes of Meeting, Committee on Aerodynamics, 4-5 October 1954. Extended quote: Appendix I, p. 2.
50 Ibid., pp. 15-18 (extended quote, pp. 17-18); Martin, "History," p. 10.
51 Martin, "History," pp. 13-14; Memorandum of Understanding, 23 December 1954 (quote, paragraph H).
52 NACA copy of signed Memorandum of Understanding with handwritten notes by Clotaire Wood; memo, Gardner to Assistant Secretary of the Navy for Air, 11 November 1954.
53 Gunston, *Fighters*, pp. 191-94; "Evaluation Report," pp. 82, 96, 99, 104-05.
54 Hallion, *Hypersonic*, p. 384 (includes quote).
55 NASA SP-2004-4236, footnote, pp. 66-67.
56 "Evaluation Report," pp. 23, 26, 99-100, 103; quote, p. 28.
57 Hallion, "American," p. 298; "Evaluation Report," pp. 44, 49, 50, 100-01 (quote, p. 49).
58 "Thirty Years" (Rocketdyne); Gibson, *Navaho*, pp. 36, 40-41.
59 "Evaluation Report," p. 65; memo from Becker, 13 June 1955.
60 NASA SP-2004-4236, p. 72 (includes quotes; see also footnotes).
61 "Evaluation Report," pp. 5, 6, 98.
62 Hallion, *Hypersonic*, p. 13.
63 "Research-Airplane-Committee Report," 1958, pp. 243-73.
64 Ibid., p. 264.
65 Ibid., pp. 151-58; Miller, *X-Planes*, p. 200. 3,500°F: Hallion, *Hypersonic*, p. 158; quote: *Time*,

27 October 1961, p. 89.
66. NASA SP-4303, p. 76.
67. "Research-Airplane-Committee Report," 1956, pp. 175-81; 1958, pp. 303-11. X-1B: NASA SP-4303, pp. 79-80, 294-95; SP-2000-4518, p. 77.
68. NASA TN D-1157, Figure 2.
69. NASA TN D-1157, TN D-1402.
70. AIAA Paper 93-0309, p. 2; Miller, *X-Planes*, p. 188. Start in 1955: NASA TM X-56008, p. 1.
71. NASA: SP-60, pp. 78-79; TN D-6208, pp. 7-8.
72. AIAA Paper 64-17, p. 1.
73. *American Heritage of Invention & Technology*, Winter 1992, pp. 19-24.
74. NASA TN D-6208, p. 10; AIAA Paper 63075, pp. 15-16.
75. *American Heritage of Invention & Technology*, Winter 1989, pp. 60-62.
76. Quotes: "Research-Airplane-Committee Report," 1956, p. 84.
77. NASA: SP-60, pp. 37-38; TN D-1159, pp. 6-7.
78. "Research-Airplane-Committee Report," 1958, pp. 107-16.
79. AIAA Paper 93-0309, p. 5; NASA SP-4303, pp. 118, 336.
80. NASA SP-4222, pp. 51, 113; quote: *Time*, 8 December 1952, 69.
81. NASA SP-4201, p. 228; "Research-Airplane-Committee Report," 1958, pp. 117-27 (5 psi: p. 117). Quote: NASA SP-4303, pp. 126-27.
82. NASA SP-4201, pp. 314-18, 402-07.
83. NASA SP-4303, pp. 315-16, 330; Miller, *X-Planes*, pp. 64, 191-192.
84. Miller, *X-Planes*, pp. 192-93.
85. Ibid., pp. 189-95; NASA SP-4303, pp. 121-22.
86. NASA CP-3105, p. 165.
87. Quote: *Raumfahrtforschung*, March-April 1969: 47.
88. Ibid., pp. 47-48; Hallion, *Hypersonic*, p. 164.
89. Hunley, "Significance," p. 4; *Astronautics & Aeronautics*, March 1964, p. 23. Quote: AIAA Paper 93-0309, p. 3.
90. AIAA Paper 93-0309, pp. 3, 9; NASA TN D-1278, pp. 8-9, 40-41; *Raumfahrtforschung*, March-April 1969, p. 49.
91. *Astronautics & Aerospace Engineering*, June 1963, p. 25; NASA TN D-1278, pp. 10-11.
92. *Astronautics & Aerospace Engineering*, June 1963, pp. 28-29.
93. NASA TN D-1278, pp. 26-28.
94. *Aviation Week*, 26 August 1968, p. 97; Yeager and Janos, *Yeager*, p. 267.
95. *Aviation Week*, 12 August 1968, p. 104; 26 August 1968, pp. 97, 100.
96. *Aviation Week*, 26 August 1968, p. 85-105 (quote, p. 101).
97. NASA SP-4303, pp. 123-25. Numbers: *Aviation Week*, 12 August 1968, pp. 104, 117.
98. Quote: *New York Times*, 13 September 1962, p. 16.

4

First Thoughts of Hypersonic Propulsion

Three new aircraft engines emerged from World War II: the turbojet, the ramjet, and the liquid rocket. The turbojet was not suitable for hypersonic flight, but the rocket and the ramjet both gave rise to related airbreathing concepts that seemed to hold promise.

Airbreathing rockets drew interest, but it was not possible to pump in outside air with a conventional compressor. Such rockets instead used liquid hydrogen fuel as a coolant, to liquefy air, with this liquid air being pumped to the engine. This arrangement wasted cooling power by also liquefying the air's nonflammable hydrogen, and so investigators sought ways to remove this nitrogen. They wanted a flow of nearly pure liquid oxygen, taken from the air, for use as the oxidizer.

Ramjets provided higher flight speeds than turbojets, but they too had limits. Antonio Ferri, one of Langley's leading researchers, took the lead in conceiving of ramjets that appeared well suited to flight at hypersonic and perhaps even orbital speeds, at least on paper. Other investigators studied combined-cycle engines. The ejector ramjet, for one, sought to integrate a rocket with a ramjet, yielding a single compact unit that might fly from a runway to orbit.

Was it possible to design a flight vehicle that in fact would do this? Ferri thought so, as did his colleague Alexander Kartveli of Republic Aviation. Air Force officials encouraged such views by sponsoring a program of feasibility studies called Aerospaceplane. Designers at several companies contributed their own ideas.

These activities unfolded within a world where work with conventional rockets was advancing vigorously.[1] In particular, liquid hydrogen was entering the mainstream of rocket engineering.[2] Ramjets also won acceptance as standard military engines, powering such missiles as Navaho, Bomarc, Talos, and the X-7. With this background, for a time some people believed that even an Aerospaceplane might prove feasible.

Ramjets As Military Engines

The ramjet and turbojet relied on fundamentally the same thermodynamic cycle. Both achieved compression of inflowing air, heated the compressed flow by burning

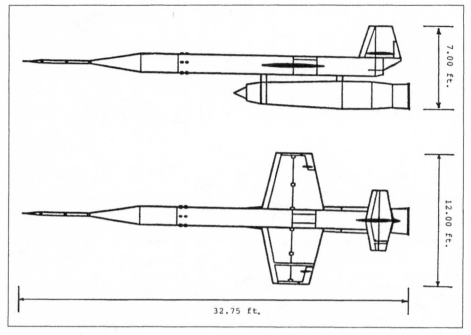

The X-7. (U.S. Air Force)

fuel, and obtained thrust by allowing the hot airflow to expand through a nozzle. The turbojet achieved compression by using a rotating turbocompressor, which inevitably imposed a requirement to tap a considerable amount of power from its propulsive jet by placing the turbine within the flow. A ramjet dispensed with this turbomachinery, compressing the incoming flow by the simple method of processing it through a normal or oblique shock. This brought the promise of higher flight speed. However, a ramjet paid for this advantage by requiring an auxiliary boost, typically with a rocket, to accelerate it to speeds at which this shock could form.[3]

The X-7 served as a testbed for development of ramjet engines as mainstream propulsion units. With an initial concept that dated to December 1946, it took shape as a three-stage flight vehicle. The first stage, a B-29 and later a B-50 bomber, played the classic role of lifting the aircraft to useful altitudes. Such bombers also served in this fashion as mother ships for the X-1 series, the X-2, and in time the X-15. For the X-7, a solid propellant rocket served as the second stage, accelerating the test aircraft to a speed high enough for sustained ramjet operation. The ramjet engine, slung beneath its fuselage, provided further acceleration along with high-speed cruise. Recovery was achieved using a parachute and a long nose spike that pierced the ground like a lance. This enabled the X-7 to remain upright, which protected the airframe and engine for possible re-use.

First Thoughts of Hypersonic Propulsion

The X-7 craft were based at Holloman Air Force Base in New Mexico, which was an early center for missile flight test. The first flight took place in April 1951, with a ramjet of 20-inch diameter built by Wright Aeronautical. The X-7 soon took on the role of supporting developmental tests of a 28-inch engine built by the Marquardt Company and intended for use with the Bomarc missile. Flights with this 28-inch design began in December 1952 and achieved a substantial success the following April. The engine burned for some 20 seconds; the vehicle reached 59,500 feet and Mach 2.6. This exceeded the Mach 2.44 of the X-1A rocket plane in December 1953, piloted by Chuck Yeager. Significantly, although the X-7 was unpiloted, it remained aerodynamically stable during this flight. By contrast, the X-1A lost stability and fell out of the sky, dropping 51,000 feet before Yeager brought it back under control.[4]

The X-7 soon became a workhorse, running off some one hundred missions between 1955 and the end of the program in July 1960. It set a number of records, including range and flight time of 134 miles and 552 seconds, respectively. Its altitude mark of 106,000 feet, achieved with an airbreathing ramjet, compared with 126,200 feet for the rocket-powered X-2 research airplane.[5]

Other achievements involved speed. The vehicle had been built of heat-treated 4130 steel, with the initial goal being Mach 3. The program achieved this—and simply kept going. On 29 August 1957 it reached Mach 3.95 with a 28-inch Marquardt engine. Following launch from a B-50 at 33,700 feet, twin solid motors mounted beneath the wings boosted the craft to Mach 2.25. These boosters fell away; the ramjet ignited, and the vehicle began to climb at a 20-degree angle before leveling out at 54,500 feet. It then went into a very shallow dive. The engine continued to operate, as it ran for a total of 91 seconds, and acceleration continued until the craft attained its peak Mach at fuel exhaustion. It was recovered through use of its parachute and nose spike, and temperature-sensitive paints showed that it had experienced heating to more than 600°F. This heating also brought internal damage to the engine.[6]

Even so, the X-7 was not yet at the limit of its capabilities. Fitted with a 36-inch ramjet, again from Marquardt, it flew to Mach 4.31 on 17 April 1958. This time the drop from the B-50 came at 28,500 feet, with the engine igniting following rocket boost to Mach 1.99. It operated for 70.9 seconds, propelling the vehicle to a peak altitude of 60,000 feet. By then it was at Mach 3.5, continuing to accelerate as it transitioned to level flight. It reached its maximum Mach—and sustained a sharp drop in thrust three seconds later, apparently due to an engine tailpipe failure. Breakup of the vehicle occurred immediately afterward, with the X-7 being demolished.[7]

This flight set a record for airbreathing propulsion that stands to this day. Its speed of 2,881 miles per hour (mph) compares with the record for an afterburning

The Bomarc. (U.S. Air Force)

turbojet of 2,193 mph, set in an SR-71 in 1976.[8] Moreover, while the X-7 was flying to glory, the Bomarc program that it supported was rolling toward operational deployment.

The name "Bomarc" derives from the contractors Boeing and the Michigan Aeronautical Research Center, which conducted early studies. It was a single-stage, ground-launched antiaircraft missile that could carry a nuclear warhead. A built-in liquid-propellant rocket provided boost; it was replaced by a solid rocket in a later version. Twin ramjets sustained cruise at Mach 2.6. Range of the initial operational model was 250 miles, later extended to 440 miles.[9]

Specifications for this missile were written in September 1950. In January 1951 an Air Force letter contract designated Boeing as the prime contractor, with Marquardt Aircraft winning a subcontract to build its ramjet. The development of this engine went forward rapidly. In July 1953 officials of the Air Force's Air Materiel Command declared that work on the 28-inch engine was essentially complete.[10]

Flight tests were soon under way. An Air Force review notes that a test vehicle "traveled 70 miles in 1.5 minutes to complete a most successful test of 17 June 1954." The missile "cruised at Mach 3+ for 15 seconds and reached an altitude of 56,000 feet." In another flight in February 1955, it reached a peak altitude of 72,000 feet as its ramjet burned for 245 seconds. This success brought a decision to order Bomarc into production. Four more test missiles flew with their ramjets later that year, with all four scoring successes.[11]

Other activity turned Bomarc from a missile into a weapon system, integrating it with the electronic Semi-Automatic Ground Environment (SAGE) that controlled air defense within North America. In October 1958, Bomarcs scored a spectacular success. Controlled remotely from a SAGE center 1,500 miles away, two missiles

First Thoughts of Hypersonic Propulsion

homed in on target drones that were more than 100 miles out to sea. The Bomarcs dived on them and made intercepts. The missiles were unarmed, but one of them actually rammed its target. A similar success occurred a year later when a Bomarc made a direct hit on a Regulus 2 supersonic target over the Gulf of Mexico. The missile first achieved operational status in September 1959. Three years later, Bomarc was in service at eight Air Force sites, with deployment of Canadian squadrons following. These missiles remained on duty until 1972.[12]

Paralleling Bomarc, the Navy pursued an independent effort that developed a ship-based antiaircraft missile named Talos, after a mythical defender of the island of Crete. It took shape at a major ramjet center, the Applied Physics Laboratory (APL) of Johns Hopkins University. Like Bomarc, Talos was nuclear-capable; *Jane's* gave its speed as Mach 2.5 and its range as 65 miles.

An initial version first flew in 1952, at New Mexico's White Sands Missile Range. A prototype of a nuclear-capable version made its own first flight in December 1953. The Korean War had sparked development of this missile, but the war ended in mid-1953 and the urgency diminished. When the Navy selected the light cruiser USS *Galveston* for the first operational deployment of Talos, the conversion of this ship became a four-year task. Nevertheless, Talos finally joined the fleet in 1958, with other cruisers installing it as well. It remained in service until 1980.[13]

The Talos. (National Archives and Records Administration)

One military ramjet project, that of Navaho, found itself broken in mid-stride. Although the ultimate version was to have intercontinental range, the emphasis during the 1950s was on an interim model with range of 2,500 miles, with the missile cruising at Mach 2.75 and 76,800 feet. The missile used a rocket-powered booster with liquid-fueled engines built by Rocketdyne. The airplane-like Navaho mounted two 51-inch ramjets from Wright Aeronautical, which gave it the capability to demonstrate long-duration supersonic cruise under ramjet power.[14]

Flight tests began in November 1956, with launches of complete missiles taking place at Cape Canaveral. The first four were flops; none even got far enough to permit ignition of the ramjets. In mid-July of 1957, three weeks after the first launch of an Atlas, the Air Force canceled Navaho. Lee Atwood, president of North American Aviation, recalls that Atlas indeed had greater promise: "Navaho would approach its target at Mach 3; a good antiaircraft missile might shoot it down. But Atlas would come in at Mach 20. There was no way that anyone would shoot *it* down."

There nevertheless was hardware for several more launches, and there was considerable interest in exercising the ramjets. Accordingly, seven more Navahos were launched during the following 18 months, with the best flight taking place in January 1958.

The Navaho. (Smithsonian Institution No. 77-10905)

The missile accelerated on rocket power and leveled off, the twin ramjet engines ignited, and it stabilized in cruise at 64,000 feet. It continued in this fashion for half an hour. Then, approaching the thousand-mile mark in range, its autopilot initiated a planned turnaround to enable this Navaho to fly back uprange. The turn was wide, and ground controllers responded by tightening it under radio control. This disturbed the airflow near the inlet of the right ramjet, which flamed out. The missile lost speed, its left engine also flamed out, and the vehicle fell into the Atlantic. It had been airborne for 42 minutes, covering 1,237 miles.[15]

Because the program had been canceled and the project staff was merely flying off its leftover hardware, there were no funds to address what clearly was a serious inlet problem. Still, Navaho at least had flown. By contrast, another project—the Air Force's XF-103 fighter, which aimed at Mach 3.7—never even reached the prototype stage.

First Thoughts of Hypersonic Propulsion

The XF-103 in artist's rendering. (U.S. Air Force)

Its engine, also from Wright Aeronautical, combined a turbojet and ramjet within a single package. The ramjet doubled as an afterburner, with internal doors closing off the ramjet's long inlet duct. Conversion to pure ramjet operation took seven seconds. This turboramjet showed considerable promise. At Arnold Engineering Development Center, an important series of ground tests was slated to require as much as six weeks. They took only two weeks, with the engine running on the first day.

Unfortunately, the XF-103 outstayed its welcome. The project dated to 1951; it took until December 1956 to carry out the AEDC tests. Much of the reason for this long delay involved the plane's highly advanced design, which made extensive use of titanium. Still, the Mach 1.8 XF-104 took less than 18 months to advance from the start of engineering design to first flight, and the XF-103 was not scheduled to fly until 1960. The Air Force canceled it in August 1957, and aviation writer Richard DeMeis pronounced its epitaph: "No matter how promising or outstanding an aircraft may be, if development takes inordinately long, the mortality rate rises in proportion."[16]

Among the five cited programs, three achieved operational status, with the X-7 actually outrunning its initial planned performance. The feasibility of Navaho was never in doubt; the inlet problem was one of engineering development, not one that would call its practicality into question. Only the XF-103 encountered serious problems of technology that lay beyond the state of the art. The ramjet of the 1950s thus was an engine whose time had come, and which had become part of mainstream design.

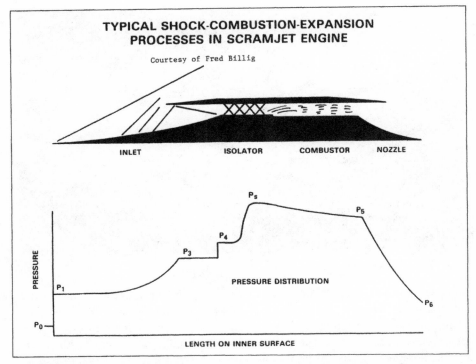

The scramjet. Oblique shocks in the isolator prevent disturbances in the combustor from propagating upstream, where they would disrupt flow in the inlet. (Courtesy of Frederick Billig)

There was a ramjet industry, featuring the firms of Marquardt Aviation and Wright Aeronautical. Facilities for developmental testing existed, not only at these companies but at NACA-Lewis and the Navy's Ordnance Aerophysics Laboratory, which had a large continuous-flow supersonic wind tunnel. With this background, a number of investigators looked ahead to engines derived from ramjets that could offer even higher performance.

Origins of the Scramjet

The airflow within a ramjet was subsonic. This resulted from its passage through one or more shocks, which slowed, compressed, and heated the flow. This was true even at high speed, with the Mach 4.31 flight of the X-7 also using a subsonic-combustion ramjet. Moreover, because shocks become stronger with increasing Mach, ramjets could achieve greater internal compression of the flow at higher speeds. This increase in compression improved the engine's efficiency.

First Thoughts of Hypersonic Propulsion

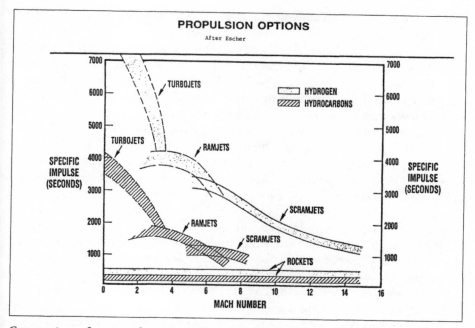

Comparative performance of scramjets and other engines. Airbreathers have very high performance because they are "energy machines," which burn fuel to heat air. Rockets have much lower performance because they are "momentum machines," which physically expel flows of mass. (Courtesy of William Escher)

Still, there were limits to a ramjet's effectiveness. Above Mach 5, designers faced increasingly difficult demands for thermal protection of an airframe and for cooling of the ramjet duct. With the internal flow being very hot, it became more difficult to add still more heat by burning fuel, without overtaxing the materials or the cooling arrangements. If the engine were to run lean to limit the temperature rise in the combustor, its thrust would fall off. At still higher Mach levels, the issue of heat addition through combustion threatened to become moot. With high internal temperatures promoting dissociation of molecules of air, combustion reactions would not go to completion and hence would cease to add heat.

A promising way around this problem involved doing away with a requirement for subsonic internal flow. Instead this airflow was to be supersonic and was to sustain combustion. Right at the outset, this approach reduced the need for internal cooling, for this airflow would not heat up excessively if it was fast enough. This relatively cool internal airflow also could continue to gain heat through combustion. It would avoid problems due to dissociation of air or failure of chemical reactions in combustion to go to completion. On paper, there now was no clear upper limit to speed. Such a vehicle might even fly to orbit.

Yet while a supersonic-combustion ramjet offered tantalizing possibilities, right at the start it posed a fundamental issue: was it feasible to burn fuel in the duct of such an engine without producing shock waves? Such shocks could produce severe internal heating, destroying the benefits of supersonic combustion by slowing the flow to subsonic speeds. Rather than seeking to achieve shock-free supersonic combustion in a duct, researchers initially bypassed this difficulty by addressing a simpler problem: demonstration of combustion in a supersonic free-stream flow.

The earliest pertinent research appears to have been performed at the Applied Physics Laboratory (APL), during or shortly after World War II. Machine gunners in aircraft were accustomed to making their streams of bullets visible by making every twentieth round a tracer, which used a pyrotechnic. They hoped that a gunner could walk his bullets into a target by watching the glow of the tracers, but experience showed that the pyrotechnic action gave these bullets trajectories of their own. The Navy then engaged two research centers to look into this. In Aberdeen, Maryland, Ballistic Research Laboratories studied the deflection of the tracer rounds themselves. Near Washington, DC, APL treated the issue as a new effect in aerodynamics and sought to make use of it.

Investigators conducted tests in a Mach 1.5 wind tunnel, burning hydrogen at the base of a shell. A round in flight experienced considerable drag at its base, but the experiments showed that this combustion set up a zone of higher pressure that canceled the drag. This work did not demonstrate supersonic combustion, for while the wind-tunnel flow was supersonic, the flow near the base was subsonic. Still, this work introduced APL to topics that later proved pertinent to supersonic-combustion ramjets (which became known as scramjets).[17]

NACA's Lewis Flight Propulsion Laboratory, the agency's center for studies of engines, emerged as an early nucleus of interest in this topic. Initial work involved theoretical studies of heat addition to a supersonic flow. As early as 1950, the Lewis investigators Irving Pinkel and John Serafini treated this problem in a two-dimensional case, as in flow over a wing or past an axisymmetric body. In 1952 they specifically treated heat addition under a supersonic wing. They suggested that this might produce more lift than could be obtained by burning the same amount of fuel in a turbojet to power an airplane.[18]

This conclusion immediately raised the question of whether it was possible to demonstrate supersonic combustion in a wind tunnel. Supersonic tunnels produced airflows having very low pressure, which added to the experimental difficulties. However, researchers at Lewis had shown that aluminum borohydride could promote the ignition of pentane fuel at air pressures as low as 0.03 atmosphere. In 1953 Robert Dorsch and Edward Fletcher launched a research program that sought to ignite pure borohydride within a supersonic flow. Two years later they declared that they had succeeded. Subsequent work showed that at Mach 3, combustion of this fuel under a wing more than doubled the lift.[19]

Also at Lewis, the aerodynamicists Richard Weber and John MacKay published the first important open-literature study of theoretical scramjet performance in 1958. Because they were working entirely with equations, they too bypassed the problem of attaining shock-free flow in a supersonic duct by simply positing that it was feasible. They treated the problem using one-dimensional gas dynamics, corresponding to flow in a duct with properties at any location being uniform across the diameter. They restricted their treatment to flow velocities from Mach 4 to 7.

They discussed the issue of maximizing the thrust and the overall engine efficiency. They also considered the merits of various types of inlet, showing that a suitable choice could give a scramjet an advantage over a conventional ramjet. Supersonic combustion failed to give substantial performance improvements or to lead to an engine of lower weight. Even so, they wrote that "the trends developed herein indicate that the [scramjet] will offer superior performance at higher hypersonic flight speeds."[20]

An independent effort proceeded along similar lines at Marquardt, where investigators again studied scramjet performance by treating the flow within an engine duct using one-dimensional gasdynamic theory. In addition, Marquardt researchers carried out their own successful demonstration of supersonic combustion in 1957. They injected hydrogen into a supersonic airflow, with the hydrogen and the air having the same velocity. This work overcame objections from skeptics, who had argued that the work at NACA-Lewis had not truly demonstrated supersonic combustion. The Marquardt experimental arrangement was simpler, and its results were less equivocal.[21]

The Navy's Applied Physics Laboratory, home of Talos, also emerged as an early center of interest in scramjets. As had been true at NACA-Lewis and at Marquardt, this group came to the concept by way of external burning under a supersonic wing. William Avery, the leader, developed an initial interest in supersonic combustion around 1955, for he saw the conventional ramjet facing increasingly stiff competition from both liquid rockets and afterburning turbojets. (Two years later such competition killed Navaho.) Avery believed that he could use supersonic combustion to extend the performance of ramjets.

His initial opportunity came early in 1956, when the Navy's Bureau of Ordnance set out to examine the technological prospects for the next 20 years. Avery took on the task of assembling APL's contribution. He picked scramjets as a topic to study, but he was well aware of an objection. In addition to questioning the fundamental feasibility of shock-free supersonic combustion in a duct, skeptics considered that a hypersonic inlet might produce large pressure losses in the flow, with consequent negation of an engine's thrust.

Avery sent this problem through Talos management to a young engineer, James Keirsey, who had helped with Talos engine tests. Keirsey knew that if a hypersonic ramjet was to produce useful thrust, it would appear as a small difference between two large quantities: gross thrust and total drag. In view of uncertainties in both these numbers, he was unable to state with confidence that such an engine would work. Still he did not rule it out, and his "maybe" gave Avery reason to pursue the topic further.

Avery decided to set up a scramjet group and to try to build an engine for test in a wind tunnel. He hired Gordon Dugger, who had worked at NACA-Lewis. Dugger's first task was to decide which of several engine layouts, both ducted and unducted, was worth pursuing. He and Avery selected an external-burning configuration with the shape of a broad upside-down triangle. The forward slope, angled downward, was to compress the incoming airflow. Fuel could be injected at the apex, with the upward slope at the rear allowing the exhaust to expand. This approach again bypassed the problem of producing shock-free flow in a duct. The use of external burning meant that this concept could produce lift as well as thrust.

Dugger soon became concerned that this layout might be too simple to be effective. Keirsey suggested placing a very short cowl at the apex, thereby easing problems of ignition and combustion. This new design lent itself to incorporation within the wings of a large aircraft of reasonably conventional configuration. At low speeds the wide triangle could retract until it was flat and flush with the wing undersurface, leaving the cowl to extend into the free stream. Following acceleration to supersonic speed, the two shapes would extend and assume their triangular shape, then function as an engine for further acceleration.

Wind-tunnel work also proceeded at APL. During 1958 this center had a Mach 5 facility under construction, and Dugger brought in a young experimentalist named Frederick Billig to work with it. His first task was to show that he too could demonstrate supersonic combustion, which he tried to achieve using hydrogen as his fuel. He tried electric ignition; an APL history states that he "generated gigantic arcs," but "to no avail." Like the NACA-Lewis investigators, he turned to fuels that ignited particularly readily. His choice, triethyl aluminum, reacts spontaneously, and violently, on contact with air.

"The results of the tests on 5 March 1959 were dramatic," the APL history continues. "A vigorous white flame erupted over the rear of [the wind-tunnel model] the instant the triethyl aluminum fuel entered the tunnel, jolting the model against its support. The pressures measured on the rear surface jumped upward." The device produced less than a pound of thrust. But it generated considerable lift, supporting calculations that had shown that external burning could increase lift. Later tests showed that much of the combustion indeed occurred within supersonic regions of the flow.[22]

By the late 1950s small scramjet groups were active at NACA-Lewis, Marquardt, and APL. There also were individual investigators, such as James Nicholls of the University of Michigan. Still it is no small thing to invent a new engine, even as an extension of an existing type such as the ramjet. The scramjet needed a really high-level advocate, to draw attention within the larger realms of aerodynamics and propulsion. The man who took on this role was Antonio Ferri.

He had headed the supersonic wind tunnel in Guidonia, Italy. Then in 1943 the Nazis took control of that country and Ferri left his research to command a band of partisans who fought the Nazis with considerable effectiveness. This made him a marked man, and it was not only Germans who wanted him. An American agent, Moe Berg, was also on his trail. Berg found him and persuaded him to come to the States. The war was still on and immigration was nearly impossible, but Berg persuaded William Donovan, the head of his agency, to seek support from President Franklin Roosevelt himself. Berg had been famous as a baseball catcher in civilian life, and when Roosevelt learned that Ferri now was in the hands of his agent, he remarked, "I see Berg is still catching pretty well."[23]

At NACA-Langley after the war, he rose in management and became director of the Gas Dynamics Branch in 1949. He wrote an important textbook, *Elements of Aerodynamics of Supersonic Flows* (Macmillan, 1949). Holding a strong fondness for the academic world, he took a professorship at Brooklyn Polytechnic Institute in 1951, where in time he became chairman of his department. He built up an aerodynamics laboratory at Brooklyn Poly and launched a new activity as a consultant. Soon he was working for major companies, drawing so many contracts that his graduate students could not keep up with them. He responded in 1956 by founding a company, General Applied Science Laboratories (GASL). With financial backing from the Rockefellers, GASL grew into a significant center for research in high-speed flight.[24]

He was a formidable man. Robert Sanator, a former student, recalls that "you had to really want to be in that course, to learn from him. He was very fast. His mind was constantly moving, redefining the problem, and you had to be fast to keep up with him. He expected people to perform quickly, rapidly." John Erdos, another ex-student, adds that "if you had been a student of his and later worked for him, you could never separate the professor-student relationship from your normal working relationship." He remained Dr. Ferri to these people, never Tony, even when they rose to leadership within their companies.[25]

He came early to the scramjet. Taking this engine as his own, he faced its technical difficulties squarely and asserted that they could be addressed, giving examples of approaches that held promise. He repeatedly emphasized that scramjets could offer performance far higher than that of rockets. He presented papers at international conferences, bringing these ideas to a wider audience. In turn, his strong professional reputation ensured that he was taken seriously. He also performed experiments as he sought to validate his claims. More than anyone else, Ferri turned the scramjet from an idea into an invention, which might be developed and made practical.

His path to the scramjet began during the 1950s, when his work as a consultant brought him into a friendship with Alexander Kartveli at Republic Aviation. Louis Nucci, Ferri's longtime colleague, recalls that the two men "made good sparks. They were both Europeans and learned men; they liked opera and history." They also complemented each other professionally, as Kartveli focused on airplane design while Ferri addressed difficult problems in aerodynamics and propulsion. The two men worked together on the XF-103 and fed off each other, each encouraging the other to think bolder thoughts. Among the boldest was a view that there were no natural limits to aircraft speed or performance. Ferri put forth this idea initially; Kartveli then supported it with more detailed studies.[26]

The key concept, again, was the scramjet. Holding a strong penchant for experimentation, Ferri conducted research at Brooklyn Poly. In September 1958, at a conference in Madrid, he declared that steady combustion, without strong shocks, had been accomplished in a supersonic airstream at Mach 3.0. This placed him midway in time between the supersonic-combustion demonstrations at Marquardt and at APL.[27]

Shock-free flow in a duct continued to loom as a major problem. The Lewis, Marquardt, and APL investigators had all bypassed this issue by treating external combustion in the supersonic flow past a wing, but Ferri did not flinch. He took the problem of shock-free flow as a point of departure, thereby turning the ducted scramjet from a wish into a serious topic for investigation.

In supersonic wind tunnels, shock-free flow was an everyday affair. However, the flow in such tunnels achieved its supersonic Mach values by expanding through a nozzle. By contrast, flow within a scramjet was to pass through a supersonic inlet and then be strongly heated within a combustor. The inlet actually had the purpose of producing a shock, an oblique one that was to slow and compress the flow while allowing it to remain supersonic. However, the combustion process was only too likely to produce unwanted shocks, which would limit an engine's thrust and performance.

Nicholls, at Michigan, proposed to make a virtue of necessity by turning a combustor shock to advantage. Such a shock would produce very strong heating of the flow. If the fuel and air had been mixed upstream, then this combustor shock could produce ignition. Ferri would have none of this. He asserted that "by using a suitable design, formation of shocks in the burner can be avoided."[28]

Specifically, he started with a statement by NACA's Weber and MacKay on combustors. These researchers had already written that the combustor needed a diverging shape, like that of a rocket nozzle, to overcome potential limits on the airflow rate due to heat addition ("thermal choking"). Ferri proposed that within such a combustor, "fuel is injected parallel to the stream to eliminate formation of shocks.... The fuel gradually mixes with the air and burns...and the combustion process can take place without the formation of shocks." Parallel injection might take place by building the combustor with a step or sudden widening. The flow could expand as it passed the step, thereby avoiding a shock, while the fuel could be injected at the step.[29]

Ferri also made an intriguing contribution in dealing with inlets, which are critical to the performance of scramjets. He did this by introducing a new concept called "thermal compression." One approaches it by appreciating that a process of heat addition can play the role of a normal shock wave. When an airflow passes through such a shock, it slows in speed and therefore diminishes in Mach, while its temperature and pressure go up. The same consequences occur when a supersonic airflow is heated. It therefore follows that a process of heat addition can substitute for a normal shock.[30]

Practical inlets use oblique shocks, which are two-dimensional. Such shocks afford good control of the aerodynamics of an inlet. If heat addition is to substitute for an oblique shock, it too must be two-dimensional. Heat addition in a duct is one-dimensional, but Ferri proposed that numerous small burners, set within a flow, could achieve the desired two-dimensionality. By turning individual burners on or off, and by regulating the strength of each one's heating, he could produce the desired pattern of heating that in fact would accomplish the substitution of heating for shock action.[31]

Why would one want to do this? The nose of a hypersonic aircraft produces a strong bow shock, an oblique shock that accomplishes initial compression of the airflow. The inlet rides well behind the nose and features an enclosing cowl. The cowl, in turn, has a lip or outer rim. For best effectiveness, the inlet should sustain a "shock-on-lip" condition. The shock should not impinge within the inlet, for only the lip is cooled in the face of shock-impingement heating. But the shock also should not ride outside the inlet, or the inlet will fail to capture all of the shock-compressed airflow.

To maintain the shock-on-lip condition across a wide Mach range, an inlet requires variable geometry. This is accomplished mechanically, using sliding seals that must not allow leakage of very hot boundary-layer air. Ferri's principle of thermal compression raised the prospect that an inlet could use fixed geometry, which was far simpler. It would do this by modulating its burners rather than by physically moving inlet hardware.

Thermal compression brought an important prospect of flexibility. At a given value of Mach, there typically was only one arrangement of a variable-geometry inlet that would produce the desired shock that would compress the flow. By contrast, the thermal-compression process might be adjusted at will simply by controlling the heating. Ferri proposed to do this by controlling the velocity of injection of the fuel. He wrote that "the heat release is controlled by the mixing process, [which] depends on the difference of velocity of the air and of the injected gas." Shock-free internal flow appeared feasible: "The fuel is injected parallel to the stream to eliminate formation of shocks [and] the combustion process can take place without the formation of shocks." He added,

> "The preliminary analysis of supersonic combustion ramjets...indicates that combustion can occur in a fixed-geometry burner-nozzle combination through a large range of Mach numbers of the air entering the combustion region. Because the Mach number entering the burner is permitted to vary with flight Mach number, the inlet and therefore the complete engine does not require variable geometry. Such an engine can operate over a large range of flight Mach numbers and, therefore, can be very attractive as an accelerating engine."[32]

There was more. As noted, the inlet was to produce a bow shock of specified character, to slow and compress the incoming air. But if the inflow was too great, the inlet would disgorge its shock. This shock, now outside the inlet, would disrupt the flow within the inlet and hence in the engine, with the drag increasing and the thrust falling off sharply. This was known as an unstart.

Supersonic turbojets, such as the Pratt & Whitney J58 that powered the SR-71 to speeds beyond Mach 3, typically were fitted with an inlet that featured a conical spike at the front, a centerbody that was supposed to translate back and forth to adjust the shock to suit the flight Mach number. Early in the program, it often did not work.[33] The test pilot James Eastham was one of the first to fly this spy plane, and he recalls what happened when one of his inlets unstarted.

> "An unstart has your full and undivided attention, right then. The airplane gives a very pronounced yaw; then you are very preoccupied with getting the inlet started again. The speed falls off; you begin to lose altitude. You follow a procedure, putting the spikes forward and opening the bypass doors. Then you would go back to the automatic positioning of the spike—which many times would unstart it again. And when you unstarted on one side, sometimes the other side would also unstart. Then you really had to give it a good massage."[34]

The SR-71 initially used a spike-positioning system from Hamilton Standard. It proved unreliable, and Eastham recalls that at one point, "unstarts were literally stopping the whole program."[35] This problem was eventually overcome through development of a more capable spike-positioning system, built by Honeywell.[36] Still, throughout the development and subsequent flight career of the SR-71, the

First Thoughts of Hypersonic Propulsion

positioning of inlet spikes was always done mechanically. In turn, the movable spike represented a prime example of variable geometry.

Scramjets faced similar issues, particularly near Mach 4. Ferri's thermal-compression principle applied here as well—and raised the prospect of an inlet that might fight against unstarts by using thermal rather than mechanical arrangements. An inlet with thermal compression then might use fixed geometry all the way to orbit, while avoiding unstarts in the bargain.

Ferri presented his thoughts publicly as early as 1960. He went on to give a far more detailed discussion in May 1964, at the Royal Aeronautical Society in London. This was the first extensive presentation on hypersonic propulsion for many in the audience, and attendees responded effusively.

One man declared that "this lecture opened up enormous possibilities. Where they had, for lack of information, been thinking of how high in flight speed they could stretch conventional subsonic burning engines, it was now clear that they should be thinking of how far down they could stretch supersonic burning engines." A. D. Baxter, a Fellow of the Society, added that Ferri "had given them an insight into the prospects and possibilities of extending the speed range of the airbreathing engine far beyond what most of them had dreamed of; in fact, assailing the field which until recently was regarded as the undisputed regime of the rocket."[37]

Not everyone embraced thermal compression. "The analytical basis was rather weak," Marquardt's Arthur Thomas commented. "It was something that he had in his head, mostly. There were those who thought it was a lot of baloney." Nor did Ferri help his cause in 1968, when he published a Mach 6 inlet that offered "much better performance" at lower Mach "because it can handle much higher flow." His paper contained not a single equation.[38]

But Fred Billig was one who accepted the merits of thermal compression and gave his own analyses. He proposed that at Mach 5, thermal compression could increase an engine's specific impulse, an important measure of its performance, by 61 percent. Years later he recalled Ferri's "great capability for visualizing, a strong physical feel. He presented a full plate of ideas, not all of which have been realized."[39]

Combined-Cycle Propulsion Systems

The scramjet used a single set of hardware and operated in two modes, sustaining supersonic combustion as well as subsonic combustion. The transition involved a process called "swallowing the shock." In the subsonic mode, the engine held a train of oblique shocks located downstream of the inlet and forward of the combustor. When the engine went over to the supersonic-combustion mode, these shocks passed through the duct and were lost. This happened automatically, when the flight

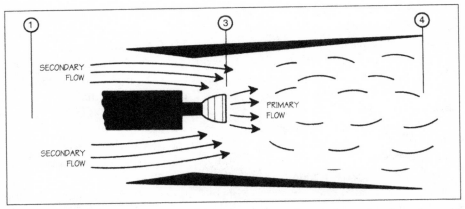

Ejector ramjet. Primary flow from a ducted rocket entrains a substantial secondary flow of external air. (U.S. Air Force)

vehicle topped a critical speed and the engine continued to burn with no diminution of thrust.

The turboramjet arrangement of the XF-103 also operated in two modes, serving both as a turbojet and as a ramjet. Here, however, the engine employed two sets of hardware, which were physically separate. They shared a common inlet and nozzle, while the ramjet also served as the turbojet's afterburner. But only one set of equipment operated at any given time. Moreover, they were mounted separately and were not extensively integrated.[40]

System integration was the key concept within a third class of prime mover: the combined-cycle engine, which sought to integrate two separate thrust-producing cycles within a single set of hardware. In contrast to the turboramjet of the XF-103, engines of this type merged their equipment rather than keeping them separate. Two important concepts that did this were the ejector ramjet, which gave thrust even when standing still, and the Liquid Air Cycle Engine (LACE), which was an airbreathing rocket.

The ejector ramjet amounted to a combined-cycle system derived from a conventional ramjet. It used the ejector principle, whereby the exhaust of a rocket motor, placed within a duct, entrains a flow of air through the duct's inlet. This increases the thrust by converting thermal energy, within the hot exhaust, to mechanical energy. The entrained air slows and cools the exhaust. The effect is much the same as when designers improve the performance of a turbojet engine by installing a fan.

Ejectors date back to the nineteenth century. Horatio Phillips, a pioneer in aeronautical research, used a steam ejector after 1870 to build a wind tunnel. His ejector was a ring of pipe pierced with holes and set within a duct with the holes facing downstream. Pressurized steam, expanding through the holes, entrained an airflow

First Thoughts of Hypersonic Propulsion

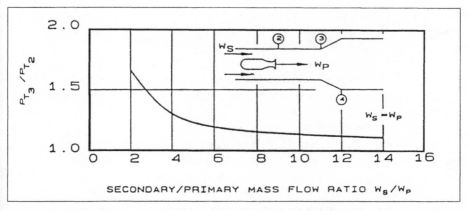

Performance of an ejector. Even with minimal flow of entrained air, the pressure ratio is much lower than that of a turbojet. A pressure ratio of 1.5 implied low efficiency. (Courtesy of William Escher)

within the duct that reached speeds of 40 mph, which served his studies.[41] Nearly a century later ejectors were used to evacuate chambers that conducted rocket-engine tests in near-vacuum, with the ejectors rapidly pumping out the rocket exhaust. Ejectors also flew, being used with both the F-104 and the SR-71. Significantly, the value of an ejector could increase with Mach. On the SR-71, for instance, it contributed only 14 percent of the total thrust at Mach 2.2, but accounted for 28 percent at Mach 3.2.[42]

Jack Charshafian of Curtiss-Wright, director of development of ramjets for Navaho, filed a patent disclosure for an ejector rocket as early as 1947. By entraining outside air, it might run fuel-rich and still burn all its fuel. Ejector concepts also proved attractive to other aerospace pioneers, with patents being awarded to Alfred Africano, a founder of the American Rocket Society; to Hans von Ohain, an inventor of the turbojet; and to Helmut Schelp, who stirred early interest in military turbojets within the Luftwaffe.[43]

A conventional ramjet needed a boost to reach speeds at which its air-ramming effect would come into play, and the hardware requirements approached those of a complete and separate flight vehicle. The turbojet of the XF-103 exemplified what was necessary, as did the large rocket-powered booster of Navaho. But after 1960 the ejector ramjet brought the prospect of a ramjet that could produce thrust even when on the runway. By placing small rocket engines in a step surrounding a duct, a designer could leave the duct free of hardware. It might even sustain supersonic combustion, with the engine converting to a scramjet.

The ejector ramjet also promised to increase the propulsive efficiency by improving the match between flight speed and exhaust speed. A large mismatch greatly reduces the effectiveness of a propulsive jet. There would be little effective thrust, for instance, if one had a jet velocity of 10,000 feet per second while flying in the atmosphere at 400 feet per second. The ejector promised to avoid this by slowing the overall flow.

The ejector ramjet thus offered the enticing concept of a unified engine that could propel a single-stage vehicle from a runway to orbit. It would take off with ejector-boosted thrust from its rocket, accelerate through the atmosphere by using the combination as an ejector-boosted ramjet and scramjet, and then go over completely to rocket propulsion for the final boost to orbit.

Yet even with help from an ejector, a rocket still had a disadvantage. A ramjet or scramjet could use air as its oxidizer, but a rocket had to carry heavy liquid oxygen in an onboard tank. Hence, there also was strong interest in airbreathing rockets. Still, it was not possible to build such a rocket through a simple extension of principles applicable to the turbojet, for there was a serious mismatch between pressures available through turbocompression and those of a rocket's thrust chamber.

In the SR-71, for instance, a combination of inlet compression and turbocompression yielded an internal pressure of approximately 20 pounds per square inch (psi) at Mach 3 and 80,000 feet. By contrast, internal pressures of rocket engines started in the high hundreds of psi and rapidly ascended into the thousands for high performance. Unless one could boost the pressure of ram air to that level, no airbreathing rocket would ever fly.[44]

The concept that overcame this difficulty was LACE. It dated to 1954, and Randolph Rae of the Garrett Corporation was the inventor. LACE used liquid hydrogen both as fuel and as a refrigerant, to liquefy air. The temperature of the liquid hydrogen was only 21 K, far below that at which air liquefies. LACE thus called for incoming ram air to pass through a heat exchanger that used liquid hydrogen as the coolant. The air would liquefy, and then a pump could raise its pressure to whatever value was desired. In this fashion, LACE bypassed the restrictions on turbocompression of gaseous air. In turn, the warmed hydrogen flowed to the combustion chamber to burn in the liquefied air.[45]

At the outset, LACE brought a problem. The limited thermal capacity of liquid hydrogen brought another mismatch, for the system needed eight times more liquid hydrogen to liquefy a given mass of air than could burn in that mass. The resulting hydrogen-rich exhaust still had a sufficiently high velocity to give LACE a prospective advantage over a hydrogen-fueled rocket using tanked oxygen. Even so, there was interest in "derichening" the fuel-air mix, by making use of some of this extra hydrogen. An ejector promised to address this issue by drawing in more air to burn the hydrogen. Such an engine was called a ramLACE or scramLACE.[46]

A complementary strategy called for removal of nitrogen from the liquefied air, yielding nearly pure liquid oxygen as the product. Nitrogen does not support combustion, constitutes some three-fourths of air by weight, and lacks the redeeming quality of low molecular weight that could increase the exhaust velocity. Moreover, a hydrogen-fueled rocket could give much better performance when using oxygen rather than air. With oxygen liquefying at 90 K while nitrogen becomes a liquid at 77 K, at atmospheric pressure the prospect existed of using this temperature difference to leave the nitrogen unliquefied. Nor would it be useless; it could flow within a precooler, an initial heat exchanger that could chill the inflowing air while reserving the much colder liquid hydrogen for the main cooler.

It did not appear feasible in practice to operate a high-capacity LACE air liquefier with the precision in temperature that could achieve this. However, a promising approach called for use of fractional distillation of liquid air, as a variant of the process used in oil refineries to obtain gasoline from petroleum. The distillation process promised fine control, allowing the nitrogen to boil off while keeping the oxygen liquid. To increase the throughput, the distillation was to take place within a rotary apparatus that could impose high g-loads, greatly enhancing the buoyancy of the gaseous nitrogen. A LACE with such an air separator was called ACES, Air Collection and Enrichment System.[47]

When liquid hydrogen chilled and liquefied the nitrogen in air, that hydrogen went only partially to waste. In effect, it transferred its coldness to the nitrogen, which used it to advantage in the precooler. Still, there was a clear prospect of greater efficiency in the heat-transfer process if one could remove the nitrogen directly from the ram air. A variant of ACES promised to do precisely this, using chemical separation of oxygen. The process relied on the existence of metal oxides that could take up additional oxygen when heated by the hot ram air and then release this oxygen when placed under reduced pressure. Only the oxygen then was liquefied. This brought the increased efficiency, for the amount of liquid hydrogen used as a coolant was reduced. This enhanced efficiency also translated into conceptual designs for chemical-separation ACES units that could be lighter in weight and smaller in size than rotary-distillation counterparts.[48]

Turboramjets, ramjets, scramjets, LACE, ramLACE and scramLACE, ACES: with all these in prospect, designers of paper engines beheld a plenitude of possibilities. They also carried a strong mutual synergism. A scramjet might use a type of turboramjet for takeoff, again with the scramjet duct also functioning as an afterburner. Alternately, it might install an internal rocket and become a scramLACE. It could use ACES for better performance, while adopting the chemical-separation process to derichen the use of hydrogen.

It did not take long before engineers rose to their new opportunities by conceiving of new types of vehicles that were to use these engines, perhaps to fly to orbit as a single stage. Everyone in aerospace was well aware that it had taken only 30 years to progress from Lindbergh in Paris to satellites in space. The studies that explored the new possibilities amounted to an assertion that this pace of technical advance was likely to continue.

Aerospaceplane

"I remember when Sputnik was launched," says Arthur Thomas, a leader in early work on scramjets at Marquardt. The date was 4 October 1957. "I was doing analysis of scramjet boosters to go into orbit. We were claiming back in those days that we could get the cost down to a hundred dollars per pound by using airbreathers." He adds that "our job was to push the frontiers. We were extremely excited and optimistic that we were really on the leading edge of something that was going to be big."[49]

At APL, other investigators proposed what may have been the first concept for a hypersonic airplane that merited consideration. In an era when the earliest jet airliners were only beginning to enter service, William Avery leaped beyond the supersonic transport to the hypersonic transport, at least in his thoughts. His colleague Eugene Pietrangeli developed a concept for a large aircraft with a wingspan of 102 feet and length of 175 feet, fitted with turbojets and with the Dugger-Keirsey external-burning scramjet, with its short cowl, under each wing. It was to accelerate to Mach 3.6 using the turbojets, then go over to scramjet propulsion and cruise at Mach 7. Carrying 130 passengers, it was to cross the country in half an hour and achieve a range of 7,000 miles. Its weight of 600,000 pounds was nearly twice that of the Boeing 707 Intercontinental, largest of that family of jetliners.[50]

Within the Air Force, an important prelude to similar concepts came in 1957 with Study Requirement 89774. It invited builders of large missiles to consider what modifications might make them reusable. It was not hard to envision that they might return to a landing on a runway by fitting them with wings and jet engines, but most such rocket stages were built of aluminum, which raised serious issues of thermal protection. Still, Convair at least had a useful point of departure. Its Atlas used stainless steel, which had considerably better heat resistance.[51]

The Convair concept envisioned a new version of this missile, fitted out as a reusable first stage for a launch vehicle. Its wings were to use the X-15's structure. A crew compartment, set atop a rounded nose, recalled that company's B-36 heavy bomber. To ease the thermal problem, designers were aware that this stage, having burned its propellants, would be light in weight. It therefore could execute a hypersonic glide while high in the atmosphere, losing speed slowly and diminishing the rate of heating.[52]

It did not take long before Convair officials began to view this reusable Atlas as merely a first step into space, for the prospect of LACE opened new vistas. Beginning late in 1957, using a combination of Air Force and in-house funding, the company launched paper studies of a new concept called Space Plane. It took shape as a large single-stage vehicle with highly-swept delta wings and a length of 235 feet. Propulsion was to feature a combination of ramjets and LACE with ACES, installed as separate engines, with the ACES being distillation type. The gross weight at takeoff, 450,000 pounds, was to include 270,000 pounds of liquid hydrogen.

First Thoughts of Hypersonic Propulsion

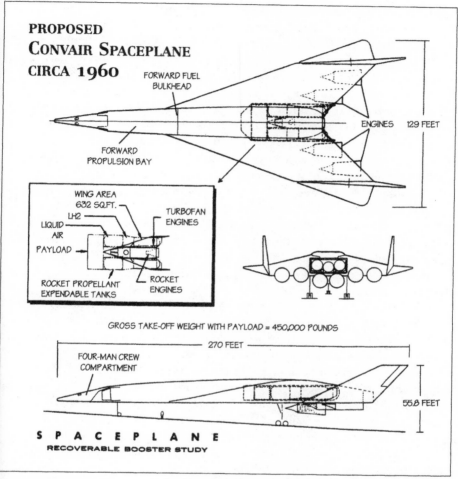

Convair's Space Plane concept. (Art by Dennis Jenkins)

Space Plane was to take off from a runway, using LACE and ACES while pumping the oxygen-rich condensate directly to the LACE combustion chambers. It would climb to 40,000 feet and Mach 3, cut off the rocket, and continue to fly using hydrogen-fueled ramjets. It was to use ACES for air collection while cruising at Mach 5.5 and 66,000 feet, trading liquid hydrogen for oxygen-rich liquid air while taking on more than 600,000 pounds of this oxidizer. Now weighing more than a million pounds, Space Plane would reach Mach 7 on its ramjets, then shut them down and go over completely to rocket power. Drawing on its stored oxidizer, it could fly to orbit while carrying a payload of 38,000 pounds.

The concept was born in exuberance. Its planners drew on estimates "that by 1970 the orbital payload accumulated annually would be somewhere between two million and 20 million pounds." Most payloads were to run near 10,000 pounds, thereby calling for a schedule of three flights per day. Still the concept lacked an important element, for if scramjets were nowhere near the state of the art, at Convair they were not even the state of the imagination.[53] Space Plane, as noted, used ramjets with subsonic combustion, installing them in pods like turbojets on a B-52. Scramjets lay beyond the thoughts of other companies as well. Thus, Northrop expected to use LACE with its Propulsive Fluid Accumulator (PROFAC) concept, which also was to cruise in the atmosphere while building up a supply of liquefied air. Like Space Plane, PROFAC also specified conventional ramjets.[54]

But Republic Aviation was home to the highly imaginative Kartveli, with Ferri being just a phone call away. Here the scramjet was very much a part of people's thinking. Like the Convair designers, Kartveli looked ahead to flight to orbit with a single stage. He also expected that this goal was too demanding to achieve in a single jump, and he anticipated that intermediate projects would lay groundwork. He presented his thoughts in August 1960 at a national meeting of the Institute of Aeronautical Sciences.[55]

The XF-103 had been dead and buried for three years, but Kartveli had crafted the F-105, which topped Mach 2 as early as 1956 and went forward into production. He now expected to continue with a Mach 2.3 fighter-bomber with enough power to lift off vertically as if levitating and to cruise at 75,000 feet. Next on the agenda was a strategic bomber powered by nuclear ramjets, which would use atomic power to heat internal airflow, with no need to burn fuel. It would match the peak speed of the X-7 by cruising at Mach 4.25, or 2,800 mph, and at 85,000 feet.[56]

Kartveli set Mach 7, or 5,000 mph, as the next goal. He anticipated achieving this speed with another bomber that was to cruise at 120,000 feet. Propulsion was to come from two turbojets and two ramjets, with this concept pressing the limits of subsonic combustion. Then for flight to orbit, his masterpiece was slated for Mach 25. It was to mount four J58 turbojets, modified to burn hydrogen, along with four scramjets. Ferri had convinced him that such engines could accelerate this craft all the way to orbit, with much of the gain in speed taking place while flying at 200,000 feet. A small rocket engine might provide a final boost into space, but Kartveli placed his trust in Ferri's scramjets, planning to use neither LACE nor ACES.[57]

These concepts drew attention, and funding, from the Aero Propulsion Laboratory at Wright-Patterson Air Force Base. Its technical director, Weldon Worth, had been closely involved with ramjets since the 1940s. Within a world that the turbojet had taken by storm, he headed a Nonrotating Engine Branch that focused on ramjets and liquid-fuel rockets. Indeed, he regarded the ramjet as holding the greater

First Thoughts of Hypersonic Propulsion

promise, taking this topic as his own while leaving the rockets to his deputy, Lieutenant Colonel Edward Hall. He launched the first Air Force studies of hypersonic propulsion as early as 1957. In October 1959 he chaired a session on scramjets at the Second USAF Symposium on Advanced Propulsion Concepts.

In the wake of this meeting, he built on the earlier SR-89774 efforts and launched a new series of studies called Aerospaceplane. It did not aim at anything so specific as a real airplane that could fly to orbit. Rather, it supported design studies and conducted basic research in advanced propulsion, seeking to develop a base for the evolution of such craft in the distant future. Marquardt and GASL became heavily involved, as did Convair, Republic, North American, GE, Lockheed, Northrop, and Douglas Aircraft.[58]

The new effort broadened the scope of the initial studies, while encouraging companies to pursue their concepts to greater depth. Convair, for one, had issued single-volume reports on Space Plane in October 1959, April 1960, and December 1960. In February 1961 it released an 11-volume set of studies, with each of them addressing a specific topic such as Aerodynamic Heating, Propulsion, Air Enrichment Systems, Structural Analysis, and Materials.[59]

Aerospaceplane proved too hot to keep under wraps, as a steady stream of disclosures presented concept summaries to the professional community and the general public. *Aviation Week*, hardly shy in these matters, ran a full-page article in October 1960:

USAF PLANS RADICAL SPACE PLANE
Studies costing $20 million sought in next budget, Earth-to-orbit vehicle would need no large booster.[60]

At the *Los Angeles Times*, the aerospace editor Marvin Miles published headlined stories of his own. The first appeared in November:

LOCKHEED WORKING ON PLANE ABLE TO GO INTO ORBIT ALONE
Air Force Interested in Project[61]

Two months later another of his articles ran as a front-page headline:

HUGE BOOSTER NOT NEEDED BY AIR FORCE SPACE PLANE
Proposed Wing Vehicle Would Take Off, Return Like Conventional Craft

It particularly cited Convair's Space Plane, with a *Times* artist presenting a view of this craft in flight.[62]

Participants in the new studies took to the work with enthusiasm matching that of Arthur Thomas at Marquardt. Robert Sanator, a colleague of Kartveli at Republic, recalls the excitement: "This one had everything. There wasn't a single thing in it that was off-the-shelf. Whatever problem there was in aerospace—propulsion, materials, cooling, aerodynamics—Aerospaceplane had it. It was a lifetime work and it had it all. I naturally jumped right in."[63]

Aerospaceplane also drew attention from the Air Force's Scientific Advisory Board, which set up an ad hoc committee to review its prospects. Its chairman, Alexander Flax, was the Air Force's chief scientist. Members specializing in propulsion included Ferri, along with Seymour Bogdonoff of Princeton University, a leading experimentalist; Perry Pratt of Pratt & Whitney, who had invented the twin-spool turbojet; NASA's Alfred Eggers; and the rocket specialist George P. Sutton. There also were hands-on program managers: Robert Widmer of Convair, builder of the Mach 2 B-58 bomber; Harrison Storms of North American, who had shaped the X-15 and the Mach 3 XB-70 bomber.[64]

This all-star group came away deeply skeptical of the prospects for Aerospaceplane. Its report, issued in December 1960, addressed a number of points and gave an overall assessment:

> The proposed designs for Aerospace Plane...appear to violate no physical principles, but the attractive performance depends on an estimated combination of optimistic assumptions for the performance of components and subsystems. There are practically no experimental data which support these assumptions.

On LACE and ACES:

> We consider the estimated LACE-ACES performance very optimistic. In several cases complete failure of the project would result from any significant performance degradation from the present estimates.... Obviously the advantages claimed for the system will not be available unless air can be condensed *and purified* very rapidly during flight. The figures reported indicate that about 0.8 ton of air per second would have to be processed. In conventional, i.e., ordinary commercial equipment, this would require a distillation column having a cross section on the order of 500 square feet.... It is proposed to increase the capacity of equipment of otherwise conventional design by using centrifugal force. This may be possible, but as far as the Committee knows this has never been accomplished.

First Thoughts of Hypersonic Propulsion

On other propulsion systems:

> When reduced to a common basis and compared with the best of current technology, all assumed large advances in the state-of-the-art.... On the basis of the best of current technology, none of the schemes could deliver useful payloads into orbits.

On vehicle design:

> We are gravely concerned that too much emphasis may be placed on the more glamorous aspects of the Aerospace Plane resulting in neglect of what appear to be more conventional problems. The achievement of low structural weight is equally important...as is the development of a highly successful propulsion system.

Regarding scramjets, the panel was not impressed with claims that supersonic combustion had been achieved in existing experiments:

> These engine ideas are based essentially upon the feasibility of diffusion deflagration flames in supersonic flows. Research should be immediately initiated using existing facilities...to substantiate the feasibility of this type of combustion.

The panelists nevertheless gave thumbs-up to the Aerospaceplane effort as a continuing program of research. Their report urged a broadening of topics, placing greater emphasis on scramjets, structures and materials, and two-stage-to-orbit configurations. The array of proposed engines were "all sufficiently interesting so that research on all of them should be continued and emphasized."[65]

As the studies went forward in the wake of this review, new propulsion concepts continued to flourish. Lockheed was in the forefront. This firm had initiated company-funded work during the spring of 1959 and had a well-considered single-stage concept two years later. An artist's rendering showed nine separate rocket nozzles at its tail. The vehicle also mounted four ramjets, set in pods beneath the wings.

Convair's Space Plane had used separated nitrogen as a propellant, heating it in the LACE precooler and allowing it to expand through a nozzle to produce thrust. Lockheed's Aerospace Plane turned this nitrogen into an important system element, with specialized nitrogen rockets delivering 125,000 pounds of thrust. This certainly did not overcome the drag produced by air collection, which would have turned the vehicle into a perpetual motion machine. However, the nitrogen rockets made a valuable contribution.[66]

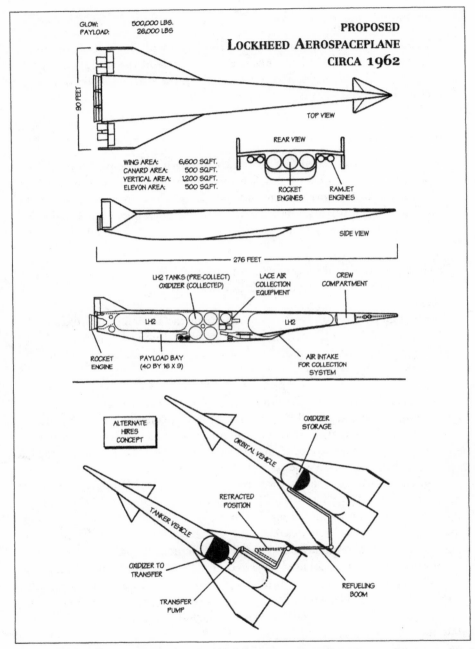

Lockheed's Aerospaceplane concept. The alternate hypersonic in-flight refueling system approach called for propellant transfer at Mach 6. (Art by Dennis Jenkins)

First Thoughts of Hypersonic Propulsion

Republic's Aerospaceplane concept showed extensive engine-airframe integration. (Republic Aviation)

For takeoff, Lockheed expected to use Turbo-LACE. This was a LACE variant that sought again to reduce the inherently hydrogen-rich operation of the basic system. Rather than cool the air until it was liquid, Turbo-Lace chilled it deeply but allowed it to remain gaseous. Being very dense, it could pass through a turbocompressor and reach pressures in the hundreds of psi. This saved hydrogen because less was needed to accomplish this cooling. The Turbo-LACE engines were to operate at chamber pressures of 200 to 250 psi, well below the internal pressure of standard rockets but high enough to produce 300,000 pounds of thrust by using turbocompressed oxygen.[67]

Republic Aviation continued to emphasize the scramjet. A new configuration broke with the practice of mounting these engines within pods, as if they were turbojets. Instead, this design introduced the important topic of engine-airframe integration by setting forth a concept that amounted to a single enormous scramjet fitted with wings and a tail. A conical forward fuselage served as an inlet spike. The inlets themselves formed a ring encircling much of the vehicle. Fuel tankage filled most of its capacious internal volume.

This design study took two views regarding the potential performance of its engines. One concept avoided the use of LACE or ACES, assuming again that this craft could scram all the way to orbit. Still, it needed engines for takeoff so turbo-ramjets were installed, with both Pratt & Whitney and General Electric providing candidate concepts. Republic thus was optimistic at high Mach but conservative at low speed.

The other design introduced LACE and ACES both for takeoff and for final ascent to orbit and made use of yet another approach to derichening the hydrogen. This was SuperLACE, a concept from Marquardt that placed slush hydrogen rather than standard liquid hydrogen in the main tank. The slush consisted of liquid that contained a considerable amount of solidified hydrogen. It therefore stood at the freezing point of hydrogen, 14 K, which was markedly lower than the 21 K of liquid hydrogen at the boiling point.[68]

SuperLACE reduced its use of hydrogen by shunting part of the flow, warmed in the LACE heat exchanger, into the tank. There it mixed with the slush, chilling again to liquid while melting some of the hydrogen ice. Careful control of this flow ensured that while the slush in the tank gradually turned to liquid and rose toward the 21 K boiling point, it did not get there until the air-collection phase of a flight was finished. As an added bonus, the slush was noticeably denser than the liquid, enabling the tank to hold more fuel.[69]

LACE and ACES remained in the forefront, but there also was much interest in conventional rocket engines. Within the Aerospaceplane effort, this approach took the name POBATO, Propellants On Board At Takeoff. These rocket-powered vehicles gave points of comparison for the more exotic types that used LACE and scramjets, but here too people used their imaginations. Some POBATO vehicles ascended vertically in a classic liftoff, but others rode rocket sleds along a track while angling sharply upward within a cradle.[70]

In Denver, the Martin Company took rocket-powered craft as its own, for this firm expected that a next-generation launch vehicle of this type could be ready far sooner than one based on advanced airbreathing engines. Its concepts used vertical liftoff, while giving an opening for the ejector rocket. Martin introduced a concept of its own called RENE, Rocket Engine Nozzle Ejector (RENE), and conducted experiments at the Arnold Engineering Development Center. These tests went forward during 1961, using a liquid rocket engine, with nozzle of 5-inch diameter set within a shroud of 17-inch width. Test conditions corresponded to flight at Mach 2 and 40,000 feet, with the shrouds or surrounding ducts having various lengths to achieve increasingly thorough mixing. The longest duct gave the best performance, increasing the rated 2,000-pound thrust of the rocket to as much as 3,100 pounds.[71]

A complementary effort at Marquardt sought to demonstrate the feasibility of LACE. The work started with tests of heat exchangers built by Garrett AiResearch that used liquid hydrogen as the working fluid. A company-made film showed dark liquid air coming down in a torrent, as seen through a porthole. Further tests used this liquefied air in a small thrust chamber. The arrangement made no attempt to derichen the hydrogen flow; even though it ran very fuel-rich, it delivered up to 275 pounds of thrust. As a final touch, Marquardt crafted a thrust chamber of 18-inch diameter and simulated LACE operation by feeding it with liquid air and gaseous

First Thoughts of Hypersonic Propulsion

hydrogen from tanks. It showed stable combustion, delivering thrust as high as 5,700 pounds.[72]

Within the Air Force, the SAB's Ad Hoc Committee on Aerospaceplane continued to provide guidance along with encouraging words. A review of July 1962 was less skeptical in tone than the one of 18 months earlier, citing "several attractive arguments for a continuation of this program at a significant level of funding":

> It will have the military advantages that accrue from rapid response times and considerable versatility in choice of landing area. It will have many of the advantages that have been demonstrated in the X-15 program, namely, a real pay-off in rapidly developing reliability and operational pace that comes from continuous re-use of the same hardware again and again. It may turn out in the long run to have a cost effectiveness attractiveness… the cost per pound may eventually be brought to low levels. Finally, the Aerospaceplane program will develop the capability for flights in the atmosphere at hypersonic speeds, a capability that may be of future use to the Defense Department and possibly to the airlines.[73]

Single-stage-to-orbit (SSTO) was on the agenda, a topic that merits separate comment. The space shuttle is a stage-and-a-half system; it uses solid boosters plus a main stage, with all engines burning at liftoff. It is a measure of progress, or its lack, in astronautics that the Soviet R-7 rocket that launched the first Sputniks was also stage-and-a-half.[74] The concept of SSTO has tantalized designers for decades, with these specialists being highly ingenious and ready to show a can-do spirit in the face of challenges.

This approach certainly is elegant. It also avoids the need to launch two rockets to do the work of one, and if the Earth's gravity field resembled that of Mars, SSTO would be the obvious way to proceed. Unfortunately, the Earth's field is considerably stronger. No SSTO has ever reached orbit, either under rocket power or by using scramjets or other airbreathers. The technical requirements have been too severe.

The SAB panel members attended three days of contractor briefings and reached a firm conclusion: "It was quite evident to the Committee from the presentation of nearly all the contractors that a single stage to orbit Aerospaceplane remains a highly speculative effort." Reaffirming a recommendation from its 1960 review, the group urged new emphasis on two-stage designs. It recommended attention to "development of hydrogen fueled turbo ramjet power plants capable of accelerating the first stage to Mach 6.0 to 10.0…. Research directed toward the second stage which will ultimately achieve orbit should be concentrated in the fields of high pressure hydrogen rockets and supersonic burning ramjets and air collection and enrichment systems."[75]

Convair, home of Space Plane, had offered single-stage configurations as early as 1960. By 1962 its managers concluded that technical requirements placed such a vehicle out of reach for at least the next 20 years. The effort shifted toward a two-stage concept that took form as the 1964 Point Design Vehicle. With a gross takeoff weight of 700,000 pounds, the baseline approach used turboramjets to reach Mach 5. It cruised at that speed while using ACES to collect liquid oxygen, then accelerated anew using ramjets and rockets. Stage separation occurred at Mach 8.6 and 176,000 feet, with the second stage reaching orbit on rocket power. The payload was 23,000 pounds with turboramjets in the first stage, increasing to 35,000 pounds with the more speculative SuperLACE.

The documentation of this 1964 Point Design, filling 16 volumes, was issued during 1963. An important advantage of the two-stage approach proved to lie in the opportunity to optimize the design of each stage for its task. The first stage was a Mach 8 aircraft that did not have to fly to orbit and that carried its heavy wings, structure, and ACES equipment only to staging velocity. The second-stage design showed strong emphasis on re-entry; it had a blunted shape along with only modest requirements for aerodynamic performance. Even so, this Point Design pushed the state of the art in materials. The first stage specified superalloys for the hot underside along with titanium for the upper surface. The second stage called for coated refractory metals on its underside, with superalloys and titanium on its upper surfaces.[76]

Although more attainable than its single-stage predecessors, the Point Design still relied on untested technologies such as ACES, while anticipating use in aircraft structures of exotic metals that had been studied merely as turbine blades, if indeed they had gone beyond the status of laboratory samples. The opportunity nevertheless existed for still greater conservatism in an airbreathing design, and the man who pursued it was Ernst Steinhoff. He had been present at the creation, having worked with Wernher von Braun on Germany's wartime V-2, where he headed up the development of that missile's guidance. After 1960 he was at the Rand Corporation, where he examined Aerospaceplane concepts and became convinced that single-stage versions would never be built. He turned to two-stage configurations and came up with an outline of a new one: ROLS, the Recoverable Orbital Launch System. During 1963 he took the post of chief scientist at Holloman Air Force Base and proceeded to direct a formal set of studies.[77]

The name of ROLS had been seen as early as 1959, in one of the studies that had grown out of SR-89774, but this concept was new. Steinhoff considered that the staging velocity could be as low as Mach 3. At once this raised the prospect that the first stage might take shape as a modest technical extension of the XB-70, a large bomber designed for flight at that speed, which at the time was being readied for flight test. ROLS was to carry a second stage, dropping it from the belly like a bomb, with that stage flying on to orbit. An ACES installation would provide the liquid oxidizer prior to separation, but to reach from Mach 3 to orbital speed, the second stage had to be simple indeed. Steinhoff envisioned a long vehicle resembling a torpedo, powered by hydrogen-burning rockets but lacking wings and thermal protection. It was not reusable and would not reenter, but it would be piloted. A project report stated, "Crew recovery is accomplished by means of a reentry capsule of the Gemini-Apollo class. The capsule forms the nose section of the vehicle and serves as the crew compartment for the entire vehicle."[78]

ROLS appears in retrospect as a mirror image of NASA's eventual space shuttle, which adopted a technically simple booster—a pair of large solid-propellant rockets—while packaging the main engines and most other costly systems within a fully-recoverable orbiter. By contrast, ROLS used a simple second stage and a highly intricate first stage, in the form of a large delta-wing airplane that mounted eight turbojet engines. Its length of 335 feet was more than twice that of a B-52. Weighing 825,000 pounds at takeoff, ROLS was to deliver a payload of 30,000 pounds to orbit.[79]

Such two-stage concepts continued to emphasize ACES, while still offering a role for LACE. Experimental test and development of these concepts therefore remained on the agenda, with Marquardt pursuing further work on LACE. The earlier tests, during 1960 and 1961, had featured an off-the-shelf thrust chamber that had seen use in previous projects. The new work involved a small LACE engine, the MA117, that was designed from the start as an integrated system.

LACE had a strong suit in its potential for a very high specific impulse, I_{sp}. This is the ratio of thrust to propellant flow rate and has dimensions of seconds. It is a key measure of performance, is equivalent to exhaust velocity, and expresses the engine's fuel economy. Pratt & Whitney's RL10, for instance, burned hydrogen and oxygen to give thrust of 15,000 pounds with an I_{sp} of 433 seconds.[80] LACE was an airbreather, and its I_{sp} could be enormously higher because it took its oxidizer from the atmosphere rather than carrying it in an onboard tank. The term "propellant flow rate" referred to tanked propellants, not to oxidizer taken from the air. For LACE this meant fuel only.

The basic LACE concept produced a very fuel-rich exhaust, but approaches such as RENE and SuperLACE promised to reduce the hydrogen flow substantially. Indeed, such concepts raised the prospect that a LACE system might use an optimized mixture ratio of hydrogen and oxidizer, with this ratio being selected to give the highest I_{sp}. The MA117 achieved this performance artificially by using a large flow of liquid hydrogen to liquefy air and a much smaller flow for the thrust chamber. Hot-fire tests took place during December 1962, and a company report stated that "the system produced 83% of the idealized theoretical air flow and 81% of the idealized thrust. These deviations are compatible with the simplifications of the idealized analysis."[81]

The best performance run delivered 0.783 pounds per second of liquid air, which burned a flow of 0.0196 pounds per second of hydrogen. Thrust was 73 pounds; I_{sp} reached 3,717 seconds, more than eight times that of the RL10. Tests of the MA117 continued during 1963, with the best measured values of I_{sp} topping 4,500 seconds.[82]

In a separate effort, the Marquardt manager Richard Knox directed the preliminary design of a much larger LACE unit, the MA116, with a planned thrust of 10,000 pounds. On paper, it achieved substantial derichening by liquefying only one-fifth of the airflow and using this liquid air in precooling, while deeply cooling the rest of the airflow without liquefaction. A turbocompressor then was to pump this chilled air into the thrust chamber. A flow of less than four pounds per second of liquid hydrogen was to serve both as fuel and as primary coolant, with the anticipated I_{sp} exceeding 3,000 seconds.[83]

New work on RENE also flourished. The Air Force had a cooperative agreement with NASA's Marshall Space Flight Center, where Fritz Pauli had developed a subscale rocket engine that burned kerosene with liquid oxygen for a thrust of 450 pounds. Twelve of these small units, mounted to form a ring, gave a basis for this new effort. The earlier work had placed the rocket motor squarely along the centerline of the duct. In the new design, the rocket units surrounded the duct, leaving it unobstructed and potentially capable of use as an ejector ramjet. The cluster was tested successfully at Marshall in September 1963 and then went to the Air Force's AEDC. As in the RENE tests of 1961, the new configuration gave a thrust increase of as much as 52 percent.[84]

While work on LACE and ejector rockets went forward, ACES stood as a particularly critical action item. Operable ACES systems were essential for the practical success of LACE. Moreover, ACES had importance distinctly its own, for it could provide oxidizer to conventional hydrogen-burning rocket engines, such as those of ROLS. As noted earlier, there were two techniques for air separation: by chemical methods and through use of a rotating fractional distillation apparatus. Both approaches went forward, each with its own contractor.

In Cambridge, Massachusetts, the small firm of Dynatech took up the challenge of chemical separation, launching its effort in May 1961. Several chemical reactions appeared plausible as candidates, with barium and cobalt offering particular promise:

$$2BaO_2 \rightleftarrows 2BaO + O_2$$

$$2Co_3O_4 \rightleftarrows 6CoO + O_2$$

First Thoughts of Hypersonic Propulsion

The double arrows indicate reversibility. The oxidation reactions were exothermic, occurring at approximately 1,600°F for barium and 1,800°F for cobalt. The reduction reactions, which released the oxygen, were endothermic, allowing the oxides to cool as they yielded this gas.

Dynatech's separator unit consisted of a long rotating drum with its interior divided into four zones using fixed partitions. A pebble bed of oxide-coated particles lined the drum interior; containment screens held the particles in place while allowing the drum to rotate past the partitions with minimal leakage. The zones exposed the oxide alternately to high-pressure ram air for oxidation and to low pressure for reduction. The separation was to take place in flight, at speeds of Mach 4 to Mach 5, but an inlet could slow the internal airflow to as little as 50 feet per second, increasing the residence time of air within a unit. The company proposed that an array of such separators weighing just under 10 tons could handle 2,000 pounds per second of airflow while producing liquid oxygen of 65 percent purity.[85]

Ten tons of equipment certainly counts within a launch vehicle, even though it included the weight of the oxygen liquefaction apparatus. Still it was vastly lighter than the alternative: the rotating distillation system. The Linde Division of Union Carbide pursued this approach. Its design called for a cylindrical tank containing the distillation apparatus, measuring nine feet long by nine feet in diameter and rotating at 570 revolutions per minute. With a weight of 9,000 pounds, it was to process 100 pounds per second of liquefied air—which made it 10 times as heavy as the Dynatech system, per pound of product. The Linde concept promised liquid oxygen of 90 percent purity, substantially better than the chemical system could offer, but the cited 9,000-pound weight left out additional weight for the LACE equipment that provided this separator with its liquefied air.[86]

A study at Convair, released in October 1963, gave a clear preference to the Dynatech concept. Returning to the single-stage Space Plane of prior years, Convair engineers considered a version with a weight at takeoff of 600,000 pounds, using either the chemical or the distillation ACES. The effort concluded that the Dynatech separator offered a payload to orbit of 35,800 using barium and 27,800 pounds with cobalt. The Linde separator reduced this payload to 9,500 pounds. Moreover, because it had less efficiency, it demanded an additional 31,000 pounds of hydrogen fuel, along with an increase in vehicle volume of 10,000 cubic feet.[87]

The turn toward feasible concepts such as ROLS, along with the new emphasis on engineering design and test, promised a bright future for Aerospaceplane studies. However, a commitment to serious research and development was another matter. Advanced test facilities were critical to such an effort, but in August 1963 the Air Force canceled plans for a large Mach 14 wind tunnel at AEDC. This decision gave a clear indication of what lay ahead.[88]

A year earlier Aerospaceplane had received a favorable review from the SAB Ad Hoc Committee. The program nevertheless had its critics, who existed particularly within the SAB's Aerospace Vehicles and Propulsion panels. In October 1963 they issued a report that dealt with proposed new bombers and vertical-takeoff-and-landing craft, as well as with Aerospaceplane, but their view was unmistakable on that topic:

> The difficulties the Air Force has encountered over the past three years in identifying an Aerospaceplane program have sprung from the facts that the requirement for a fully recoverable space launcher is at present only vaguely defined, that today's state-of-the-art is inadequate to support any real hardware development, and the cost of any such undertaking will be extremely large.... [T]he so-called Aerospaceplane program has had such an erratic history, has involved so many clearly infeasible factors, and has been subject to so much ridicule that from now on this name should be dropped. It is also recommended that the Air Force increase the vigilance that no new program achieves such a difficult position.[89]

Aerospaceplane lost still more of its rationale in December, as Defense Secretary Robert McNamara canceled Dyna-Soar. This program was building a mini-space shuttle that was to fly to orbit atop a Titan III launch vehicle. This craft was well along in development at Boeing, but program reviews within the Pentagon had failed to find a compelling purpose. McNamara thus disposed of it.[90]

Prior to this action, it had been possible to view Dyna-Soar as a prelude to operational vehicles of that general type, which might take shape as Aerospaceplanes. The cancellation of Dyna-Soar turned the Aerospaceplane concept into an orphan, a long-term effort with no clear relation to anything currently under way. In the wake of McNamara's decision, Congress deleted funds for further Aerospaceplane studies, and Defense Department officials declined to press for its restoration within the FY 1964 budget, which was under consideration at that time. The Air Force carried forward with new conceptual studies of vehicles for both launch and hypersonic cruise, but these lacked the focus on advanced airbreathing propulsion that had characterized Aerospaceplane.[91]

There nevertheless was real merit to some of the new work, for this more realistic and conservative direction pointed out a path that led in time toward NASA's space shuttle. The Martin Company made a particular contribution. It had designed no Aerospaceplanes; rather, using company funding, its technical staff had examined concepts called Astrorockets, with the name indicating the propulsion mode. Scramjets and LACE won little attention at Martin, but all-rocket vehicles were another matter. A concept of 1964 had a planned liftoff weight of 1,250 tons, making it intermediate in size between the Saturn I-B and Saturn V. It was a two-stage fully-reusable configuration, with both stages having delta wings and flat undersides. These undersides fitted together at liftoff, belly to belly.

First Thoughts of Hypersonic Propulsion

Martin's Astrorocket. (U.S. Air Force)

The design concepts of that era were meant to offer glimpses of possible futures, but for this Astrorocket, the future was only seven years off. It clearly foreshadowed a class of two-stage fully reusable space shuttles, fitted with delta wings, that came to the forefront in NASA-sponsored studies of 1971. The designers at Martin were not clairvoyant; they drew on the background of Dyna-Soar and on studies at NASA-Ames of winged re-entry vehicles. Still, this concept demonstrated that some design exercises were returning to the mainstream.[92]

Further work on ACES also proceeded, amid unfortunate results at Dynatech. That company's chemical separation processes had depended for success on having a very large area of reacting surface within the pebble-bed air separators. This appeared achievable through such means as using finely divided oxide powders or porous particles impregnated with oxide. But the research of several years showed that the oxide tended to sinter at high temperatures, markedly diminishing the reacting surface area. This did not make chemical separation impossible, but it sharply increased the size and weight of the equipment, which robbed this approach of its initially strong advantage over the Linde distillation system. This led to abandonment of Dynatech's approach.[93]

Linde's system was heavy and drastically less elegant than Dynatech's alternative, but it amounted largely to a new exercise in mechanical engineering and went forward to successful completion. A prototype operated in test during 1966, and while limits to the company's installed power capacity prevented the device from processing the rated flow of 100 pounds of air per second, it handled 77 pounds per second, yielding a product stream of oxygen that was up to 94 percent pure. Studies of lighter-weight designs also proceeded. In 1969 Linde proposed to build a distillation air separator, rated again at 100 pounds per second, weighing 4,360 pounds. This was only half the weight allowance of the earlier configuration.[94]

In the end, though, Aerospaceplane failed to identify new propulsion concepts that held promise and that could be marked for mainstream development. The program's initial burst of enthusiasm had drawn on a view that the means were in hand, or soon would be, to leap beyond the liquid-fuel rocket as the standard launch vehicle and to pursue access to orbit using methods that were far more advanced. The advent of the turbojet, which had swiftly eclipsed the piston engine, was on everyone's mind. Yet for all the ingenuity behind the new engine concepts, they failed to deliver. What was worse, serious technical review gave no reason to believe that they could deliver.

In time it would become clear that hypersonics faced a technical wall. Only limited gains were achievable in airbreathing propulsion, with single-stage-to-orbit remaining out of reach and no easy way at hand to break through to the really advanced performance for which people hoped.

First Thoughts of Hypersonic Propulsion

1. See, for instance, Emme, *History*; Ley, *Rockets*; McDougall, *Heavens*; Neufeld, *Ballistic*; NASA SP-4206; and Heppenheimer, *Countdown*.
2. NASA SP-4404; SP-4221, pp. 105-08, 423-25.
3. *Johns Hopkins APL Technical Digest*, Vol. 13, No. 1 (1992), p. 57; Lay, *Thermodynamics*, pp. 574-76.
4. Miller, *X-Planes*, ch. 11.
5. Ibid., pp. 116-17. X-2 record: NASA SP-4303, p. 316.
6. Peterson, "Evaluation." 4130 steel: Miller, *X-Planes*, p. 119.
7. Ritchie, "Evaluation."
8. Miller, *X-Planes*, p. 117; Crickmore, *SR-71*, pp. 154-55.
9. "Standard Missile Characteristics: IM-99A Bomarc." Its name: Cornett, "Overview," p. 109. Extended range: *Jane's*, 1967-1968, p. 490.
10. "Development of Bomarc," pp. 10-12; Bagwell, "History," pp. 12-13.
11. Bagwell, "History," p. 17 (includes quotes); Pfeifer, "Bomarc," p. 1; Report Boeing D-11532 (Boeing), pp. 43-44.
12. *Jane's*, 1960-61, p. 447; Cornett, "Overview," pp. 118, 124-25.
13. *Johns Hopkins APL Technical Digest*, Vol. 3, No. 2 (1982), pp. 117-22; *Jane's*, 1960-61, p. 446.
14. "Standard Missile Characteristics: XSM-64 Navaho"; Gibson, *Navaho*, pp. 35-36, 45-48.
15. Gibson, *Navaho*, pp. 63-77; "Development of the SM-64," 1954-58, pp. 100-07. Author interview, J. Leland Atwood, 18 July 1988. Folder 18649, NASA Historical Reference Collection, NASA History Division, Washington, D.C. 20546.
16. *Air Enthusiast*, July-September 1978, pp. 198-213 (quote, p. 213). XF-103, XF-104: Gunston, *Fighters*, pp. 121-22, 193.
17. Author interview, Frederick Billig, 27 June 1987. Folder 18649, NASA Historical Reference Collection, NASA History Division, Washington, D.C. 20546. See also CASI 88N-12321, p. 12-6.
18. Fletcher, "Supersonic"; NACA: TN 2206, RM E51K26.
19. Fletcher, "Supersonic."
20. NACA TN 4386 (quote, p. 22).
21. Fletcher, "Supersonic," p. 737.
22. *Johns Hopkins APL Technical Digest*, Vol. 11, Nos. 3 and 4 (1990), pp. 319-35 (quotes, pp. 328-29).
23. "Statement Regarding the Military Service of Antonio Ferri"; Kaufman et al., *Moe Berg*, pp. 182-88 (quote, p. 186).
24. *Journal of Aircraft*, January-February 1968, p. 3; *Journal of the Royal Aeronautical Society*, September 1964, p. 575; *Johns Hopkins Magazine*, December 1988 p. 53; author interview, Louis Nucci, 24 June 1987.
25. Author interviews: John Erdos, 22 July 1985. Folder 18649, NASA Historical Reference Collection, NASA History Division, Washington, D.C. 20546; Robert Sanator, 1 February 1988. NASA Historical Reference Collection, NASA History Division, Washington, D.C. 20546.

26 Author interview, Louis Nucci, 24 June 1988 (includes quote). Folder 18649, NASA Historical Reference Collection, NASA History Division, Washington, D.C. 20546.

27 ISABE Paper 97-7004, p. 1.

28 Quote: *Journal of the Royal Aeronautical Society,* September 1964, p. 577.

29 ISABE Paper 97-7004, p. 2. Quote: *Astronautics & Aeronautics,* August 1964, p. 33.

30 Shapiro, *Compressible,* pp. 198-99.

31 Ferri and Fox, "Analysis."

32 Quotes: *Astronautics & Aeronautics,* August 1964, p. 33; *Journal of the Royal Aeronautical Society,* September 1964, p. 577.

33 Shapiro, *Compressible,* pp. 147-51; *Lockheed Horizons,* Winter 1981/1982, 11-12; Crickmore, *SR-71,* pp. 26, 94-95.

34 Author interview, James Eastham, 1 May 1988. Folder 18649, NASA Historical Reference Collection, NASA History Division, Washington, D.C. 20546.

35 Ibid.

36 Crickmore, *SR-71,* pp. 125-26.

37 Ferri, "Possible Directions"; *Journal of the Royal Aeronautical Society,* September 1964, pp. 575-97 (quotes, pp. 595, 597).

38 Ferri and Fox, "Analysis" (quotes, p. 1111). Arthur Thomas quotes: Arthur Thomas, author interview, 24 September 1987. Folder 18649, NASA Historical Reference Collection, History Division, Washington, D.C. 20546.

39 *Journal of Spacecraft and Rockets,* September 1968, pp. 1076-81. Quote: Fred Billig interview, 16 October 1987. Folder 18649, NASA Historical Reference Collection, NASA History Division, Washington, D.C. 20546.

40 *Air Enthusiast,* July-September 1978, p. 210.

41 Heppenheimer, *First Flight,* pp. 63, 142.

42 Gunston, *Fighters,* p. 121; Crickmore, *SR-71,* pp. 94-95.

43 U.S. Patents 2,735,263 (Charshafian), 2,883,829 (Africano), 3,172,253 (Schelp), 3,525,474 (Von Ohain). For Ohain and Schelp, see Schlaifer and Heron, *Development,* pp. 377-79, 383-86.

44 Crickmore, *SR-71,* pp. 94-95.

45 U.S. Patents 2,922,286, 3,040,519, 3,040,520 (all to Rae).

46 AIAA Paper 86-1680.

47 AIAA Paper 92-3499.

48 DTIC AD-351239.

49 Author interview, Arthur Thomas, 24 September 1987. Folder 18649, NASA Historical Reference Collection, NASA History Division, Washington, D.C. 20546.

50 *Johns Hopkins APL Technical Digest, Vol.* 11, Nos. 3 and 4 (1990), pp. 332-33; Boeing 707s: *Pedigree,* pp. 55-57.

51 Jenkins, *Space Shuttle,* p. 52; DTIC AD-372320.

52 "Reusable Space Launch Vehicle Systems Study" (Convair); Convair Report GD/C-DCJ 65-004.

53 Jenkins, *Space Shuttle*, p. 54; Convair ZP-M-095 (quote, p. ix).
54 *Los Angeles Times*, 15 January 1961, pp. 1, 3.
55 *Republic Aviation News*, 9 September 1960, pp. 1, 5.
56 Ibid. F-105: Gunston, *Fighters*, pp. 194-200.
57 *Republic Aviation News*, 9 September 1960, pp. 1, 5.
58 Author interview, Colonel Edward Hall, 25 January 1989. Folder 18649, NASA Historical Reference Collection, NASA History Division, Washington, D.C. 20546. Also Hallion, *Hypersonic*, p. 751.
59 DTIC AD-342992, p. i. Nine of these 11 reports are available from DTIC: AD-342992 through AD-343000.
60 *Aviation Week*, 31 October 1960, p. 26.
61 *Los Angeles Times*, 3 November 1960, p. 3A.
62 *Los Angeles Times*, 15 January 1961, pp. 1, 3.
63 Author interview, Robert Sanator, February 1988. Folder 18649, NASA Historical Reference Collection, NASA History Division, Washington, D.C. 20546.
64 "Memorandum of the Scientific Advisory Board Ad Hoc Committee on Aerospace Plane," December 1960.
65 Ibid. (includes all quotes.)
66 *Los Angeles Times*, 3 November 1960, p. 3A; Lockheed LAC 571375, cover and chart 6.
67 Lockheed LAC 571375, chart 6. For Turbo-LACE, see Heppenheimer, *Hypersonic*, p. 165.
68 Jenkins, *Space Shuttle*, pp. 55-57; DTIC ADC-053100, AD-351039 through AD-351041, AD-351581 through AD-351584.
69 AIAA Paper 86-1680.
70 Jenkins, *Space Shuttle*, p. 56.
71 Martin M-63-1 (RENE, pp. IV-2 to IV-5).
72 DTIC AD-318628, AD-322199; film, "Liquid Air Cycle Engine" (Marquardt, March 1961).
73 "Report of the Scientific Advisory Board Ad Hoc Committee on Aerospaceplane," 23-25 July 1962.
74 NASA SP-2000-4408, p. 131; Heppenheimer, *Countdown*, pp. 110, 119.
75 "Report of the Scientific Advisory Board Ad Hoc Committee on Aerospaceplane," 23-25 July 1962 (includes quotes).
76 AIAA Paper 92-3499, p. 4.
77 Neufeld, *Rocket*, p. 101; Cornett, "History," pp. 1-3.
78 Hallion, *Hypersonic*, p. 949; Cornett, "History," pp. 2, 7-8. Quote: DTIC AD-359545, p. 11.
79 Cornett, "History," pp. 7, 11.
80 For RL10, see NASA SP-4206, p. 139.
81 DTIC AD-335212, pp. 5-7 (quote, p. 1).
82 Ibid., pp. 18-19; CASI 75N-75668, pp. ii, 15.
83 DTIC AD-350952, AD-384302 (performance, p. 2).

84 AIAA Paper 99-4896, p. 4; DTIC AD-359763, p. II-2.
85 DTIC AD-336169, pp. 1, 13, 26-28.
86 DTIC AD-351207, pp. iii, 3-4.
87 DTIC AD-345178, pp. 22, 397.
88 Cornett, "History," p. 2.
89 "Report of the USAF Scientific Advisory Board Aerospace Vehicles/Propulsion Panels on Aerospaceplane, VTOL, and Strategic Manned Aircraft," 24 October 1963; extended quote also in Hallion, *Hypersonic*, p. 951.
90 NASA SP-4221, pp. 49-54. For general overviews of Dyna-Soar, see Hallion, *Hypersonic*, Case II; Miller, *X-Planes*, ch. 24; DTIC ADA-303832.
91 Hallion, *Hypersonic*, pp. 951-52; Jenkins, *Space Shuttle*, pp. 66-67; Cornett, "History," p. 4.
92 Hallion, *Hypersonic*, pp. 952-54; Jenkins, *Space Shuttle*, p. 62.
93 DTIC: AD-353898, AD-461220.
94 1966 work: AIAA Paper 92-3499, pp. 5-6; DTIC AD-381504, p. 10. 1969 proposal: DTIC AD-500489, p. iv.

5

WIDENING PROSPECTS FOR RE-ENTRY

The classic spaceship has wings, and throughout much of the 1950s both NACA and the Air Force struggled to invent such a craft. Design studies addressed issues as fundamental as whether this hypersonic rocket plane should have one particular wing-body configuration, or whether it should be upside down. The focus of the work was Dyna-Soar, a small version of the space shuttle that was to ride to orbit atop a Titan III. It brought remarkable engineering advances, but Pentagon policy makers, led by Defense Secretary Robert McNamara, saw it as offering little more than technical development, with no mission that could offer a military justification. In December 1963 he canceled it.

Better prospects attended NASA's effort in manned spaceflight, which culminated in the Apollo piloted flights to the Moon. Apollo used no wings; rather, it relied on a simple cone that used the Allen-Eggers blunt-body principle. Still, its demands were stringent. It had to re-enter successfully with twice the energy of an entry from Earth orbit. Then it had to navigate a corridor, a narrow range of altitudes, to bleed off energy without either skipping back into space or encountering g-forces that were too severe. By doing these things, it showed that hypersonics was ready for this challenge.

WINGED SPACECRAFT AND DYNA-SOAR

Boost-glide rockets, with wings, entered the realm of advanced conceptual design with postwar studies at Bell Aircraft called Bomi, Bomber Missile. The director of the work, Walter Dornberger, had headed Germany's wartime rocket development program and had been in charge of the V-2. The new effort involved feasibility studies that sought to learn what might be done with foreseeable technology, but Bomi was a little too advanced for some of Dornberger's colleagues. Historian Roy Houchin writes that when Dornberger faced "abusive and insulting remarks" from an Air Force audience, he responded by declaring that his Bomi would be receiving more respect if he had had the chance to fly it against the United States during the war. In Houchin's words, "The silence was deafening."[1]

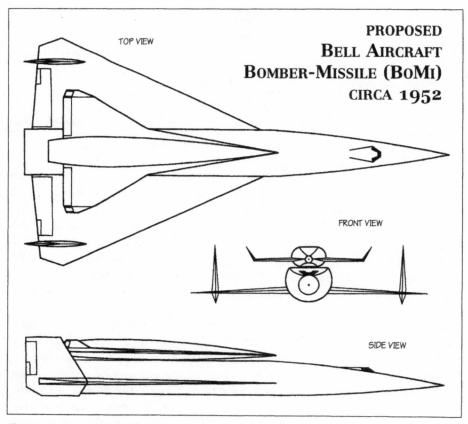

The Bomi concept. (Art by Dennis Jenkins)

The initial Bomi concept, dating back to 1951, took form as an in-house effort. It called for a two-stage rocket, with both stages being piloted and fitted with delta wings. The lower stage was mostly of aluminum, with titanium leading edges and nose; the upper stage was entirely of titanium and used radiative cooling. With an initial range of 3,500 miles, it was to come over the target above 100,000 feet and at speeds greater than Mach 4. Operational concepts called for bases in England or Spain, targets in the western Soviet Union, and a landing site in northern Africa.[2]

During the spring of 1952, Bell officials sought funds for further study from Wright Air Development Center (WADC). A year passed, and WADC responded with a firm no. The range was too short. Thermal protection and onboard cooling raised unanswered questions. Values assumed for L/D appeared highly optimistic, and no information was available on stability, control, or aerodynamic flutter at the proposed speeds. Bell responded by offering to consider higher speeds and greater range. Basic feasibility then lay even farther in the future, but the Air Force's interest in the Atlas ICBM meant that it wanted missiles of longer range, even though shorter-range designs could be available sooner. An intercontinental Bomi at least could be evaluated as a potential alternative to Atlas, and it might find additional roles such as strategic reconnaissance.[3]

In April 1954, with that ICBM very much in the ascendancy, WADC awarded Bell its desired study contract. Bomi now had an Air Force designation, MX-2276. Bell examined versions of its two-stage concept with 4,000- and 6,000-mile ranges while introducing a new three-stage configuration with the stages mounted belly-to-back. Liftoff thrust was to be 1.2 million pounds, compared with 360,000 for the three-engine Atlas. Bomi was to use a mix of liquid oxygen and liquid fluorine, the latter being highly corrosive and hazardous, whereas Atlas needed only liquid oxygen, which was much safer. The new Bomi was to reach 22,000 feet per second, slightly less than Atlas, but promised a truly global glide range of 12,000 miles. Even so, Atlas clearly was preferable.[4]

But the need for reconnaissance brought new life to the Bell studies. At WADC, in parallel with initiatives that were sparking interest in unpiloted reconnaissance satellites, officials defined requirements for Special Reconnaissance System 118P. These called initially for a range of 3,500 miles at altitudes above 100,000 feet. Bell won funding in September 1955, as a follow-on to its recently completed MX-2276 activity, and proposed a two-stage vehicle with a Mach 15 glider. In March 1956 the company won a new study contract for what now was called Brass Bell. It took shape as a fairly standard advanced concept of the mid-1950s, with a liquid-fueled expendable first stage boosting a piloted craft that showed sharply swept delta wings. The lower stage was conventional in design, burning Atlas propellants with uprated Atlas engines, but the glider retained the company's preference for fluorine. Officials at Bell were well aware of its perils, but John Sloop at NACA-Lewis was successfully testing a fluorine rocket engine with 20,000 pounds of thrust, and this gave hope.[5]

The Brass Bell study contract went into force at a moment when prospects for boost-glide were taking a sharp step upward. In February 1956 General Thomas Power, head of the Air Research and Development Command (ARDC), stated that the Air Force should stop merely considering such radical concepts and begin developing them. High on his list was a weapon called Robo, Rocket Bomber, for which several firms were already conducting in-house work as a prelude to funded study contracts. Robo sought to advance beyond Brass Bell, for it was to circle the globe and hence required near-orbital speed. In June ARDC Headquarters set forth System Requirement 126 that defined the scope of the studies. Convair, Douglas, and North American won the initial awards, with Martin, Bell, and Lockheed later participating as well.

The X-15 by then was well along in design, but it clearly was inadequate for the performance requirements of Brass Bell and Robo. This raised the prospect of a new and even more advanced experimental airplane. At ARDC Headquarters, Major George Colchagoff took the initiative in pursuing studies of such a craft, which took the name HYWARDS: Hypersonic Weapons Research and Development Supporting System. In November 1956 the ARDC issued System Requirement 131, thereby placing this new X-plane on the agenda as well.[6]

The initial HYWARDS concept called for a flight speed of Mach 12. However, in December Bell Aircraft raised the speed of Brass Bell to Mach 18. This increased the boost-glide range to 6,300 miles, but it opened a large gap between the performance of the two craft, inviting questions as to the applicability of HYWARDS results. In January a group at NACA-Langley, headed by John Becker, weighed in with a report stating that Mach 18, or 18,000 feet per second, was appropriate for HYWARDS. The reason was that "at this speed boost gliders approached their peak heating environment. The rapidly increasing flight altitudes at speeds above Mach 18 caused a reduction in the heating rates."[7]

With the prospect now strong that Brass Bell and HYWARDS would have the same flight speed, there was clear reason not to pursue them as separate projects but to consolidate them into a single program. A decision at Air Force Headquarters, made in March 1957, accomplished this and recognized their complementary characters. They still had different goals, with HYWARDS conducting flight research and Brass Bell being the operational reconnaissance system, but HYWARDS now was to stand as a true testbed.[8]

Robo still was a separate project, but events during 1957 brought it into the fold as well. In June an ad hoc review group, which included members from ARDC and WADC, looked at Robo concepts from contractors. Robert Graham, a NACA attendee, noted that most proposals called for "a boost-glide vehicle which would fly at Mach 20-25 at an altitude above 150,000 feet." This was well beyond the state of the art, but the panel concluded that with several years of research, an experimental craft could enter flight test in 1965, an operational hypersonic glider in 1968, and Robo in 1974.[9]

On 10 October—less than a week after the Soviets launched their first Sputnik—ARDC endorsed this three-part plan by issuing a lengthy set of reports, "Abbreviated Systems Development Plan, System 464L—Hypersonic Strategic Weapon System." It looked ahead to a research vehicle capable of 18,000 feet per second and 350,000 feet, to be followed by Brass Bell with the same speed and 170,000 feet, and finally Robo, rated at 25,000 feet per second and 300,000 feet but capable of orbital flight.

The ARDC's Lieutenant Colonel Carleton Strathy, a division chief and a strong advocate of program consolidation, took the proposed plan to Air Force Headquarters. He won endorsement from Brigadier General Don Zimmerman, Deputy

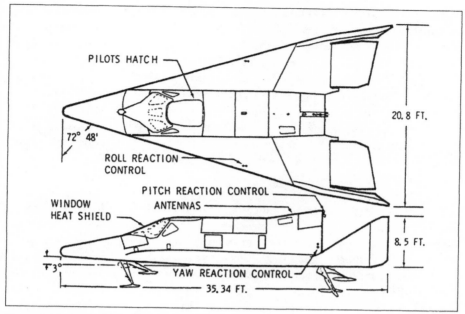

Top and side views of Dyna-Soar. (U.S. Air Force)

Director of Development Planning, and from Brigadier General Homer Boushey, Deputy Director of Research and Development. NACA's John Crowley, Associate Director for Research, gave strong approval to the proposed test vehicle, viewing it as a logical step beyond the X-15. On 25 November, having secured support from his superiors, Boushey issued Development Directive 94, allocating $3 million to proceed with more detailed studies following a selection of contractors.[10]

The new concept represented another step in the sequence that included Eugen Sänger's *Silbervogel*, his suborbital skipping vehicle, and among live rocket craft, the X-15. It was widely viewed as a tribute to Sänger, who was still living. It took the name Dyna-Soar, which drew on "dynamic soaring," Sänger's name for his skipping technique, and which also stood for "dynamic ascent and soaring flight," or boost-glide. Boeing and Martin emerged as the finalists in June 1958, with their roles being defined in November 1959. Boeing was to take responsibility for the winged spacecraft. Martin, described as the associate contractor, was to provide the Titan missile that would serve as the launch vehicle.[11]

The program now demanded definition of flight modes, configuration, structure, and materials. The name of Sänger was on everyone's lips, but his skipping flight path had already proven to be uncompetitive. He and his colleague Bredt had treated its dynamics, but they had not discussed the heating. That task fell to NACA's Allen and Eggers, along with their colleague Stanford Neice.

In 1954, following their classic analysis of ballistic re-entry, Eggers and Allen turned their attention to comparison of this mode with boost-glide and skipping entries. They assumed the use of active cooling and found that boost-glide held the advantage:

> The glide vehicle developing lift-drag ratios in the neighborhood of 4 is far superior to the ballistic vehicle in ability to convert velocity into range. It has the disadvantage of having far more heat convected to it; however, it has the compensating advantage that this heat can in the main be radiated back to the atmosphere. Consequently, the mass of coolant material may be kept relatively low.

A skip vehicle offered greater range than the alternatives, in line with Sänger's advocacy of this flight mode. But it encountered more severe heating, along with high aerodynamic loads that necessitated a structurally strong and therefore heavy vehicle. Extra weight meant extra coolant, with the authors noting that "ultimately the coolant is being added to cool coolant. This situation must obviously be avoided." They concluded that "the skip vehicle is thought to be the least promising of the three types of hypervelocity vehicle considered here."[12]

Following this comparative assessment of flight modes, Eggers worked with his colleague Clarence Syvertson to address the issue of optimum configuration. This issue had been addressed for the X-15; it was a mid-wing airplane that generally resembled the high-performance fighters of its era. In treating Dyna-Soar, following the Robo review of mid-1957, NACA's Robert Graham wrote that "high-wing, mid-wing and low-wing configurations were proposed. All had a highly swept wing, and a small angle cone as the fuselage or body." This meant that while there was agreement on designing the fuselage, there was no standard way to design the wing.[13]

Eggers and Syvertson proceeded by treating the design problem entirely as an exercise in aerodynamics. They concluded that the highest values of L/D were attainable by using a high-wing concept with the fuselage mounted below as a slender half-cone and the wing forming a flat top. Large fins at the wing tips, canted sharply downward, directed the airflow under the wings downward and increased the lift. Working with a hypersonic wind tunnel at NACA-Ames, they measured a maximum L/D of 6.65 at Mach 5, in good agreement with a calculated value of 6.85.[14]

This configuration had attractive features, not the least of which was that the base of its half-cone could readily accommodate a rocket engine. Still, it was not long before other specialists began to argue that it was upside down. Instead of having a flat top with the fuselage below, it was to be flipped to place the wing below the fuselage, giving it a flat bottom. This assertion came to the forefront during Becker's HYWARDS study, which identified its preferred velocity as 18,000 feet per second. His colleague Peter Korycinski worked with Becker to develop heating analyses of flat-top and flat-bottom candidates, with Roger Anderson and others within Langley's Structures Division providing estimates for the weight of thermal protection.

A simple pair of curves, plotted on graph paper, showed that under specified assumptions the flat-bottom weight at that velocity was 21,400 pounds and was increasing at a modest rate at higher speeds. The flat-top weight was 27,600 pounds and was rising steeply. Becker wrote that the flat-bottom craft placed its fuselage "in the relatively cool shielded region on the top or lee side of the wing—i.e., the wing was used in effect as a partial heat shield for the fuselage.... This 'flat-bottomed' design had the least possible critical heating area...and this translated into least circulating coolant, least area of radiative heat shields, and least total thermal protection in flight."[15]

These approaches—flat-top at Ames, flat-bottom at Langley—brought a debate between these centers that continued through 1957. At Ames, the continuing strong interest in high L/D reflected an ongoing emphasis on excellent supersonic aerodynamics for military aircraft, which needed high L/D as a matter of course. To ease the heating problem, Ames held for a time to a proposed speed of 11,000 feet per second, slower than the Langley concept but lighter in weight and more attainable in technology while still offering a considerable leap beyond the X-15. Officials at NACA diplomatically described the Ames and Langley HYWARDS concepts respectively as "high L/D" and "low heating," but while the debate continued, there remained no standard approach to the design of wings for a hypersonic glider.[16]

There was a general expectation that such a craft would require active cooling. Bell Aircraft, which had been studying Bomi, Brass Bell, and lately Robo, had the most experience in the conceptual design of such arrangements. Its Brass Bell of 1957, designed to enter its glide at 18,000 feet per second and 170,000 feet in altitude, featured an actively cooled insulated hot structure. The primary or load-bearing structure was of aluminum and relied on cooling in a closed-loop arrangement that used water-glycol as the coolant. Wing leading edges had their own closed-loop cooling system that relied on a mix of sodium and potassium metals. Liquid hydrogen, pumped initially to 1,000 pounds per square inch, flowed first through a heat exchanger and cooled the heated water-glycol, then proceeded to a second heat exchanger to cool the hot sodium-potassium. In an alternate design concept, this gas cooled the wing leading edges directly, with no intermediate liquid-metal coolant loop. The warmed hydrogen ran a turbine within an onboard auxiliary power unit and then was exhausted overboard. The leading edges reached a maximum temperature of 1,400°F, for which Inconel X was a suitable material.[17]

During August of that year Becker and Korycinski launched a new series of studies that further examined the heating and thermal protection of their flat-bottom glider. They found that for a glider of global range, flying with angle of attack of 45 degrees, an entry trajectory near the upper limit of permissible altitudes gave peak uncooled skin temperatures of 2,000°F. This appeared achievable with improved metallic or ceramic hot structures. Accordingly, no coolant at all was required![18]

Artist's rendering showing Dyna-Soar boosted by a Titan III launch vehicle. (Boeing Company archives)

This conclusion, published early in 1959, influenced the configuration of subsequent boost-glide vehicles—Dyna-Soar, the space shuttle—much as the Eggers-Allen paper of 1953 had defined the blunt-body shape for ballistic entry. Preliminary and unpublished results were in hand more than a year prior to publication, and when the prospect emerged of eliminating active cooling, the concepts that could do this were swept into prominence. They were of the flat-bottom type, with Dyna-Soar being the first to proceed into mainstream development.

This uncooled configuration proved robust enough to accommodate substantial increases in flight speed and performance. In April 1959 Herbert York, the Defense Director of Research and Engineering, stated that Dyna-Soar was to fly at 15,000 miles per hour. This was well above the planned speed of Brass Bell but still below orbital velocity.

During subsequent years the booster changed from Martin's Titan I to the more capable Titan II and then to the powerful Titan III-C, which could easily boost it to orbit. A new plan, approved in December 1961, dropped suborbital missions and called for "the early attainment of orbital flight." Subsequent planning anticipated that Dyna-Soar would reach orbit with the Titan III upper stage, execute several circuits of the Earth, and then come down from orbit by using this stage as a retrorocket.[19]

After that, though, advancing technical capabilities ran up against increasingly stringent operational requirements. The Dyna-Soar concept had grown out of HYWARDS, being intended initially to serve as a testbed for the reconnaissance

Widening Prospects for Re-entry

Full-scale model of Dyna-Soar, on display at an Air Force exhibition in 1962. The scalloped pattern on the base was intended to suggest Sänger's skipping entry. (Boeing Company archives)

boost-glider Brass Bell and for the manned rocket-powered bomber Robo. But the rationale for both projects became increasingly questionable during the early 1960s. The hypersonic Brass Bell gave way to a new concept, the Manned Orbiting Laboratory (MOL), which was to fly in orbit as a small space station while astronauts took reconnaissance photos. Robo fell out of the picture completely, for the success of the Minuteman ICBM, which used solid propellant, established such missiles as the nation's prime strategic force. Some people pursued new concepts that continued to hold out hope for Dyna-Soar applications, with satellite interception standing in the forefront. The Air Force addressed this with studies of its Saint project, but Dyna-Soar proved unsuitable for such a mission.[20]

Dyna-Soar was a potentially superb technology demonstrator, but Defense Secretary Robert McNamara took the view that it had to serve a military role in its own right or lead to a follow-on program with clear military application. The cost of Dyna-Soar was approaching a billion dollars, and in October 1963 he declared that he could not justify spending such a sum if it was a dead-end program with no ultimate purpose. He canceled it on 10 December, noting that it was not to serve as a cargo rocket, could not carry substantial payloads, and could not stay in orbit for

Artist's rendering showing Dyna-Soar in orbit. (Boeing Company archives)

long durations. He approved MOL as a new program, thereby giving the Air Force continuing reason to hope that it would place astronauts in orbit, but stated that Dyna-Soar would serve only "a very narrow objective."[21]

At that moment the program called for production of 10 flight vehicles, and Boeing had completed some 42 percent of the necessary tasks. McNamara's decision therefore was controversial, particularly because the program still had high-level supporters. These included Eugene Zuckert, Air Force Secretary; Alexander Flax, Assistant Secretary for Research and Development; and Brockway McMillan, Zuckert's Under Secretary and Flax's predecessor as Assistant Secretary. Still, McNamara gave more attention to Harold Brown, the Defense Director of Research and Engineering, who made the specific proposal that McNamara accepted: to cancel Dyna-Soar and proceed instead with MOL.[22]

Dyna-Soar never flew. The program had expended $410 million when canceled, but the schedule still called for another $373 million, and the vehicle was still some two and a half years away from its first flight. Even so, its technology remained available for further development, contributing to the widening prospects for reentry that marked the era.[23]

THE TECHNOLOGY OF DYNA-SOAR

Its thermal environment during re-entry was less severe than that of an ICBM nose cone, allowing designers to avoid not only active structural cooling but ablative thermal protection as well. This meant that it could be reusable; it did not have to change out its thermal protection after every flight. Even so, its environment imposed temperatures and heat loads that pervaded the choice of engineering solutions throughout the vehicle.

Dyna-Soar used radiatively-cooled hot structure, with the primary or load-bearing structure being of Rene 41. Trusses formed the primary structure of the wings and fuselage, with many of their beams meeting at joints that were pinned rather than welded. Thermal gradients, imposing differential expansion on separate beams, caused these members to rotate at the pins. This accommodated the gradients without imposing thermal stress.

Rene 41 was selected as a commercially available superalloy that had the best available combination of oxidation resistance and high-temperature strength. Its yield strength, 130,000 psi at room temperature, fell off only slightly at 1,200°F and retained useful values at 1,800°F. It could be processed as sheet, strip, wire, tubes, and forgings. Used as the primary structure of Dyna-Soar, it supported a design specification that indeed called for reusability. The craft was to withstand at least four re-entries under the most severe conditions permitted.

As an alloy, Rene 41 had a standard composition of 19 percent chromium, 11 percent cobalt, 10 percent molybdenum, 3 percent titanium, and 1.5 percent aluminum, along with 0.09 percent carbon and 0.006 percent boron, with the balance being nickel. It gained strength through age hardening, with the titanium and aluminum precipitating within the nickel as an intermetallic compound. Age-hardening weldments initially showed susceptibility to cracking, which occurred in parts that had been strained through welding or cold working. A new heat-treatment process permitted full aging without cracking, with the fabricated assemblies showing no significant tendency to develop cracks.[24]

As a structural material, the relatively mature state of Rene 41 reflected the fact that it had already seen use in jet engines. It nevertheless lacked the temperature resistance necessary for use in the metallic shingles or panels that were to form the outer skin of the vehicle, reradiating the heat while withstanding temperatures as high as 3,000°F. Here there was far less existing art, and investigators at Boeing had to find their way through a somewhat roundabout path.

Four refractory or temperature-resistant metals initially stood out: tantalum, tungsten, molybdenum, and columbium. Tantalum was too heavy, and tungsten was not available commercially as sheet. Columbium also appeared to be ruled out for it required an antioxidation coating, but vendors were unable to coat it without rendering it brittle. Molybdenum alloys also faced embrittlement due to recrystallization produced by a prolonged soak at high temperature in the course of coating formation. A promising alloy, Mo-0.5Ti, overcame this difficulty through addition of 0.07 percent zirconium. The alloy that resulted, Mo-0.5Ti-0.07Zr, was called TZM. For a time it appeared as a highly promising candidate for all the other panels.[25]

Wing design also promoted its use, for the craft mounted a delta wing with a leading-edge sweep of 73 degrees. Though built for hypersonic re-entry from orbit, it resembled the supersonic delta wings of contemporary aircraft such as the B-58 bomber. However, this wing was designed using the Eggers-Allen blunt-body principle, with the leading edge being curved or blunted to reduce the rate of heating. The wing sweep then reduced equilibrium temperatures along the leading edge to levels compatible with the use of TZM.[26]

Boeing's metallurgists nevertheless held an ongoing interest in columbium because in uncoated form it showed superior ease of fabrication and lack of brittleness. A new Boeing-developed coating method eliminated embrittlement, putting columbium back in the running. A survey of its alloys showed that they all lacked the hot strength of TZM. Columbium nevertheless retained its attractiveness because it promised less weight. Based on coatability, oxidation resistance, and thermal emissivity, the preferred alloy was Cb-10Ti-5Zr, called D-36. It replaced TZM in many areas of the vehicle but proved to lack strength against creep at the highest temperatures. Moreover, coated TZM gave more of a margin against oxidation than coated D-36, again at the most extreme temperatures. D-36 indeed was chosen to cover most of the vehicle, including the flat underside of the wing. But TZM retained its advantage for such hot areas as the wing leading edges.[27]

The vehicle had some 140 running feet of leading edges and 140 square feet of associated area. This included leading edges of the vertical fins and elevons as well as of the wings. In general, D-36 served where temperatures during re-entry did not exceed 2,700°F, while TZM was used for temperatures between 2,700 and 3,000°F. In accordance with the Stefan-Boltzmann law, all surfaces radiated heat at a rate proportional to the fourth power of the temperature. Hence for equal emissivities, a surface at 3,000°F radiated 43 percent more heat than one at 2,700°F.[28]

Panels of both TZM and D-36 demanded antioxidation coatings. These coatings were formed directly on the surfaces as metallic silicides (silicon compounds), using a two-step process that employed iodine as a chemical intermediary. Boeing introduced a fluidized-bed method for application of the coatings that cut the time for preparation while enhancing uniformity and reliability. In addition, a thin layer of silicon carbide, applied to the surface, gave the vehicle its distinctive black color. It enhanced the emissivity, lowering temperatures by as much as 200°F.

Development testing featured use of an oxyacetylene torch, operated with excess oxygen, which heated small samples of coated refractory sheet to temperatures as high as 3,000°F, measured by optical pyrometer. Test durations ran as long as four hours, with a published review noting that failures of specimens "were easily detected by visual observation as soon as they occurred." This work showed that although TZM had better oxidation resistance than D-36, both coated alloys could resist oxidation for more than two hours at 3,000°F. This exceeded design requirements. Similar tests applied stress to hot samples by hanging weights from them, thereby demonstrating their ability to withstand stress of 3,100 psi, again at 3,000°F.[29]

Other tests showed that complete panels could withstand aerodynamic flutter. This issue was important; a report of the Aerospace Vehicles Panel of the Air Force Scientific Advisory Board (SAB)—a panel on panels, as it were—came out in April 1962 and singled out the problem of flutter, citing it as one that called for critical attention. The test program used two NASA wind tunnels: the 4 by 4-foot Unitary facility at Langley that covered a range of Mach 1.6 to 2.8 and the 11 by 11-foot Unitary installation at Ames for Mach 1.2 to 1.4. Heaters warmed test samples to 840°F as investigators started with steel panels and progressed to versions fabricated from Rene nickel alloy.

"Flutter testing in wind tunnels is inherently dangerous," a Boeing review declared. "To carry the test to the actual flutter point is to risk destruction of the test specimen. Under such circumstances, the safety of the wind tunnel itself is jeopardized." Panels under test were as large as 24 by 45 inches; actual flutter could easily have brought failure through fatigue, with parts of a specimen being blown through the tunnel at supersonic speed. The work therefore proceeded by starting at modest dynamic pressures, 400 and 500 pounds per square foot, and advancing over 18 months to levels that exceeded the design requirement of close to 1,400 pounds per square foot. The Boeing report concluded that the success of this test program, which ran through mid-1962, "indicates that an adequate panel flutter capability has been achieved."[30]

Between the outer panels and the inner primary structure, a corrugated skin of Rene 41 served as a substructure. On the upper wing surface and upper fuselage, where temperatures were no higher than 2,000°F, the thermal-protection panels were also of Rene 41 rather than of a refractory. Measuring 12 by 45 inches, these panels were spot-welded directly to the corrugations of the substructure. For the wing undersurface, and for other areas that were hotter than 2,000°F, designers specified an insulated structure. Standoff clips, each with four legs, were riveted to the underlying corrugations and supported the refractory panels, which also were 12 by 45 inches in size.

The space between the panels and the substructure was to be filled with insulation. A survey of candidate materials showed that most of them exhibited a strong tendency to shrink at high temperatures. This was undesirable; it increased the rate of heat transfer and could create uninsulated gaps at seams and corners. Q-felt, a silica fiber from Johns-Manville, also showed shrinkage. However, nearly all of it occurred at 2,000°F and below; above 2,000°F, further shrinkage was negligible. This meant that Q-felt could be "pre-shrunk" through exposure to temperatures above 2,000°F for several hours. The insulation that resulted had density no greater than 6.2 pounds per cubic foot, one-tenth that of water. In addition, it withstood temperatures as high as 3,000°F.[31]

TZM outer panels, insulated with Q-felt, proved suitable for wing leading edges. These were designed to withstand equilibrium temperatures of 2,825°F and short-duration overtemperatures of 2,900°F. However, the nose cap faced temperatures of 3,680°F, along with a peak heat flux of 143 BTU per square foot-second. This cap had a radius of curvature of 7.5 inches, making it far less blunt than the Project Mercury heat shield that had a radius of 120 inches.[32] Its heating was correspondingly more severe. Reliable thermal protection of the nose was essential, and so the program conducted two independent development efforts that used separate approaches. The firm of Chance Vought pursued the main line of activity, while Boeing also devised its own nose-cap design.

The work at Vought began with a survey of materials that paralleled Boeing's review of refractory metals for the thermal-protection panels. Molybdenum and columbium had no strength to speak of at the pertinent temperatures, but tungsten retained useful strength even at 4,000°F. However, this metal could not be welded, while no known coating could protect it against oxidation. Attention then turned to nonmetallic materials, including ceramics.

WIDENING PROSPECTS FOR RE-ENTRY

Ceramics of interest existed as oxides such as silica and magnesia, which meant that they could not undergo further oxidation. Magnesia proved to be unsuitable because it had low thermal emittance, while silica lacked strength. However, carbon in the form of graphite showed clear promise. It held considerable industrial experience; it was light in weight, while its strength actually increased with temperature. It oxidized readily but could be protected up to 3,000°F by treating it with silicon, in a vacuum and at high temperatures, to form a thin protective layer of silicon carbide. Near the stagnation point, the temperatures during re-entry would exceed that level. This brought the concept of a nose cap with siliconized graphite as the primary material, with an insulating layer of a temperature-resistant ceramic covering its forward area. With graphite having good properties as a heat sink, it would rise in temperature uniformly and relatively slowly, while remaining below the 3,000°F limit through the full time of re-entry.

Suitable grades of graphite proved to be available commercially from the firm of National Carbon. Candidate insulators included hafnia, thoria, magnesia, ceria, yttria, beryllia, and zirconia. Thoria was the most refractory but was very dense and showed poor resistance to thermal shock. Hafnia brought problems of availability and of reproducibility of properties. Zirconia stood out. Zirconium, its parent metal, had found use in nuclear reactors; the ceramic was available from the Zirconium Corporation of America. It had a melting point above 4,500°F, was chemically stable and compatible with siliconized graphite, offered high emittance with low thermal conductivity, provided adequate resistance to thermal shock and thermal stress, and lent itself to fabrication.[33]

For developmental testing, Vought used two in-house facilities that simulated the flight environment, particularly during re-entry. A ramjet, fueled with JP-4 and running with air from a wind tunnel, produced an exhaust with velocity up to 4,500 feet per second and temperature up to 3,500°F. It also generated acoustic levels above 170 decibels, reproducing the roar of a Titan III booster and showing that samples under test could withstand the resulting stresses without cracking. A separate installation, built specifically for the Dyna-Soar program, used an array of propane burners to test full-size nose caps.

The final Vought design used a monolithic shell of siliconized graphite that was covered over its full surface by zirconia tiles held in place using thick zirconia pins. This arrangement relieved thermal stresses by permitting mechanical movement of the tiles. A heat shield stood behind the graphite, fabricated as a thick disk-shaped container made of coated TZM sheet metal and filled with Q-felt. The nose cap attached to the vehicle with a forged ring and clamp that also were of coated TZM. The cap as a whole relied on radiative cooling. It was designed to be reusable; like the primary structure, it was to withstand four re-entries under the most severe conditions permitted.[34]

The backup Boeing effort drew on that company's own test equipment. Study of samples used the Plasma Jet Subsonic Splash Facility, which created a jet with temperature as high as 8,000°F that splashed over the face of a test specimen. Full-scale nose caps went into the Rocket Test Chamber, which burned gasoline to produce a nozzle exit velocity of 5,800 feet per second and an acoustic level of 154 decibels. Both installations were capable of long-duration testing, reproducing conditions during re-entries that could last for 30 minutes.[35]

The Boeing concept used a monolithic zirconia nose cap that was reinforced against cracking with two screens of platinum-rhodium wire. The surface of the cap was grooved to relieve thermal stress. Like its counterpart from Vought, this design also installed a heat shield that used Q-felt insulation. However, there was no heat sink behind the zirconia cap. This cap alone provided thermal protection at the nose through radiative cooling. Lacking both pinned tiles and an inner shell, its design was simpler than that of Vought.[36]

Its fabrication bore comparison to the age-old work of potters, who shape wet clay on a rotating wheel and fire the resulting form in a kiln. Instead of using a potter's wheel, Boeing technicians worked with a steel die with an interior in the shape of a bowl. A paper honeycomb, reinforced with Elmer's Glue and laid in place, defined the pattern of stress-relieving grooves within the nose cap surface. The working material was not moist clay, but a mix of zirconia powder with binders, internal lubricants, and wetting agents.

With the honeycomb in position against the inner face of the die, a specialist loaded the die by hand, filling the honeycomb with the damp mix and forming layers of mix that alternated with the wire screens. The finished layup, still in its die, went into a hydraulic press. A pressure of 27,000 psi compacted the form, reducing its porosity for greater strength and less susceptibility to cracks. The cap was dried at 200°F, removed from its die, dried further, and then fired at 3,300°F for 10 hours. The paper honeycomb burned out in the course of the firing. Following visual and x-ray inspection, the finished zirconia cap was ready for machining to shape in the attachment area, where the TZM ring-and-clamp arrangement was to anchor it to the fuselage.[37]

The nose cap, outer panels, and primary structure all were built to limit their temperatures through passive methods: radiation, insulation. Active cooling also played a role, reducing temperatures within the pilot's compartment and two equipment bays. These used a "water wall," which mounted absorbent material between sheet-metal panels to hold a mix of water and a gel. The gel retarded flow of this fluid, while the absorbent wicking kept it distributed uniformly to prevent hot spots.

During reentry, heat reached the water walls as it penetrated into the vehicle. Some of the moisture evaporated as steam, transferring heat to a set of redundant water-glycol cooling loops resembling those proposed for Brass Bell of 1957. In Dyna-Soar, liquid hydrogen from an onboard supply flowed through heat exchangers and cooled these loops. Brass Bell had called for its warmed hydrogen to flow through a turbine, operating the onboard Auxiliary Power Unit. Dyna-Soar used an arrangement that differed only slightly: a catalytic bed to combine the stream of warm hydrogen with oxygen that again came from an onboard supply. This produced gas that drove the turbine of the Dyna-Soar APU, which provided both hydraulic and electric power.

A cooled hydraulic system also was necessary to move the control surfaces as on a conventional aircraft. The hydraulic fluid operating temperature was limited to 400°F by using the fluid itself as an initial heat-transfer medium. It flowed through an intermediate water-glycol loop that removed its heat by cooling with hydrogen. Major hydraulic system components, including pumps, were mounted within an actively cooled compartment. Control-surface actuators, along with their associated valves and plumbing, were insulated using inch-thick blankets of Q-felt. Through this combination of passive and active cooling methods, the Dyna-Soar program avoided a need to attempt to develop truly high-temperature hydraulic arrangements, remaining instead within the state of the art.[38]

Specific vehicle parts and components brought their own thermal problems. Bearings, both ball and antifriction, needed strength to carry mechanical loads at high temperatures. For ball bearings, the cobalt-base superalloy Stellite 19 was known to be acceptable up to 1,200°F. Investigation showed that it could perform under high load for short durations at 1,350°F. However, Dyna-Soar needed ball bearings qualified for 1,600°F and obtained them as spheres of Rene 41 plated with gold. The vehicle also needed antifriction bearings as hinges for control surfaces, and here there was far less existing art. The best available bearings used stainless steel and were suitable only to 600°F, whereas Dyna-Soar again faced a requirement of 1,600°F. A survey of 35 candidate materials led to selection of titanium carbide with nickel as a binder.[39]

Antenna windows demanded transparency to radio waves at similarly high temperatures. A separate program of materials evaluation led to selection of alumina, with the best grade being available from the Coors Porcelain Company. Its emittance had the low value of 0.4 at 2,500°F, which meant that waveguides beneath these windows faced thermal damage even though they were made of columbium alloy. A mix of oxides of cobalt, aluminum, and nickel gave a suitable coating when fired at 3,000°F, raising the emittance to approximately 0.8.[40]

The pilot needed his own windows. The three main ones, facing forward, were the largest yet planned for a manned spacecraft. They had double panes of fused silica, with infrared-reflecting coatings on all surfaces except the outermost. This inhibited the inward flow of heat by radiation, reducing the load on the active cooling of the pilot's compartment. The window frames expanded when hot; to hold the panes in position, the frames were fitted with springs of Rene 41. The windows also needed thermal protection, and so they were covered with a shield of D-36.

The cockpit was supposed to be jettisoned following re-entry, around Mach 5, but this raised a question: what if it remained attached? The cockpit had two other windows, one on each side, which faced a less severe environment and were to be left unshielded throughout a flight. The test pilot Neil Armstrong flew approaches and landings with a modified Douglas F5D fighter and showed that it was possible to land Dyna-Soar safely with side vision only.[41]

The vehicle was to touch down at 220 knots. It lacked wheeled landing gear, for inflated rubber tires would have demanded their own cooled compartments. For the same reason, it was not possible to use a conventional oil-filled strut as a shock absorber. The craft therefore deployed tricycle landing skids. The two main skids, from Goodyear, were of Waspaloy nickel steel and mounted wire bristles of Rene 41. These gave a high coefficient of friction, enabling the vehicle to skid to a stop in a planned length of 5,000 feet while accommodating runway irregularities. In place of the usual oleo strut, a long rod of Inconel stretched at the moment of touchdown and took up the energy of impact, thereby serving as a shock absorber. The nose skid, from Bendix, was forged from Rene 41 and had an undercoat of tungsten carbide to resist wear. Fitted with its own energy-absorbing Inconel rod, the front skid had a reduced coefficient of friction, which helped to keep the craft pointing straight ahead during slideout.[42]

Through such means, the Dyna-Soar program took long strides toward establishing hot structures as a technology suitable for operational use during re-entry from orbit. The X-15 had introduced heat sink fabricated from Inconel X, a nickel steel. Dyna-Soar went considerably further, developing radiation-cooled insulated structures fabricated from Rene 41 superalloy and from refractory materials. A chart from Boeing made the point that in 1958, prior to Dyna-Soar, the state of the art for advanced aircraft structures involved titanium and stainless steel, with temperature limits of 600°F. The X-15 with its Inconel X could withstand temperatures above 1,200°F. Against this background, Dyna-Soar brought substantial advances in the temperature limits of aircraft structures:[43]

TEMPERATURE LIMITS BEFORE AND AFTER DYNA-SOAR (in °F)

Element	1958	1963
Nose cap	3,200	4,300
Surface panels	1,200	2,750
Primary structure	1,200	1,800
Leading edges	1,200	3,000
Control surfaces	1,200	1,800
Bearings	1,200	1,800

Meanwhile, while Dyna-Soar was going forward within the Air Force, NASA had its own approaches to putting man in space.

WIDENING PROSPECTS FOR RE-ENTRY

HEAT SHIELDS FOR MERCURY AND CORONA

In November 1957, a month after the first Sputnik reached orbit, the Soviets again startled the world by placing a much larger satellite into space, which held the dog Laika as a passenger. This clearly presaged the flight of cosmonauts, and the question then was how the United States would respond. No plans were ready at the moment, but whatever America did, it would have to be done quickly.

HYWARDS, the nascent Dyna-Soar, was proceeding smartly. In addition, at North American Aviation the company's chief engineer, Harrison Storms, was in Washington, DC, with a concept designated X-15B. Fitted with thermal protection for return from orbit, it was to fly into space atop a cluster of three liquid-fueled boosters for an advanced Navaho, each with thrust of 415,000 pounds.[44] However, neither HYWARDS nor the X-15B could be ready soon. Into this breach stepped Maxime Faget of NACA-Langley, who had already shown a talent for conceptual design during the 1954 feasibility study that led to the original X-15.

In 1958 he was a branch chief within Langley's Pilotless Aircraft Research Division. Working on speculation, amid full awareness that the Army or Air Force might win the man-in-space assignment, he initiated a series of paper calculations and wind-tunnel tests of what he described as a "simple nonlifting satellite vehicle which follows a ballistic path in reentering the atmosphere." He noted that an "attractive feature of such a vehicle is that the research and production experiences of the ballistic-missile programs are applicable to its design and construction," and "since it follows a ballistic path, there is a minimum requirement for autopilot, guidance, or control equipment."[45]

In seeking a suitable shape, Faget started with the heat shield. Invoking the Allen-Eggers principle, he at first considered a flat face. However, it proved to trap heat by interfering with the rapid airflow that could carry this heat away. This meant that there was an optimum bluntness, as measured by radius of curvature.

Calculating thermal loads and heat-transfer rates using theories of Lees and of Fay and Riddell, and supplementing these estimates with experimental data from his colleague William Stoney, he considered a series of shapes. The least blunt was a cone with a rounded tip that faced the airflow. It had the highest heat input and the highest peak heating rate. A sphere gave better results in both areas, while the best estimates came with a gently rounded surface that faced the flow. It had only two-thirds the total heat input of the rounded cone—and less than one-third the peak heating rate. It also was the bluntest shape of those considered, and it was selected.[46]

With a candidate heat-shield shape in hand, he turned his attention to the complete manned capsule. An initial concept had the shape of a squat dome that was recessed slightly from the edge of the shield, like a circular Bundt cake that does not quite extend to the rim of its plate. The lip of this heat shield was supposed to produce separated flow over the afterbody to reduce its heating. When tested in a wind tunnel, however, it proved to be unstable at subsonic speeds.

Faget's group eliminated the open lip and exchanged the domed afterbody for a tall cone with a rounded tip that was to re-enter with its base end forward. It proved to be stable in this attitude, but tests in the 11-inch Langley hypersonic wind tunnel showed that it transferred too much heat to the afterbody. Moreover, its forward tip did not give enough room for its parachutes. This brought a return to the domed afterbody, which now was somewhat longer and had a cylinder on top to stow the chutes. Further work evolved the domed shape into a funnel, a conic frustum that retained the cylinder. This configuration provided a basis for design of the Mercury and later of the Gemini capsules, both of which were built by the firm of McDonnell Aircraft.[47]

Choice of thermal protection quickly emerged as a critical issue. Fortunately, the thermal environment of a re-entering satellite proved to be markedly less demanding than that of an ICBM. The two vehicles were similar in speed and kinetic energy, but an ICBM was to slam back into the atmosphere at a steep angle, decelerating rapidly due to drag and encountering heating that was brief but very severe. Re-entry from orbit was far easier, taking place over a number of minutes. Indeed, experimental work showed that little if any ablation was to be expected under the relatively mild conditions of satellite entry.

But satellite entry involved high total heat input, while its prolonged duration imposed a new requirement for good materials properties as insulators. They also had to stay cool through radiation. It thus became possible to critique the usefulness of ICBM nose-cone ablators for the prospective new role of satellite reentry.[48]

Heat of ablation, in BTU per pound, had been a standard figure of merit. For satellite entry, however, with little energy being carried away by ablation, it could be irrelevant. Phenolic glass, a fine ICBM material with a measured heat of 9,600 BTU per pound, was unusable for a satellite because it had an unacceptably high thermal conductivity. This meant that the prolonged thermal soak of re-entry could have time enough to fry a spacecraft. Teflon, by contrast, had a measured heat only one-third as large. It nevertheless made a superb candidate because of its excellent properties as an insulator.[49]

Such results showed that it was not necessary to reopen the problem of thermal protection for satellite entry. With appropriate caveats, the experience and research techniques of the ICBM problem could carry over to this new realm. This background made it possible for the Central Intelligence Agency to build operational orbital re-entry vehicles at a time when nose cones for Atlas were still in flight test.

This happened beginning in 1958, when Richard Bissell, a senior manager within the CIA, launched a highly classified reconnaissance program called Corona. General Electric, which was building nose cones for Atlas, won a contract to build the film-return capsule. The company selected ablation as the thermal-protection method, with phenolic nylon as the ablative material.[50]

WIDENING PROSPECTS FOR RE-ENTRY

The second Corona launch, in April 1959, flew successfully and became the world's first craft to return safely from orbit. It was supposed to come down near Hawaii, and a ground controller transmitted a command to have the capsule begin re-entry at a particular time. However, he forgot to press a certain button. The director of the recovery effort, Lieutenant Colonel Charles "Moose" Mathison, then learned that it would actually come down near the Norwegian island of Spitzbergen.

Mathison telephoned a friend in Norway's air force, Major General Tufte Johnsen, and told him to watch for a small spacecraft that was likely to be descending by parachute. Johnsen then phoned a mining company executive on the island and had him send out ski patrols. A three-man patrol soon returned with news: They had seen the orange parachute as the capsule drifted downward near the village of Barentsburg. That was not good because its residents were expatriate Russians. General Nathan Twining, Chairman of the Joint Chiefs, summarized the craft's fate in a memo: "From concentric circular tracks found in the snow at the suspected impact point and leading to one of the Soviet mining concessions on the island, we strongly suspect that the Soviets are in possession of the capsule."[51]

Meanwhile, NASA's Maxime Faget was making decisions concerning thermal protection for his own program, which now had the name Project Mercury. He was well aware of ablation but preferred heat sink. It was heavier, but he doubted that industrial contractors could fabricate an ablative heat shield that had adequate reliability.[52]

The suitability of ablation could not be tested by flying a subscale heat shield atop a high-speed rocket. Nothing less would do than to conduct a full-scale test using an Atlas ICBM as a booster. This missile was still in development, but in December 1958 the Air Force Ballistic Missile Division agreed to provide one Atlas C within six months, along with eight Atlas Ds over the next several years. This made it possible to test an ablative heat shield for Mercury as early as September 1959.[53]

The contractor for this shield was General Electric. The ablative material, phenolic-fiberglass, lacked the excellent insulating properties of Teflon or phenolic-nylon. Still, it had flown successfully as a ballistic-missile nose cone. The project engineer Aleck Bond adds that "there was more knowledge and experience with fiberglass-phenolic than with other materials. A great deal of ground-test information was available.... There was considerable background and experience in the fabrication, curing, and machining of assemblies made of Fiberglass." These could be laid up and cured in an autoclave.[54]

The flight test was called Big Joe, and it showed conservatism. The shield was heavy, with a density of 108 pounds per cubic foot, but designers added a large safety factor by specifying that it was to be twice as thick as calculations showed to be necessary. The flight was to be suborbital, with range of 1,800 miles but was to simulate a re-entry from orbit that was relatively steep and therefore demanding, producing higher temperatures on the face of the shield and on the afterbody.[55]

Liftoff came after 3 a.m., a time chosen to coincide with dawn in the landing area so as to give ample daylight for search and recovery. "The night sky lit up and the beach trembled with the roar of the Rocketdyne engines," notes NASA's history of Project Mercury. Two of those engines were to fall away during ascent, but they remained as part of the Atlas, increasing its weight and reducing its peak velocity by some 3,000 feet per second. What was more, the capsule failed to separate. It had an onboard attitude-control system that was to use spurts of compressed nitrogen gas to turn it around, to enter the atmosphere blunt end first. But this system used up all its nitrogen trying fruitlessly to swing the big Atlas that remained attached. Separation finally occurred at an altitude of 345,000 feet, while people waited to learn what would happen.[56]

The capsule performed better than planned. Even without effective attitude control, its shape and distribution of weights gave it enough inherent stability to turn itself around entirely through atmospheric drag. Its reduced speed at re-entry meant that its heat load was only 42 percent of the planned value of 7,100 BTU per square foot. But a particularly steep flight-path angle gave a peak heating rate of 77 percent of the intended value, thereby subjecting the heat shield to a usefully severe test. The capsule came down safely in the Atlantic, some 500 miles short of the planned impact area, but the destroyer USS *Strong* was not far away and picked it up a few hours later.

Subsequent examination showed that the heating had been uniform over the face of the heat shield. This shield had been built as an ablating laminate with a thickness of 1.075 inches, supported by a structural laminate half as thick. However, charred regions extended only to a depth of 0.20 inch, with further discoloration reaching to 0.35 inch. Weight loss due to ablation came to only six pounds, in line with experimental findings that had shown that little ablation indeed would occur.[57]

The heat shield not only showed fine thermal performance, it also sustained no damage on striking the water. This validated the manufacturing techniques used in its construction. The overall results from this flight test were sufficiently satisfactory to justify the choice of ablation for Mercury. This made it possible to drop heat sink from consideration and to go over completely to ablation, not only for Mercury but for Gemini, which followed.[58]

GEMINI AND APOLLO

An Apollo spacecraft, returning from the Moon, had twice the kinetic energy of a flight in low orbit and an aerodynamic environment that was nearly three times as severe. Its trajectory also had to thread a needle in its accuracy. Too steep a return

WIDENING PROSPECTS FOR RE-ENTRY

would subject its astronauts to excessive g-forces. Too shallow a re-entry meant that it would show insufficient loss of speed within the upper atmosphere and would fly back into space, to make a final entry and then land at an unplanned location. For a simple ballistic trajectory, this "corridor" was as little as seven miles wide, from top to bottom.[59]

At the outset, these issues raised two problems that were to be addressed in flight test. The heat shield had to be qualified, in tests that resembled those of the X-17 but took place at much higher velocity. In addition, it was necessary to show that a re-entering spacecraft could maneuver with some precision. It was vital to broaden the corridor, and the only way to do this was to use lift. This meant demonstrating successful maneuvers that had to be planned in advance, using data from tests in ground facilities at near-orbital speeds, when such facilities were most prone to error.

Apollo's Command Module, which was to execute the re-entry, lacked wings. Still, spacecraft of this general type could show lift-to-drag ratios of 0.1 or 0.2 by flying at a nonzero angle of attack, thereby tilting the heat shield and turning it into a lifting surface. Such values were far below those achievable with wings, but they brought useful flexibility during re-entry by permitting maneuver, thereby achieving a more accurate splashdown.

As early as 1958, Faget and his colleagues had noted three methods for trimming a capsule to a nonzero angle. Continuous thrust from a reaction-control system could do this, tilting the craft from its equilibrium attitude. A drag flap could do it as well by producing a modest amount of additional air resistance on one side of the vehicle. The simplest method required no onboard mechanism that might fail in flight and that expended no reaction-control propellant. It called for nothing more than a nonsymmetrical distribution of weight within the spacecraft, creating an offset in the location of the center of gravity. During re-entry, this offset would trim the craft to a tilted attitude, again automatically, due to the extra weight on one side. An astronaut could steer his capsule by using attitude control to roll it about its long axis, thereby controlling the orientation of the lift vector.[60]

This center-of-gravity offset went into the Gemini capsules that followed those of Project Mercury. The first manned Gemini flight carried the astronauts Virgil "Gus" Grissom and John Young on a three-orbit mission in March 1965. Following re-entry, they splashed down 60 miles short of the carrier USS *Intrepid*, which was on the aim point. This raised questions as to the adequacy of the preflight hypersonic wind-tunnel tests that had provided estimates of the spacecraft L/D used in mission planning.

The pertinent data had come from only two facilities. The Langley 11-inch tunnel had given points near Mach 7, while an industrial hotshot installation covered Mach 15 to 22, which was close to orbital speed. The latter facility lacked instruments of adequate precision and had produced data points that showed a large scatter. Researchers had averaged and curve-fit the measurements, but it was clear that this work had introduced inaccuracies.[61]

During that year flight data became available from the Grissom-Young mission and from three others, yielding direct measurements of flight angle of attack and L/D. To resolve the discrepancies, investigators at the Air Force's Arnold Engineering Development Center undertook further studies using two additional facilities. Tunnel F, a hotshot, had a 100-inch-diameter test section and reached Mach 20, heating nitrogen with an electric arc and achieving run times of 0.05 to 0.1 seconds. Tunnel L was a low-density, continuous-flow installation that also used arc-heated nitrogen. The Langley 11-inch data was viewed as valid and was retained in the reanalysis.

This work gave an opportunity to benchmark data from continuous-flow and hotshot tunnels against flight data, at very high Mach numbers. Size did not matter, for the big Tunnel F accommodated a model at one-fifteenth scale that incorporated much detail, whereas Tunnel L used models at scales of 1/120 and 1/180, the latter being nearly small enough to fit on a tie tack. Even so, the flight data points gave a good fit to curves derived using both tunnels. Billy Griffith, supervising the tests, concluded: "Generally, excellent agreement exists" between data from these sources.

The preflight data had brought estimated values of L/D that were too high by 60 percent. This led to a specification for the re-entry trim angle that proved to be off by 4.7 degrees, which produced the miss at splashdown. Julius Lukasiewicz, longtime head of the Von Karman Gas Dynamics Facility at AEDC, later added that if AEDC data had been available prior to the Grissom-Young flight, "the impact point would have been predicted to within ± 10 miles."[62]

The same need for good data reappeared during Apollo. The first of its orbital missions took place during 1966, flying atop the Saturn I-B. The initial launch, designated AS-201, flew suborbitally and covered 5,000 miles. A failure in the reaction controls produced uncontrolled lift during entry, but the craft splashed down 38 miles from its recovery ship. AS-202, six months later, was also suborbital. It executed a proper lifting entry—and undershot its designated aim point by 205 miles. This showed that its L/D had also been mispredicted.[63]

Estimates of the Apollo L/D had relied largely on experimental data taken during 1962 at Cornell Aeronautical Laboratory and Mach 15.8, and at AEDC and Mach 18.7. Again these measurements lacked accuracy, and once more Billy Griffith of AEDC stepped forward to direct a comprehensive set of new measurements. In addition to Tunnels F and L, used previously, the new work used Tunnels A, B, and C, which with the other facilities covered a range from Mach 3 to 20. To account for effects due to model supports in the wind tunnels, investigators also used a gun range that fired small models as free-flight projectiles, at Mach 6.0 to 8.5.

Widening Prospects for Re-entry

The 1962 estimates of Apollo L/D proved to be off by 20 percent, with the trim angle being in error by 3 degrees.[64] As with the Gemini data, these results showed anew that one could not obtain reliable data by working with a limited range of facilities. But when investigators broadened their reach to use more facilities, and sought accuracy through such methods as elimination of model-support errors, they indeed obtained results that matched flight test. This happened twice, with both Gemini and Apollo, with researchers finally getting the accurate estimates they needed.

These studies dealt with aerodynamic data at hypervelocity. In a separate series, other flights sought data on the re-entry environment that could narrow the range of acceptable theories of hypervelocity heating. Two such launches constituted Project Fire, which flew spacecraft that were approximately two feet across and had the general shape of Apollo's Command Module. Three layers of beryllium served as calorimeters, with measured temperature rises corresponding to total absorbed heat. Three layers of phenolic-asbestos alternated with those layers to provide thermal protection. Windows of fused quartz, which is both heat-resistant and transparent over a broad range of optical wavelengths, permitted radiometers to directly observe the heat flux due to radiation, at selected locations. These included the nose, where heating was most intense.

The Fire spacecraft rode atop Atlas boosters, with flights taking place in April 1964 and May 1965. Following cutoff of the Atlas, an Antares solid-fuel booster, modified from the standard third stage of the Scout booster, gave the craft an additional 17,000 feet per second and propelled it into the atmosphere at an angle of nearly 15 degrees, considerably steeper than the range of angles that were acceptable for an Apollo re-entry. This increased the rate of heating and enhanced the contribution from radiation. Each beryllium calorimeter gave useful data until its outer surface began to melt, which took only 2.5 seconds as the heating approached its maximum. When decelerations due to drag reached specified levels, an onboard controller ejected the remnants of each calorimeter in turn, along with its underlying layer of phenolic-asbestos. Because these layers served as insulation, each ejection exposed a cool beryllium surface as well as a clean set of quartz windows.

Fire 1 entered the atmosphere at 38,000 feet per second, markedly faster than the 35,000 feet per second of operational Apollo missions. Existing theories gave a range in estimates of total peak heating rate from 790 to 1,200 BTU per square foot-second. The returned data fell neatly in the middle of this range. Fire 2 did much the same, re-entering at 37,250 feet per second and giving a measured peak heating rate of just over 1,000 BTU per square foot-second. Radiative heating indeed was significant, amounting to some 40 percent of this total. But the measured values, obtained by radiometer, were at or below the minimum estimates obtained using existing theories.[65]

Earlier work had also shown that radiative heating was no source of concern. The new work also validated the estimates of total heating that had been used in designing the Apollo heat shield. A separate flight test, in August 1964, placed a small vehicle—the R-4—atop a five-stage version of the Scout. As with the X-17, this fifth stage ignited relatively late in the flight, accelerating the test vehicle to its peak speed when it was deep in the upper atmosphere. This speed, 28,000 feet per second, was considerably below that of an Apollo entry. But the increased air density subjected this craft to a particularly high heating rate.[66]

This was a materials-testing flight. The firm of Avco had been developing ablators of lower and lower weight and had come up with its 5026-39 series. They used epoxy-novolac as the resin, with phenolic microballoons added to the silica-fiber filler of an earlier series. Used with a structural honeycomb made of phenolic reinforced with fiberglass, it cut the density to 35 pounds per cubic foot and, with subsequent improvements, to as little as 31 pounds per cubic foot. This was less than three-tenths the density of the ancestral phenolic-fiberglass of Mercury—which merely orbited the Earth and did not fly back from the Moon.[67]

The new material had the designation Avcoat 5026-39G. The new flight sought to qualify it under its most severe design conditions, corresponding to re-entry at the bottom of the corridor with deceleration of 20 g. The peak aerodynamic load occurred at Mach 16.4 and 102,000 feet. Observed ablation rates proved to be much higher than expected. In fact, the ablative heat shield eroded away completely! This caused serious concern, for if that were to happen during a manned mission, the spacecraft would burn up in the atmosphere and would kill its astronauts.[68]

The relatively high air pressure had subjected the heat shield to dynamic pressures three times higher than those of an Apollo re-entry. Those intense dynamic pressures corresponded to a hypersonic wind that had blown away the ablative char soon after it had formed. This char was important; it protected the underlying virgin ablator, and when it was severely thinned or removed, the erosion rate on the test heat shield increased markedly.

Much the same happened in October 1965, when another subscale heat shield underwent flight test atop another multistage solid rocket, the Pacemaker, that accelerated its test vehicle to Mach 10.6 at 67,500 feet. These results showed that failure to duplicate the true re-entry environment in flight test could introduce unwarranted concern, causing what analysts James Pavlosky and Leslie St. Leger described as "unnecessary anxiety and work."[69]

An additional Project Fire flight could indeed have qualified the heat shield under fully realistic re-entry conditions, but NASA officials had gained confidence through their ability to understand the quasi-failure of the R-4. Rather than conduct further ad hoc heat-shield flight tests, they chose to merge its qualification with unmanned flights of complete Apollo spacecraft. Following three shots aboard the

WIDENING PROSPECTS FOR RE-ENTRY

Saturn I-B that went no further than earth orbit, and which included AS-201 and -202, the next flight lifted off in November 1967. It used a Saturn V to simulate a true lunar return.

No larger rocket had ever flown. This one was immense, standing 36 stories tall. The anchorman Walter Cronkite gave commentary from a nearby CBS News studio, and as this behemoth thundered upward atop a dazzling pillar of yellow-white flame, Cronkite shouted, "Oh, my God, our building is shaking! Part of the roof has come in here!" The roar was as loud as a major volcanic eruption. People saw the ascent in Jacksonville, 150 miles away.[70]

Heat-shield qualification stood as a major goal. The upper stages operated in sequence, thrusting the spacecraft to an apogee of 11,242 miles. It spent several hours coasting, oriented with the heat shield in the cold soak of shadow to achieve the largest possible thermal gradient around the shield. Re-ignition of the main engine pushed the spacecraft into re-entry at 35,220 feet per second relative to the atmosphere of the rotating Earth. Flying with an effective L/D of 0.365, it came down 10 miles from the aim point and only six miles from the recovery ship, close enough for news photos that showed a capsule in the water with one of its chutes still billowing.

The heat shield now was ready for the Moon, for it had survived a peak heating rate of 425 BTU per square foot-second and a total heat load of 37,522 BTU per pound. Operational lunar flights imposed loads and heating rates that were markedly less demanding. In the words of Pavlosky and St. Leger, "the thermal protection subsystem was overdesigned."[71]

A 1968 review took something of an offhand view of what once had been seen as an extraordinarily difficult problem. This report stated that thermal performance of ablative material "is one of the lesser criteria in developing a TPS." Significant changes had been made to enhance access for inspection, relief of thermal stress, manufacturability, performance near windows and other penetrations, and control of the center of gravity to achieve design values of L/D, "but never to obtain better thermal performance of the basic ablator."[72]

Thus, on the eve of the first lunar landing, specialists in hypersonics could look at a technology of re-entry whose prospects had widened significantly. A suite of materials now existed that were suitable for re-entry from orbit, having high emissivity to keep the temperature down, along with low thermal conductivity to prevent overheating during the prolonged heat soak. Experience had shown how careful research in ground facilities could produce reliable results and could permit maneuvering entry with accuracy in aim. This had been proven to be feasible for missions as demanding as lunar return.

Dyna-Soar had not flown, but it introduced metallic hot structures that brought the prospect of reusability. It also introduced wings for high L/D and particular freedom during maneuver. Indeed, by 1970 there was only one major frontier in re-entry: the development of a lightweight heat shield that was simpler than the hot structure of Dyna-Soar and was reusable. This topic was held over for the following decade, amid the development of the space shuttle.

1. Quotes: DTIC ADA-303832, p. 78.
2. Hallion, *Hypersonic*, p. 189; *Spaceflight*, July-August 1980, pp. 270-272.
3. Hallion, *Hypersonic*, pp. 189-91; DTIC ADA-303832, pp. 82-87.
4. Hallion, *Hypersonic*, p. 192; *Spaceflight*, July-August 1980, p. 270; Bell Aircraft D143-981-009.
5. DTIC ADA-303832, pp. 93-100; Hallion, *Hypersonic*, pp. 193-95; Bell Aircraft D143-945-055; Sloop, *Astronautics & Aeronautics*, October 1972, pp. 52-57.
6. Hallion, *Hypersonic*, pp. 195-97; DTIC ADA-303832, pp. 106-08.
7. Hallion, *Hypersonic*, pp. 195, 388-90 (quote, p. 390); DTIC ADA-303832, pp. 122-23.
8. Hallion, *Hypersonic*, p. 198; DTIC ADA-303832, pp. 126-28.
9. Hallion, *Hypersonic*, pp. 199-201; DTIC ADA-303832, pp. 128-30. Quote: Memo, Graham to Associate Director, 3 July 1957.
10. Hallion, *Hypersonic*, pp. 201-204; Houchin, *Air Power History*, Fall 1999, pp. 15-16; DTIC ADA-303832, pp. 134-40.
11. Jenkins, *Space Shuttle*, pp. 22-25; Sunday and London, "X-20," pp. 259, 262.
12. NACA TN 4046 (quotes, pp. 1, 29).
13. Quote: Memo, Graham to Associate Director, 3 July 1957.
14. NACA RM A55L05.
15. NASA SP-4305, pp. 367-69; Hallion, *Hypersonic*, pp. 388-91 (quote, p. 390).
16. NASA SP-4305, pp. 369-73; Hallion, *Hypersonic*, pp. 392-400.
17. Bell Aircraft D143-945-055.
18. Hallion, *Hypersonic*, pp. 406-07.
19. Sunday and London, "X-20," pp. 260-61, 264-69 (quote, p. 266).
20. NASA SP-4221, p. 203; Jenkins, *Space Shuttle*, p. 26.
21. Geiger, "Termination"; Sunday and London, "X-20," pp. 274-75 (includes quote). Cancellation documents: McNamara news briefing, 10 December 1963; telex, AFCVC-1918/63, HQ USAF to All Commands, 10 December 1963.
22. Sunday and London, "X-20," p. 275; Hallion, *Hypersonic*, pp. 294-310. Technical issues: DTIC AD-449685, p. 61.
23. Hallion, *Hypersonic*, p. II-xvi.
24. DTIC AD-346912, pp. III-3-1-2 to -5, III-3-1-18 to -23, III-4-2-2 to -8. Rene 41: AD-449685, pp. 28, 41.
25. DTIC AD-609169; AD-449685, pp. 3-5, 20; AD-346912, pp. III-3-1-7, III-4-1-2 to -3.
26. DTIC AD-346912, p. III-3-1-8; Hallion, *Hypersonic*, p. II-xv; NACA RM A55E26.
27. DTIC AD-609169; AD-449685, pp. 3-5, 20; AD-346912, pp. III-3-1-7, III-4-1-2 to -3.
28. DTIC AD-346912, p. III-3-1-8; AD-449685, p. iii.
29. DTIC AD-441740; AD-449685, pp. 5-7, 20 (quote, p. 7).
30. "Scientific Advisory Board Memo Report of the Aerospace Vehicles Panel on Dyna Soar Panel Flutter," 20 April 1962; DTIC AD-346912, pp. III-3-6-2 to -15 (quotes, pp. -8, -11).
31. DTIC AD-449685, pp. 49-50; AD-609169, pp. 13, 29; AD-346912, pp. III-3-1-6 to -7, III-

3-6-2, III-3-6-13 to -14.

32 DTIC AD-346912, pp. III-3-1-8, III-3-4-4; *Aviation Week*, 21 September 1959, p. 53.
33 DTIC AD-346912, pp. III-4-5-3 to -6, III-4-5-16.
34 Ibid., pp. III-4-5-13 to -18, III-4-5-35.
35 Ibid., pp. III-3-4-4 to -6, III-3-4-20.
36 Ibid., p. III-3-4-4; Hallion, *Hypersonic*, p. 357.
37 DTIC AD-346912, pp. III-4-6-5 to -7.
38 Ibid., pp. III-3-1-10 to -12; Hallion, *Hypersonic*, pp. 361-368.
39 DTIC AD-346912, pp. III-4-6-9 to -11; AD-449685, pp. 63-65.
40 DTIC AD-346912, pp. III-4-6-7 to -8; AD-449685, pp. 57-59.
41 DTIC AD-346912, pp. III-3-1-10 to -11; Jenkins, *Space Shuttle*, p. 28.
42 DTIC AD-346912, pp. III-3-1-14 to -15; Hallion, *Hypersonic*, pp. 347-49, 360-61.
43 Hallion, *Hypersonic*, pp. 344-46; Sunday and London, "X-20," p. 278. X-15: Schleicher, *Journal of the Royal Aeronautical Society*, October 1963, pp. 618-36.
44 Wolfe, *Right Stuff*, p. 56.
45 Faget: NASA SP-4201, p. 86. Quotes: NACA RM L58E07a, p. 1.
46 NACA RM L58E07a, Table I.
47 *Aviation Week*, 21 September 1959, pp. 52-59; NASA SP-4201, pp. 94-95.
48 Riddell and Teare, "Differences," p. 184.
49 Steg, *ARS Journal*, September 1960, pp. 815-22.
50 Day et al., *Eye*, pp. 112-14; Greer, "Corona," pp. 5-6; DTIC AD-328815.
51 *Time*, 27 April 1959, pp. 16, 65; Richelson: *Espionage*, p. 184; *Secret Eyes*, pp. 34-35. Twining quote: Day et al., *Eye*, p. 53.
52 NASA SP-4201, pp. 64-65, 95, 127-28.
53 NASA SP-4201, pp. 123, 125-26, 207.
54 Bond, "Big Joe" (quote, pp. 8-9).
55 Bond, "Big Joe"; memo with attachment, Grimwood to Everline, 3 July 1963; NASA TM X-490.
56 Ibid; pp. 16-17; NASA SP-4201, pp. 203-205 (quote, p. 203).
57 NASA TM X-490.
58 Bond, "Big Joe"; Erb and Stephens, "Project Mercury."
59 NASA TN D-7564; AIAA Paper 68-1142.
60 Lukasiewicz, *Experimental*, pp. 2-3; NASA RM L58E07a.
61 Lukasiewicz, *Experimental*, pp. 3-4.
62 Ibid., pp. 4-6 (quote, p. 6); AIAA Paper 67-166 (Griffith quote, p. 10).
63 NASA SP-4206, p. 340; Ley, *Rockets*, p. 495; Lukasiewicz, *Experimental*, p. 6.
64 AIAA Paper 68-1142; Lukasiewicz, *Experimental*, pp. 6-11; *Journal of Spacecraft and Rockets*, July 1968, pp. 843-48.
65 NASA TN D-2996, TN D-7564, TM X-1053, TM X-1120, TM X-1222, TM X-1305.

66 NASA TN D-7564, TM X-1182; Langley Working Paper LWP-54.
67 AIAA Paper 69-98; NASA TN D-7564, TM X-490, TM X-1407.
68 NASA TN D-7564, TM X-1182; Langley Working Paper LWP-54.
69 AIAA Paper 68-1142; NASA TN D-4713, TN D-7564 (quote, p. 16).
70 *Time*, 17 November 1967, 84-85 (includes quote).
71 NASA TN D-7564 (quote, p. 19).
72 AIAA Paper 68-1142 (quotes, p. 11).

6

Hypersonics and the Space Shuttle

During the mid-1960s, two advanced flight projects sought to lay technical groundwork for an eventual reusable space shuttle. ASSET, which flew first, progressed beyond Dyna-Soar by operating as a flight vehicle that used a hot structure, placing particular emphasis on studies of aerodynamic flutter. PRIME, which followed, had a wingless and teardrop-shaped configuration known as a lifting body. Its flight tests exercised this craft in maneuvering entries. Separate flights, using piloted lifting bodies, were conducted for landings and to give insight into their handling qualities.

From the perspective of ASSET and PRIME then, one would have readily concluded that the eventual shuttle would be built as a hot structure and would have the aerodynamic configuration of a lifting body. Indeed, initial shuttle design studies, late in the 1960s, followed these choices. However, they were not adopted in the final design.

The advent of a highly innovative type of thermal protection, Lockheed's reusable "tiles," completely changed the game in both the design and the thermal areas. Now, instead of building the shuttle with the complexities of a hot structure, it could be assembled as an aluminum airplane of conventional type, protected by the tiles. Lifting bodies also fell by the wayside, with the shuttle having wings. The Air Force insisted that these be delta wings that would allow the shuttle to fly long distances to the side of a trajectory. While NASA at first preferred simple straight wings, in time it agreed.

The shuttle relied on carbon-carbon for thermal protection in the hottest areas. It was structurally weak, but this caused no problem for more than 100 missions. Then in 2003, damage to a wing leading edge led to the loss of *Columbia*. It was the first space disaster to bring the death of astronauts due to failure of a thermal protection system.

Preludes: Asset and Lifting Bodies

At the end of the 1950s, ablatives stood out both for the ICBM and for return from space. Insulated hot structures, as on Dyna-Soar, promised reusability and lighter weight but were less developed. As early as August 1959, the Flight Dynamics Laboratory at Wright-Patterson Air Force Base launched an in-house study of a small recoverable boost-glide vehicle that was to test hot structures during re-entry. From the outset there was strong interest in problems of aerodynamic flutter. This was reflected in the concept name: ASSET or Aerothermodynamic/elastic Structural Systems Environmental Tests.

ASSET won approval as a program late in January 1961. In April of that year the firm of McDonnell Aircraft, which was already building Mercury capsules, won a contract to develop the ASSET flight vehicles. Initial thought had called for use of the solid-fuel Scout as the booster. Soon, however, it became clear that the program could use the Thor for greater power. The Air Force

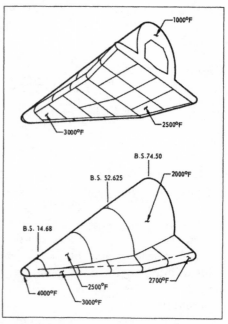

ASSET, showing peak temperatures.
(U.S. Air Force)

had deployed these missiles in England. When they came home, during 1963, they became available for use as launch vehicles.

ASSET took shape as a flat-bottomed wing-body craft that used the low-wing configuration recommended by NASA-Langley. It had a length of 59 inches and a span of 55 inches. Its bill of materials closely resembled that of Dyna-Soar, for it used TZM to withstand 3,000°F on the forward lower heat shield, graphite for similar temperatures on the leading edges, and zirconia rods for the nose cap, which was rated at 4,000°F. But ASSET avoided the use of Rene 41, with cobalt and columbium alloys being employed instead.[1]

ASSET was built in two varieties: the Aerothermodynamic Structural Vehicle (ASV), weighing 1,130 pounds, and the Aerothermodynamic Elastic Vehicle (AEV), at 1,225 pounds. The AEVs were to study panel flutter along with the behavior of a trailing-edge flap, which represented an aerodynamic control surface in hypersonic flight. These vehicles did not demand the highest possible flight speeds and hence flew with single-stage Thors as the boosters. But the ASVs were built to study materials and structures in the re-entry environment, while taking data on temperatures, pressures, and heat fluxes. Such missions demanded higher speeds. These boost-glide craft therefore used the two-stage Thor-Delta launch vehicle, which resembled the Thor-Able that had conducted nose-cone tests at intercontinental range as early as 1958.[2]

The program conducted six flights, which had the following planned values of range and of altitude and velocity at release:

ASSET FLIGHT TESTS

Date	Vehicle	Booster	Velocity, feet/second	Altitude, feet	Range, nautical miles
18 September 1963	ASV-1	Thor	16,000	205,000	987
24 March 1964	ASV-2	Thor-Delta	18,000	195,000	1800
22 July 1964	ASV-3	Thor-Delta	19,500	225,000	1830
27 October 1964	AEV-1	Thor	13,000	168,000	830
8 December 1964	AEV-2	Thor	13,000	187,000	620
23 February 1965	ASV-4	Thor-Delta	19,500	206,000	2300

Source: Hallion, *Hypersonic*, pp. 505, 510-519.

Several of these craft were to be recovered. Following standard practice, their launches were scheduled for the early morning, to give downrange recovery crews the maximum hours of daylight. This did not help ASV-1, the first flight in the program, which sank into the sea. Still, it flew successfully and returned good data. In addition, this flight set a milestone. In the words of historian Richard Hallion, "for the first time in aerospace history, a lifting reentry spacecraft had successfully returned from space."[3]

ASV-2 followed, using the two-stage Thor-Delta, but it failed when the second stage did not ignite. The next one carried ASV-3, with this mission scoring a double achievement. It not only made a good flight downrange but was successfully recovered. It carried a liquid-cooled double-wall test panel from Bell Aircraft, along with a molybdenum heat-shield panel from Boeing, home of Dyna-Soar. ASV-3 also had a new nose cap. The standard ASSET type used zirconia dowels, 1.5 inches long by 0.5 inch in diameter, that were bonded together with a zirconia cement. The new cap, from International Harvester, had a tungsten base covered with thorium oxide and was reinforced with tungsten.

A company advertisement stated that it withstood re-entry so well that it "could have been used again," and this was true for the craft as a whole. Hallion writes that "overall, it was in excellent condition. Water damage...caused some problems, but not so serious that McDonnell could not have refurbished and reflown the vehicle." The Boeing and Bell panels came through re-entry without damage, and the importance of physical recovery was emphasized when columbium aft leading edges showed significant deterioration. They were redesigned, with the new versions going into subsequent ASV and AEV spacecraft.[4]

The next two flights were AEVs, each of which carried a flutter test panel and a test flap. AEV-1 returned only one high-Mach data point, at Mach 11.88, but this sufficed to indicate that its panel was probably too stiff to undergo flutter. Engineers made it thinner and flew a new one on AEV-2, where it returned good data until it failed at Mach 10. The flap experiment also showed value. It had an electric motor that deflected it into the airstream, with potentiometers measuring the force required to move it, and it enabled aerodynamicists to critique their theories. Thus, one treatment gave pressures that were in good agreement with observations, whereas another did not.

ASV-4, the final flight, returned "the highest quality data of the ASSET program," according to the flight test report. The peak speed of 19,400 feet per second, Mach 18.4, was the highest in the series and was well above the design speed of 18,000 feet per second. The long hypersonic glide covered 2,300 nautical miles and prolonged the data return, which presented pressures at 29 locations on the vehicle and temperatures at 39. An onboard system transferred mercury ballast to trim the angle of attack, increasing L/D from its average of 1.2 to 1.4 and extending the trajectory. The only important problem came when the recovery parachute failed to deploy properly and ripped away, dooming ASV-4 to follow ASV-1 into the depths of the Atlantic.[5]

On the whole, ASSET nevertheless scored a host of successes. It showed that insulated hot structures could be built and flown without producing unpleasant surprises, at speeds up to three-fourths of orbital velocity. It dealt with such practical issues of design as fabrication, fasteners, and coatings. In hypersonic aerodynamics, ASSET contributed to understanding of flutter and of the use of movable control surfaces. The program also developed and successfully used a reaction control system built for a lifting re-entry vehicle. Only one flight vehicle was recovered in four attempts, but it complemented the returned data by permitting a close look at a hot structure that had survived its trial by fire.

A separate prelude to the space shuttle took form during the 1960s as NASA and the Air Force pursued a burgeoning interest in lifting bodies. The initial concept represented one more legacy of the blunt-body principle of H. Julian Allen and Alfred Eggers at NACA's Ames Aeronautical Laboratory. After developing this principle, they considered that a re-entering body, while remaining blunt to reduce its heat load, might produce lift and thus gain the ability to maneuver at hypersonic

Hypersonics and the Space Shuttle

speeds. An early configuration, the M-1 of 1957, featured a blunt-nosed cone with a flattened top. It showed some capacity for hypersonic maneuver but could not glide subsonically or land on a runway. A new shape, the M-2, appeared as a slender half-cone with its flat side up. Its hypersonic L/D of 1.4 was nearly triple that of the M-1. Fitted with two large vertical fins for stability, it emerged as a basic configuration that was suitable for further research.[6]

Dale Reed, an engineer at NASA's Flight Research Center, developed a strong interest in the bathtub-like shape of the M-2. He was a sailplane enthusiast and a builder of radio-controlled model aircraft. With support from the local community of airplane model builders, he proceeded to craft the M-2 as a piloted glider. Designating it as the M2-F1, he built it of plywood over a tubular steel frame. Completed early in 1963, it was 20 feet long and 13 feet across.

It needed a vehicle that could tow it into the air for initial tests. However, it produced too much drag for NASA's usual vans and trucks, and Reed needed a tow car with more power. He and his friends bought a stripped-down Pontiac with a big engine and a four-barrel carburetor that reached speeds of 110 miles per hour. They took it to a funny-car shop in Long Beach for modification. Like any other flightline vehicle, it was painted yellow with "National Aeronautics and Space Administration" on its side. Early tow tests showed enough success to allow the project to use a C-47, called the Gooney Bird, for true aerial flights. During these tests the Gooney Bird towed the M2-F1 above 10,000 feet and then set it loose to glide to an Edwards AFB lakebed. Beginning in August 1963, the test pilot Milt Thompson did this repeatedly. Reed thus showed that although the basic M-2 shape had been crafted for hypersonic re-entry, it could glide to a safe landing.

As he pursued this work, he won support from Paul Bikle, the director of NASA Flight Research Center. As early as April 1963, Bikle alerted NASA Headquarters that "the lifting-body concept looks even better to us as we get more into it." The success of the M2-F1 sparked interest within the Air Force as well. Some of its officials, along with their NASA counterparts, went on to pursue lifting-body programs that called for more than plywood and funny cars. An initial effort went beyond the M2-F1 by broadening the range of lifting-body shapes while working to develop satisfactory landing qualities.[7]

NASA contracted with the firm of Northrop to build two such aircraft: the M2-F2 and HL-10. The M2-F2 amounted to an M2-F1 built to NASA standards; the HL-10 drew on an alternate lifting-body design by Eugene Love of NASA-Langley. This meant that both Langley and Ames now had a project. The Air Force effort, the X-24A, went to the Martin Company. It used a design of Frederick Raymes at the Aerospace Corporation that resembled a teardrop fitted with two large fins.

All three flew initially as gliders, with a B-52 rather than a C-47 as the mother ship. The lifting bodies mounted small rocket engines for acceleration to supersonic speeds, thereby enabling tests of stability and handling qualities in transonic flight. The HL-10 set records for lifting bodies by making safe approaches and landings at Edwards from speeds up to Mach 1.86 and altitudes of 90,000 feet.[8]

Acceptable handling qualities were not easy to achieve. Under the best of circumstances, a lifting body flew like a brick at low speeds. Lowering the landing gear made the problem worse by adding drag, and test pilots delayed this deployment as long as possible. In May 1967 the pilot Bruce Peterson, flying the M2-F2, failed to get his gear down in time. The aircraft hit the lakebed at more than 250 mph, rolled over six times, and then came to rest on its back minus its cockpit canopy, main landing gear, and right vertical fin. Peterson, who might have died in the crash, got away with a skull fracture, a mangled face, and the loss of an eye. While surgeons reconstructed his face and returned him to active duty, the M2-F2 underwent surgery as well. Back at Northrop, engineers installed a center fin and a roll-control system that used reaction jets, while redistributing the internal weights. Gerauld Gentry, an Air Force test pilot, said that these changes turned "something I really did not enjoy flying at all into something that was quite pleasant to fly."[9]

The manned lifting-body program sought to turn these hypersonic shapes into aircraft that could land on runways, but the Air Force was not about to overlook the need for tests of their hypersonic performance during re-entry. The program that addressed this issue took shape with the name PRIME, Precision Recovery Including Maneuvering Entry. Martin Marietta, builder of the X-24A, also developed the PRIME flight vehicle, the SV-5D that later was referred to as the X-23. Although it was only seven feet in length, it faithfully duplicated the shape of the X-24A, even including a small bubble-like protrusion near the front that represented the cockpit canopy.

PRIME complemented ASSET, with both programs conducting flight tests of boost-glide vehicles. However, while ASSET pushed the state of the art in materials and hot structures, PRIME used ablative thermal protection for a more straightforward design and emphasized flight performance. Accelerated to near-orbital velocities by Atlas launch vehicles, the PRIME missions called for boost-glide flight from Vandenberg AFB to locations in the western Pacific near Kwajalein Atoll. The SV-5D had higher L/D than Gemini or Apollo, and as with those NASA programs, it was to demonstrate precision re-entry. The plans called for crossrange, with the vehicle flying up to 710 nautical miles to the side of a ballistic trajectory and then arriving within 10 miles of its recovery point.[10]

The X-24A was built of aluminum. The SV-5D used this material as well, for both the skin and primary structure. It mounted both aerodynamic and reaction controls, with the former taking shape as right and left body-mounted flaps set well aft. Used together, they controlled pitch; used individually, they produced yaw and roll. These flaps were beryllium plates that provided thermal heat sink. The fins were of steel honeycomb with surfaces of beryllium sheet.

HYPERSONICS AND THE SPACE SHUTTLE

Lifting bodies. Left to right: the X-24A, the M2-F3 which was modified from the M2-F2, and the HL-10. (NASA)

Landing a lifting body. The wingless X-24B required a particularly high angle of attack. (NASA)

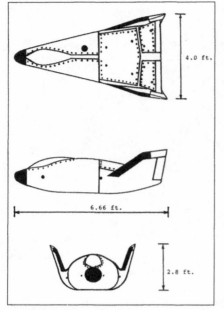

Martin SV-5D, which became the X-23. (U.S. Air Force)

Mission of the SV-5D. (U.S. Air Force)

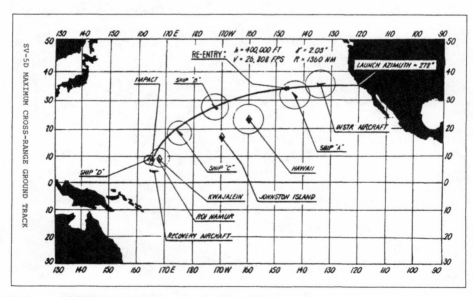

Trajectory of the SV-5D, showing crossrange. (U.S. Air Force)

Most of the vehicle surface obtained thermal protection from ESA 3560 HF, a flexible ablative blanket of phenolic fiberglass honeycomb that used a silicone elastomer as the filler, with fibers of nylon and silica holding the ablative char in place during re-entry. ESA 5500 HF, a high-density form of this ablator, gave added protection in hotter areas. The nose cap and the beryllium flaps used a different material: a carbon-phenolic composite. At the nose, its thickness reached 3.5 inches.[11]

The PRIME program made three flights, which took place between December 1966 and April 1967. All returned data successfully, with the third flight vehicle also being recovered. The first mission reached 25,300 feet per second and flew 4,300 miles downrange, missing its target by only 900 feet. The vehicle executed pitch maneuvers but made no attempt at crossrange. The next two flights indeed achieved crossrange, of 500 and 800 nautical miles, and the precision again was impressive. Flight 2 missed its aim point by less than two miles. Flight 3 missed by more than four miles, but this still was within the allowed limit. Moreover, the terminal guidance radar had been inoperative, which probably contributed to the lack of absolute accuracy.[12]

By demonstrating both crossrange and high accuracy during maneuvering entry, PRIME broadened the range of hypersonic aircraft configurations and completed a line of development that dated to 1953. In December of that year the test pilot Chuck Yeager had nearly been killed when his X-1A fell out of the sky at Mach 2.44 because it lacked tail surfaces that could produce aerodynamic stability. The X-15 was to fly to Mach 6, and Charles McLellan of NACA-Langley showed that it could use vertical fins of reasonable size if they were wedge-shaped in cross section. Meanwhile, Allen and Eggers were introducing their blunt-body principle. This led to missile nose cones with rounded tips, designed both as cones and as blunted cylinders that had stabilizing afterbodies in the shape of conic frustums.

For manned flight, Langley's Maxime Faget introduced the general shape of a cone with its base forward, protected by an ablative heat shield. Langley's John Becker entered the realm of winged re-entry configurations with his low-wing flat-bottom shapes that showed advantage over the high-wing flat-top concepts of NACA-Ames. The advent of the lifting body then raised the prospect of a structurally efficient shape that lacked wings, demanded thermal protection and added weight, and yet could land on a runway. Faget's designs had found application in Mercury, Gemini, and Apollo, while Becker's winged vehicle had provided a basis for Dyna-Soar. As NASA looked to the future, both winged designs and lifting bodies were in the forefront.[13]

Reusable Surface Insulation

As PRIME and the lifting bodies broadened the choices of hypersonic shape, work at Lockheed made similar contributions in the field of thermal protection. Ablatives were unrivaled for once-only use, but during the 1960s the hot structure continued to stand out as the preferred approach for reusable craft such as Dyna-Soar. As noted, it used an insulated primary or load-bearing structure with a skin of outer panels. These emitted heat by radiation, maintaining a temperature that was high but steady. Metal fittings supported these panels, and while the insulation could be high in quality, these fittings unavoidably leaked heat to the underlying structure. This raised difficulties in crafting this structure of aluminum or even of titanium, which had greater heat resistance. On Dyna-Soar, only Rene 41 would do.[14]

Ablatives avoided such heat leaks, while being sufficiently capable as insulators to permit the use of aluminum, as on the SV-5D of PRIME. In principle, a third approach combined the best features of hot structure and ablatives. It called for the use of temperature-resistant tiles, made perhaps of ceramic, that could cover the vehicle skin. Like hot-structure panels, they would radiate heat while remaining cool enough to avoid thermal damage. In addition, they were to be reusable. They also were to offer the excellent insulating properties of good ablators, preventing heat from reaching the underlying structure—which once more might be of aluminum. This concept, known as reusable surface insulation (RSI), gave rise in time to the thermal protection of the shuttle.

RSI grew out of ongoing work with ceramics for thermal protection. Ceramics had excellent temperature resistance, light weight, and good insulating properties. But they were brittle, and they cracked rather than stretched in response to the flexing under load of an underlying metal primary structure. Ceramics also were sensitive to thermal shock, as when heated glass breaks when plunged into cold water. This thermal shock resulted from rapid temperature changes during re-entry.[15]

Monolithic blocks of the ceramic zirconia had been specified for the nose cap of Dyna-Soar, but a different point of departure used mats of ceramic fiber in lieu of the solid blocks. The background to the shuttle's tiles lay in work with such mats that dated to the early 1960s at Lockheed Missiles and Space Company. Key people included R. M. Beasley, Ronald Banas, Douglas Izu, and Wilson Schramm. A Lockheed patent disclosure of December 1960 gave the first presentation of a reusable insulation made of ceramic fibers for use as a heat shield. Initial research dealt with casting fibrous layers from a slurry and bonding the fibers together.

Related work involved filament-wound structures that used long continuous strands. Silica fibers showed promise and led to an early success: a conical radome of 32-inch diameter built for Apollo in 1962. Designed for re-entry, it had a filament-wound external shell and a lightweight layer of internal insulation cast from short fibers of silica. The two sections were densified with a colloid of silica particles and sintered into a composite. This resulted in a non-ablative structure of silica composite, reinforced with fiber. It never flew, as design requirements changed during the development of Apollo. Even so, it introduced silica fiber into the realm of re-entry design.

Another early research effort, Lockheat, fabricated test versions of fibrous mats that had controlled porosity and microstructure. These were impregnated with organic fillers such as Plexiglas (methyl methacrylate). These composites resembled ablative materials, although the filler did not char. Instead it evaporated or volatilized, producing an outward flow of cool gas that protected the heat shield at high heat-transfer rates. The Lockheat studies investigated a range of fibers that included silica, alumina, and boria. Researchers constructed multilayer composite structures of filament-wound and short-fiber materials that resembled the Apollo radome. Impregnated densities were 40 to 60 pounds per cubic foot, the higher number being close to the density of water. Thicknesses of no more than an inch resulted in acceptably low back-face temperatures during simulations of re-entry.

This work with silica-fiber ceramics was well under way during 1962. Three years later a specific formulation of bonded silica fibers was ready for further development. Known as LI-1500, it was 89 percent porous and had a density of 15 pounds per cubic foot, one-fourth that of water. Its external surface was impregnated with filler to a predetermined depth, again to provide additional protection during the most severe re-entry heating. By the time this filler was depleted, the heat shield was to have entered a zone of more moderate heating, where the fibrous insulation alone could provide protection.

Initial versions of LI-1500, with impregnant, were intended for use with small space vehicles, similar to Dyna-Soar, that had high heating rates. Space shuttle concepts were already attracting attention—the January 1964 issue of the trade journal *Astronautics & Aeronautics* presents the thinking of the day—and in 1965 a Lockheed specialist, Max Hunter, introduced an influential configuration called Star Clipper. His design called for LI-1500 as the thermal protection.

Like other shuttle concepts, Star Clipper was to fly repeatedly, but the need for an impregnant in LI-1500 compromised its reusability. In contrast to earlier entry vehicle concepts, Star Clipper was large, offering exposed surfaces that were sufficiently blunt to benefit from the Allen-Eggers principle. They had lower temperatures and heating rates,

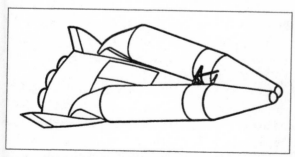

Star Clipper concept. (Art by Dan Gautier)

which made it possible to dispense with the impregnant. An unfilled version of LI-1500, which was inherently reusable, now could serve.

Here was the first concept of a flight vehicle with reusable insulation, bonded to the skin, that could reradiate heat in the fashion of a hot structure. However, the matted silica by itself was white and had low thermal emissivity, making it a poor radiator of heat. This brought excessive surface temperatures that called for thick layers of the silica insulation, adding weight. To reduce the temperatures and the thickness, the silica needed a coating that could turn it black, for high emissivity. It then would radiate well and remain cooler.

The selected coating was a borosilicate glass, initially with an admixture of chromium oxide and later with silicon carbide, which further raised the emissivity. The glass coating and silica substrate were both silicon dioxide; this assured a match of their coefficients of thermal expansion, to prevent the coating from developing cracks under the temperature changes of re-entry. The glass coating could soften at very high temperatures to heal minor nicks or scratches. It also offered true reusability, surviving repeated cycles to 2,500°F. A flight test came in 1968, as NASA-Langley investigators mounted a panel of LI-1500 to a Pacemaker re-entry test vehicle, along with several candidate ablators. This vehicle carried instruments, and it was recovered. Its trajectory reproduced the peak heating rates and temperatures of a re-entering Star Clipper. The LI-1500 test panel reached 2,300°F and did not crack, melt, or shrink. This proof-of-concept test gave further support to the concept of high-emittance reradiative tiles of coated silica for thermal protection.[16]

Lockheed conducted further studies at its Palo Alto Research Center. Investigators cut the weight of RSI by raising its porosity from the 89 percent of LI-1500 to 93 percent. The material that resulted, LI-900, weighed only nine pounds per cubic foot, one-seventh the density of water.[17] There also was much fundamental work on materials. Silica exists in three crystalline forms: quartz, cristobalite, tridymite. These not only have high coefficients of thermal expansion but also show sudden expansion or contraction with temperature due to solid-state phase changes. Cristobalite is particularly noteworthy; above 400°F it expands by more than 1 percent as it transforms from one phase to another. Silica fibers for RSI were to be glass, an amorphous rather than crystalline state having a very low coefficient of thermal expansion and absence of phase changes. The glassy form thus offered superb resistance to thermal stress and thermal shock, which would recur repeatedly during each return from orbit.[18]

The raw silica fiber came from Johns-Manville, which produced it from high-purity sand. At elevated temperatures it tended to undergo "devitrification," transforming from a glass into a crystalline state. Then, when cooling, it passed through phase-change temperatures and the fiber suddenly shrank, producing large internal tensile stresses. Some fibers broke, giving rise to internal cracking within the RSI and degradation of its properties. These problems threatened to grow worse during subsequent cycles of re-entry heating.

To prevent devitrification, Lockheed worked to remove impurities from the raw fiber. Company specialists raised the purity of the silica to 99.9 percent while reducing contaminating alkalis to as low as six parts per million. Lockheed did these things not only at the laboratory level but also in a pilot plant. This plant took the silica from raw material to finished tile, applying 140 process controls along the way. Established in 1970, the pilot plant was expanded in 1971 to attain a true manufacturing capability. Within this facility, Lockheed produced tiles of LI-1500 and LI-900 for use in extensive programs of test and evaluation. In turn, the increasing availability of these tiles encouraged their selection for shuttle thermal protection, in lieu of a hot-structure approach.[19]

General Electric also became actively involved, studying types of RSI made from zirconia and from mullite, as well as from silica. The raw fibers were commercial grade, with the zirconia coming from Union Carbide and the mullite from Babcock and Wilcox. Devitrification was a problem, but whereas Lockheed had addressed it by purifying its fiber, GE took the raw silica from Johns-Manville and tried to use it with little change. The basic fiber, the Q-felt of Dyna-Soar, also had served as insulation on the X-15. It contained 19 different elements as impurities. Some were present at a few parts per million, but others—aluminum, calcium, copper, lead, magnesium, potassium, sodium—ran from 100 to 1000 parts per million. In total, up to 0.3 percent was impurity. General Electric treated this fiber with a silicone resin that served as a binder, pyrolyzing the resin and causing it to break down at high temperatures. This transformed the fiber into a composite, sheathing each strand with a layer of amorphous silica that had a purity of 99.98 percent and higher. This high purity resulted from that of the resin. The amorphous silica bound the fibers together while inhibiting their devitrification. General Electric's RSI had a density of 11.5 pounds per cubic foot, midway between that of LI-900 and LI-1500.[20]

In January 1972, President Richard Nixon gave his approval to the space shuttle program, thereby raising it to the level of a presidential initiative. Within days, NASA's Dale Myers spoke to a lunar science conference in Houston and stated that the agency had made the basic decision to use RSI. Requests for proposal soon went out, inviting leading aerospace corporations to bid for the prime contract on the shuttle orbiter, and North American won this $2.6-billion prize in July. It specified mullite RSI for the undersurface and forward fuselage, a design feature that had been held over from the fully-reusable orbiter of the previous year.

Most of the primary structure was aluminum, but that of the nose was titanium, with insulation of zirconia lining the nose cap. The wing and fuselage upper surfaces, which had been titanium hot structure, now went over to an elastomeric RSI consisting of a foamed methylphenyl silicone, bonded to the orbiter in panel sizes as large as 36 inches. This RSI gave protection to 650°F.[21]

Still, was mullite RSI truly the one to choose? It came from General Electric and had lower emissivity than the silica RSI of Lockheed but could withstand higher temperatures. Yet the true basis for selection lay in the ability to withstand a hundred re-entries, as simulated in ground test. NASA conducted these tests during the last five months of 1972, using facilities at its Ames, Johnson, and Kennedy centers, with support from Battelle Memorial Institute.

The main series of tests ran from August to November and gave a clear advantage to Lockheed. That firm's LI-900 and LI-1500 went through 100 cycles to 2,300°F and met specified requirements for maintenance of low back-face temperatures and minimal thermal conductivity. The mullite showed excessive back-face temperatures and higher thermal conductivity, particularly at elevated temperatures. As test conditions increased in severity, the mullite also developed coating cracks and gave indications of substrate failure.

The tests then introduced acoustic loads, with each cycle of the simulation now subjecting the RSI to loud roars of rocket flight along with the heating of re-entry. LI-1500 continued to show promise. By mid-November it demonstrated the equivalent of 20 cycles to 160 decibels, the acoustic level of a large launch vehicle, and 2,300°F. A month later NASA conducted what Lockheed describes as a "sudden death shootout": a new series of thermal-acoustic tests, in which the contending materials went into a single large 24-tile array at NASA-Johnson. After 20 cycles, only Lockheed's LI-900 and LI-1500 remained intact. In separate tests, LI-1500 withstood 100 cycles to 2,500°F and survived a thermal overshoot to 3,000°F as well as an acoustic overshoot to 174 decibels. Clearly, this was the material NASA wanted.[22]

As insulation, they were astonishing. You could heat a tile in a furnace until it was white-hot, remove it, allow its surface to cool for a couple of minutes—and pick it up at its edges using your fingers, with its interior still at white heat. Lockheed won the thermal-protection subcontract in 1973, with NASA specifying LI-900 as the baseline RSI. The firm responded with preparations for a full-scale production facility in Sunnyvale, California. With this, tiles entered the mainstream of thermal protection.

Designing the Shuttle

In its overall technologies, the space shuttle demanded advances in a host of areas: rocket propulsion, fuel cells and other onboard systems, electronics and computers, and astronaut life support. As an exercise in hypersonics, two issues stood out: configuration and thermal protection. The Air Force supported some of the early studies, which grew seamlessly out of earlier work on Aerospaceplane. At

Douglas Aircraft, for instance, Melvin Root had his two-stage Astro, a fully-reusable rocket-powered concept with both stages shaped as lifting bodies. It was to carry a payload of 37,150 pounds, and Root expected that a fleet of such craft would fly 240 times per year. The contemporary Astrorocket of Martin Marietta, in turn, looked like two flat-bottom Dyna-Soar craft set belly to belly, foreshadowing fully-reusable space shuttle concepts of several years later.[23]

These concepts definitely belonged to the Aerospaceplane era. Astro dated to 1963, whereas Martin's Astrorocket studies went forward from 1961 to 1965. By mid-decade, though, the name "Aerospaceplane" was in bad odor within the Air Force. The new concepts were rocket-powered, whereas Aerospaceplanes generally had called for scramjets or LACE, and officials referred to these rocket craft as Integrated Launch and Re-entry Vehicles (ILRVs).[24]

Early contractor studies showed a definite preference for lifting bodies, generally with small foldout wings for use when landing. At Lockheed, Hunter's Star Clipper introduced the stage-and-a-half configuration that mounted expendable propellant tanks to a reusable core vehicle. The core carried the flight crew and payload along with the engines and onboard systems. It had a triangular planform and fitted neatly into a large inverted V formed by the tanks. The McDonnell Tip Tank concept was broadly similar; it also mounted expendable tanks to a lifting-body core.[25]

At Convair, people took the view that a single airframe could serve both as a core and, when fitted with internal tankage, as a reusable carrier of propellants. This led to the Triamese concept, whereby a triplet of such vehicles was to form a single ILRV that could rise into the sky. All three were to have thermal protection and would re-enter, flying to a runway and deploying their extendable wings. The concept was excessively hopeful; the differing requirements of core and tankage vehicles proved to militate strongly against a one-size-fits-all approach to airframe design. Still, the Triamese approach showed anew that designers were ready to use their imaginations.[26]

NASA became actively involved in the ongoing ILRV studies during 1968. George Mueller, the Associate Administrator for Manned Space Flight, took a particular interest and introduced the term "space shuttle" by which such craft came to be known. He had an in-house design leader, Maxime Faget of the Manner Spacecraft Center, who was quite strong-willed and had definite ideas of his own as to how a shuttle should look. Faget particularly saw lifting bodies as offering no more than mixed blessings: "You avoid wing-body interference," which brings problems of aerodynamics. "You have a simple structure. And you avoid the weight of wings." He nevertheless saw difficulties that appeared severe enough to rule out lifting bodies for a practical design.

They had low lift and high drag, which meant a dangerously high landing speed. As he put it, "I don't think it's charming to come in at 250 knots." Deployable wings could not be trusted; they might fail to extend. Lifting bodies also posed serious difficulties in development, for they required a fuselage that could do the work of a wing. This ruled out straightforward solutions to aerodynamic problems; the attempted solutions would ramify throughout the entire design. "They are very difficult to develop," he added, "because when you're trying to solve one problem, you're creating another problem somewhere else."[27] His colleague Milton Silveira, who went on to head the Shuttle Engineering Office at MSC, held a similar view:

> "If we had a problem with the aerodynamics on the vehicle, where the body was so tightly coupled to the aerodynamics, you couldn't simply go out and change the wing. You had to change the whole damn vehicle, so if you make a mistake, being able to correct it was a very difficult thing to do."[28]

Faget proposed instead to design his shuttle as a two-stage fully-reusable vehicle, with each stage being a winged airplane having low wings and a thermally-protected flat bottom. The configuration broadly resembled the X-15, and like that craft, it was to re-enter with its nose high and with its underside acting as the heat shield.

Faget wrote that "the vehicle would remain in this flight attitude throughout the entire descent to approximately 40,000 feet, where the velocity will have dropped to less than 300 feet per second. At this point, the nose gets pushed down, and the vehicle dives until it reaches adequate velocity for level flight." The craft then was to approach a runway and land at a moderate 130 knots, half the landing speed of a lifting body.[29]

During 1969 NASA sponsored a round of contractor studies that examined anew the range of alternatives. In June the agency issued a directive that ruled out the use of expendable boosters such as the Saturn V first stage, which was quite costly. Then in August, a new order called for the contractors to consider only two-stage fully reusable concepts and to eliminate partially-reusable designs such as Star Clipper and Tip Tank. This decision also was based in economics, for a fully-reusable shuttle could offer the lowest cost per flight. But it also delivered a new blow to the lifting bodies.[30]

There was a strong mismatch between lifting-body shapes, which were dictated by aerodynamics, and the cylindrical shapes of propellant tanks. Such tanks had to be cylindrical, both for ease in manufacturing and to hold internal pressure. This pressure was unavoidable; it resulted from boiloff of cryogenic propellants, and it served such useful purposes as stiffening the tanks' structures and delivering propellants to the turbopumps. However, tanks did not fit well within the internal volume of a lifting body; in Faget's words, "the lifting body is a damn poor container." The Lockheed and McDonnell designers had bypassed that problem by mounting their tanks externally, with no provision for reuse, but the new requirement of full reusability meant that internal installation now was mandatory. Yet although lifting bodies made poor containers, Faget's wing-body concept was an excellent one. Its fuselage could readily be cylindrical, being given over almost entirely to propellant tankage.[31]

The design exercises of 1969 covered thermal protection as well as configuration. McDonnell Douglas introduced orbiter designs derived from the HL-10 lifting body, examining 13 candidate configurations for the complete two-stage vehicle. The orbiter had a titanium primary structure, obtaining thermal protection from a hot-structure approach that used external panels of columbium, nickel-chromium, and Rene 41. This study also considered the use of tiles made of a "hardened compacted fiber," which was unrelated to Lockheed's RSI. However, the company did not recommend this. Those tiles were heavier than panels or shingles of refractory alloy and less durable.[32]

North American Rockwell took Faget's two-stage airplane as its preferred approach. It also used a titanium primary structure, with a titanium hot structure protecting the top of the orbiter, which faced a relatively mild thermal environment. For the thermally-protected bottom, North American adopted the work of Lockheed and specified LI-1500 tiles. The design also called for copious use of fiberglass insulation, which gave internal protection to the crew compartment and the cryogenic propellant tanks.[33]

Lockheed turned the Star Clipper core into a reusable second stage that retained its shape as a lifting body. Its structure was aluminum, as in a conventional airplane. The company was home to LI-1500, and designers considered its use for thermal protection. They concluded, though, that this carried high risk. They recommended instead a hot-structure approach that used corrugated Rene 41 along with shingles of nickel-chromium and columbium. The Lockheed work was independent of that at McDonnell Douglas, but engineers at the two firms reached similar conclusions.[34]

Convair, home of the Triamese concept, came in with new variants. These included a triplet launch vehicle with a core vehicle that was noticeably smaller than the two propellant carriers that flanked it. Another configuration placed the orbiter on the back of a single booster that continued to mount retractable wings. The orbiter had a primary structure of aluminum, with titanium for the heat-shield supports on the vehicle underside. Again this craft used a hot structure, with shingles of cobalt superalloy on the bottom and of titanium alloy on the top and side surfaces.

Significantly, these concepts were not designs that the companies were prepared to send to the shop floor and build immediately. They were paper vehicles that would take years to develop and prepare for flight. Yet despite this emphasis on the future, and notwithstanding the optimism that often pervades such preliminary design exercises, only North American was willing to recommend RSI as the baseline. Even Lockheed, its center of development, gave it no more than a highly equivocal recommendation. It lacked maturity, with hot structures standing as the approach that held greater promise.[35]

In the wake of the 1969 studies, NASA officials turned away from lifting bodies. Lockheed continued to study new versions of the Star Clipper, but the lifting body now was merely an alternative. The mainstream lay with Faget's two-stage fully-reusable approach, showing rocket-powered stages that looked like airplanes. Very soon, though, the shape of the wings changed anew, as a result of problems in Congress.

The space shuttle was a political program, funded by federal appropriations, and it had to make its way within the environment of Washington. On Capitol Hill, an influential viewpoint held that the shuttle was to go forward only if it was a national program, capable of meeting the needs of military as well as civilian users. NASA's shuttle studies had addressed the agency's requirements, but this proved not to be the way to proceed. Matters came to a head in the mid-1970 as Congressman Joseph Karth, a longtime NASA supporter, declared that the shuttle was merely the first step on a very costly road to Mars. He opposed funding for the shuttle in committee, and when he did not prevail, he made a motion from the floor of the House to strike the funds from NASA's budget. Other congressmen assured their colleagues that the shuttle had nothing to do with Mars, and Karth's measure went down to defeat—but by the narrowest possible margin: a tie vote of 53 to 53. In the Senate, NASA's support was only slightly greater.[36]

Such victories were likely to leave NASA undone, and the agency responded by seeking support for the shuttle from the Air Force. That service had tried and failed to build Dyna-Soar only a few years earlier; now it found NASA offering a much larger and more capable space shuttle on a silver platter. However, the Air Force was quite happy with its Titan III launch vehicles and made clear that it would work with NASA only if the shuttle was redesigned to meet the needs of the Pentagon. In particular, NASA was urged to take note of the views of the Air Force Flight Dynamics Laboratory (FDL), where specialists had been engaged in a running debate with Faget since early 1969.

The FDL had sponsored ILRV studies in parallel with the shuttle studies of NASA and had investigated such concepts as Lockheed's Star Clipper. One of its managers, Charles Cosenza, had directed the ASSET program. Another FDL scientist, Alfred Draper, had taken the lead in questioning Faget's approach. Faget wanted his shuttle stages to come in nose-high and then dive through 15,000 feet to pick up flying speed. With the nose so high, these airplanes would be fully stalled, and the Air Force disliked both stalls and dives, regarding them as preludes to an out-of-control crash. Draper wanted the shuttle to enter its glide while still supersonic, thereby maintaining much better control.

If the shuttle was to glide across a broad Mach range, from supersonic to subsonic, then it would face an important aerodynamic problem: a shift in the wing's center of lift. A wing generates lift across its entire lower surface, but one may regard this lift as concentrated at a point, the center of lift. At supersonic speeds, this center is located midway between the wing's leading and trailing edges. At subsonic speeds, this center moves forward and is much closer to the leading edge. To keep an airplane in proper balance, it requires an aerodynamic force that can compensate for this shift.

The Air Force had extensive experience with supersonic fighters and bombers that had successfully addressed this problem, maintaining good control and satisfactory handling qualities from Mach 3 to touchdown. Particularly for large aircraft—the B-58 and XB-70 bombers and the SR-71 spy plane—the preferred solution was a delta wing, triangular in shape. Delta wings typically ran along much of the length of the fuselage, extending nearly to the tail. Such aircraft dispensed with horizontal stabilizers at the tail and relied instead on elevons, control surfaces resembling ailerons that were set at the wing's trailing edge. Small deflections of these elevons then compensated for the shift in the center of lift, maintaining proper trim and balance without imposing excessive drag. Draper therefore proposed that both stages of Faget's shuttle have delta wings.[37]

Faget would have none of this. He wrote that because the only real flying was to take place during the landing approach, a wing design "can be selected solely on the basis of optimization for subsonic cruise and landing." The wing best suited to this limited purpose would be straight and unswept, like those of fighter planes in World War II. A tail would provide directional stability, as on a conventional airplane, enabling the shuttle to land in standard fashion. He was well aware of the center-of-lift shift but expected to avoid it by avoiding reliance on his wings until the craft was well below the speed of sound. He also believed that the delta would lose on its design merits. To achieve a suitably low landing speed, he argued that the delta would need a large wingspan. A straight wing, narrow in distance between its leading and trailing edges, would be light and would offer relatively little area demanding thermal protection. A delta of the same span, necessary for a moderate landing speed, would have a much larger area than the straight wing. This would add a great deal of weight, while substantially increasing the area that needed thermal protection.[38]

Draper responded with his own view. He believed that Faget's straight-wing design would be barred on grounds of safety from executing its maneuver of stall, dive, and recovery. Hence, it would have to glide from supersonic speeds through the transonic zone and could not avoid the center-of-lift problem. To deal with it, a good engineering solution called for installation of canards, small wings set well forward on the fuselage that would deflect to give the desired control. Canards produce lift and would tend to push the main wings farther to the back. They would be well aft from the outset, for they were to support an airplane that was empty of fuel but that had heavy rocket engines at the tail, placing the craft's center of gravity far to the rear. The wings' center of lift was to coincide closely with this center of gravity.

Draper wrote that the addition of canards "will move the wings aft and tend to close the gap between the tail and the wing." The wing shape that fills this gap is the delta, and Draper added that "the swept delta would most likely evolve."[39]

Faget had other critics, while Draper had supporters within NASA. Faget's thoughts indeed faced considerable resistance within NASA, particularly among the highly skilled test and research pilots at the Flight Research Center. Their spokesman, Milton Thompson, was certainly a man who knew how to fly airplanes, for he was an X-15 veteran and had been slated to fly Dyna-Soar as well. But in addition, these aerodynamic issues involved matters of policy, which drove the Air Force strongly toward the delta. The reason was that a delta could achieve high crossrange, whereas Faget's straight wing could not.

Crossrange was essential for single-orbit missions, launched from Vandenberg AFB on the California coast, which were to fly in polar orbit. The orbit of a spacecraft is essentially fixed with respect to distant stars, but the Earth rotates. In the course of a 90-minute shuttle orbit, this rotation carries the Vandenberg site eastward by 1,100 nautical miles. The shuttle therefore needed enough crossrange to cover that distance.

The Air Force had operational reasons for wanting once-around missions. A key example was rapid-response satellite reconnaissance. In addition, the Air Force was well aware that problems following launch could force a shuttle to come down at the earliest opportunity, executing a "once-around abort." NASA's Leroy Day, a senior shuttle manager, emphasized this point: "If you were making a polar-type launch out of Vandenberg, and you had Max's straight-wing vehicle, there was no place you could go. You'd be in the water when you came back. You've got to go crossrange quite a few hundred miles in order to make land."[40]

By contrast, NASA had little need for crossrange. It too had to be ready for once-around abort, but it expected to launch the shuttle from Florida's Kennedy Space Center on trajectories that ran almost due east. Near the end of its single orbit, the shuttle was to fly across the United States and could easily land at an emergency base. A 1969 baseline program document, "Desirable System Characteristics," stated that the agency needed only 250 to 400 nautical miles of crossrange, which Faget's straight wing could deliver with straightforward modifications.[41]

Faget's shuttle had a hypersonic L/D of about 0.5. Draper's delta-wing design was to achieve an L/D of 1.7, and the difference in the associated re-entry trajectories increased the weight penalty for the delta. A delta orbiter in any case needed a heavier wing and a larger area of thermal protection, and there was more. The straight-wing craft was to have a relatively brief re-entry and a modest heating rate. The delta orbiter was to achieve its crossrange by gliding hypersonically, executing a hypersonic glide that was to produce more lift and less drag. It also would increase both the rate of heating and its duration. Hence, its thermal protection had to be

Hypersonics and the Space Shuttle

more robust and therefore heavier still. In turn, the extra weight ramified throughout the entire two-stage shuttle vehicle, making it larger and more costly.[42]

NASA's key officials included the acting administrator, George Low, and the Associate Administrator for Manned Space Flight, Dale Myers. They would willingly have embraced Faget's shuttle. But on the military side, the Undersecretary of the Air Force for Research and Development, Michael Yarymovich, had close knowledge of the requirements of the National Reconnaissance Office. He played a key role in emphasizing that only a delta would do.

The denouement came at a meeting in Williamsburg, Virginia, in January 1971. At nearby Yorktown, in 1781, Britain's Lord Charles Cornwallis had surrendered to General George Washington, thereby ending America's war of independence. One hundred and ninety years later NASA surrendered to the Air Force, agreeing particularly to build a delta-wing shuttle with full military crossrange of 1,100 miles. In return, though, NASA indeed won the support from the Pentagon that it needed. Opposition faded on Capitol Hill, and the shuttle program went forward on a much stronger political foundation.[43]

The design studies of 1969 had counted as Phase A and were preliminary in character. In 1970 the agency launched Phase B, conducting studies in greater depth, with North American Rockwell and McDonnell Douglas as the contractors. Initially they considered both straight-wing and delta designs, but the Williamsburg decision meant that during 1971 they were to emphasize the deltas. These remained as two-stage fully-reusable configurations, which were openly presented at an AIAA meeting in July of that year.

In the primary structure and outer skin of the wings and fuselage, both contractors proposed to use titanium freely. They differed, however, in their approaches to thermal protection. McDonnell Douglas continued to favor hot structures. Most of the underside of the orbiter was covered with shingles of Hastelloy-X nickel superalloy. The wing leading edges called for a load-bearing structure of columbium, with shingles of coated columbium protecting these leading edges as well as other areas that were too hot for the Hastelloy. A nose cap of carbon-carbon completed the orbiter's ensemble.[44]

North American had its own interest in titanium hot structures, specifying them as well for the upper wing surfaces and the upper fuselage. Everywhere else possible, the design called for applying mullite RSI directly to a skin of aluminum. Such tiles were to cover the entire underside of the wings and fuselage, along with much of the fuselage forward of the wings. The nose and leading edges, both of the wings, and the vertical fin used carbon-carbon. In turn, the fin was designed as a hot structure with a skin of Inconel 718 nickel alloy.[45]

By mid-1971, though, hot structures were in trouble. The new Office of Management and Budget had made clear that it expected to impose stringent limits on funding for the shuttle, which brought a demand for new configurations that could cut the cost of development. Within weeks, the contractors did a major turnabout. They went over to primary structures of aluminum. They also abandoned hot structures and embraced RSI. Managers were aware that it might take time to develop for operational use, but they were prepared to use ablatives for interim thermal protection, switching to RSI once it was ready.[46]

What brought this dramatic change? The advent of RSI production at Lockheed was critical. This drew attention from Faget, who had kept his hand in the field of shuttle design, offering a succession of conceptual configurations that had helped to guide the work of the contractors. His most important concept, designated MSC-040, came out in September 1971 and served as a point of reference. It used aluminum and RSI.[47]

"My history has always been to take the most conservative approach," Faget declared. Everyone knew how to work with aluminum, for it was the most familiar of materials, but titanium was literally a black art. Much of the pertinent shop-floor experience had been gained within the SR-71 program and was classified. Few machine shops had the pertinent background, for only Lockheed had constructed an airplane—the SR-71—that used titanium hot structure. The situation was worse for columbium and the superalloys because these metals had been used mostly in turbine blades. Lockheed had encountered serious difficulties as its machinists and metallurgists wrestled with the use of titanium. With the shuttle facing cost constraints, no one wanted to risk an overrun while machinists struggled with the problems of other new materials.[48]

NASA-Langley had worked to build a columbium heat shield for the shuttle and had gained a particularly clear view of its difficulties. It was heavier than RSI but offered no advantage in temperature resistance. In addition, coatings posed serious problems. Silicides showed promise of reusability and long life, but they were fragile and damaged easily. A localized loss of coating could result in rapid oxygen embrittlement at high temperatures. Unprotected columbium oxidized readily, and above the melting point of its oxide, 2,730°F, it could burst into flame.[49] "The least little scratch in the coating, the shingle would be destroyed during re-entry," said Faget. Charles Donlan, the shuttle program manager at NASA Headquarters, placed this in a broader perspective in 1983:

> "Phase B was the first really extensive effort to put together studies related to the completely reusable vehicle. As we went along, it became increasingly evident that there were some problems. And then as we looked at the development problems, they became pretty expensive. We learned also that the metallic heat shield, of which the wings were to be made, was by no means ready for use. The slightest scratch and you are in trouble."[50]

Hypersonics and the Space Shuttle

Other refractory metals offered alternatives to columbium, but even when proposing to use them, the complexity of a hot structure also militated against their selection. As a mechanical installation, it called for large numbers of clips, brackets, stand-offs, frames, beams, and fasteners. Structural analysis loomed as a formidable task. Each of many panel geometries needed its own analysis, to show with confidence that the panels would not fail through creep, buckling, flutter, or stress under load. Yet this confidence might be fragile, for hot structures had limited ability to resist overtemperatures. They also faced the continuing issue of sealing panel edges against ingestion of hot gas during re-entry.[51]

In this fashion, having taken a long look at hot structures, NASA did an about-face as it turned toward the RSI that Lockheed's Max Hunter had recommended as early as 1965. The choice of aluminum for the primary structure reflected the excellent insulating properties of RSI, but there was more. Titanium offered a potential advantage because of its temperature resistance; hence, its thermal protection might be lighter. However, the apparent weight saving was largely lost due to a need for extra insulation to protect the crew cabin, payload bay, and onboard systems. Aluminum could compensate for its lack of heat resistance because it had higher thermal conductivity than titanium. Hence, it could more readily spread its heat throughout the entire volume of the primary structure.

Designers expected to install RSI tiles by bonding them to the skin, and for this, aluminum had a strong advantage. Both metals form thin layers of oxide when exposed to air, but that of aluminum is more strongly bound. Adhesive, applied to aluminum, therefore held tightly. The bond with titanium was considerably weaker and appeared likely to fail in operational use at approximately 500°F. This was not much higher than the limit for aluminum, 350°F, which showed that the temperature resistance of titanium did not lend itself to operational use.[52]

The move toward RSI and aluminum simplified the design and cut the development cost. Substantially larger cost savings came into view as well, as NASA moved away from full reusability of its two-stage concepts. The emphasis now was on partial reusability, which the prescient Max Hunter had advocated as far back as 1965 when he placed the liquid hydrogen of Star Clipper in expendable external tanks. The new designs kept propellants within the booster, but they too called for the use of external tankage for the orbiter. This led to reduced sizes for both stages and had a dramatic impact on the problem of providing thermal protection for the shuttle's booster.

On paper, a shuttle booster amounted to the world's largest airplane, combining the size of a Boeing 747 or C-5A with performance exceeding that of the X-15. It was to re-enter at speeds well below orbital velocity, but still it needed thermal protection, and the reduced entry velocities did not simplify the design. North American, for one, specified RSI for its Phase B orbiter, but the company also had to show

Facing the Heat Barrier: A History of Hypersonics

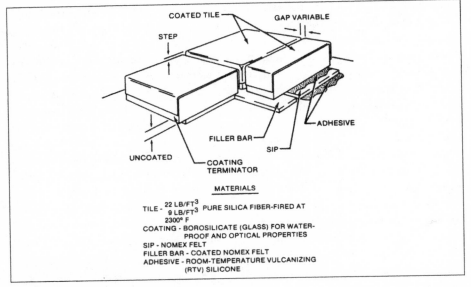

Thermal-protection tiles for the space shuttle. (NASA)

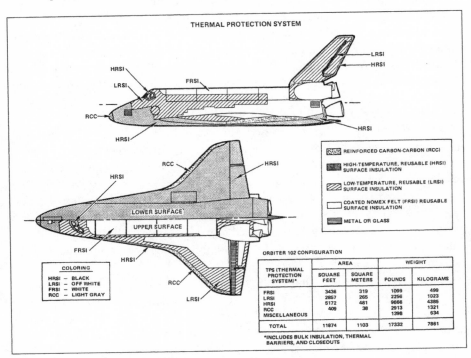

Thermal-protection system for the space shuttle. (NASA)

that it understood hot structures. These went into the booster, which protected its hot areas with titanium, Inconel 718, carbon-carbon, Rene 41, Haynes 188 steel—and coated columbium.

The move toward external tankage brought several advantages, the most important of which was a reduction in staging velocity. When designing a two-stage rocket, standard methods exist for dividing the work load between the two stages so as to achieve the lowest total weight. These methods give an optimum staging velocity. A higher value makes the first stage excessively large and heavy; a lower velocity means more size and weight for the orbiter. Ground rules set at NASA's Marshall Space Flight Center, based on such optimization, placed this staging velocity close to 10,000 feet per second.

But by offloading propellants into external tankage, the orbiter could shrink considerably in size and weight. The tanks did not need thermal protection or heavy internal structure; they might be simple aluminum shells stiffened with internal pressure. With the orbiter being lighter, and being little affected by a change in staging velocity, a recalculation of the optimum value showed advantage in making the tanks larger so that they could carry more propellant. This meant that the orbiter was to gain more speed in flight—and the booster would gain less. Hence, the booster also could shrink in size. Better yet, the reduction in staging velocity eased the problem of booster thermal protection.

Grumman was the first company to pursue this line of thought, as it studied alternative concepts alongside the mainstream fully reusable designs of McDonnell Douglas and North American. Grumman gave a fully-reusable concept of its own, for purposes of comparison, but emphasized a partially-reusable orbiter that put the liquid hydrogen in two external tanks. The liquid oxygen, which was dense and compact, remained aboard this vehicle, but the low density of the hydrogen meant that its tanks could be bulky while remaining light in weight.

The fully-reusable design followed the NASA Marshall Space Flight Center ground rules and showed a staging velocity of 9,750 feet per second. The external-tank configuration cut this to 7,000 feet per second. Boeing, which was teamed with Grumman, found that this substantially reduced the need for thermal protection as such. The booster now needed neither tiles nor exotic metals. Instead, like the X-15, it was to use its structure as a heat sink. During re-entry, it would experience a sharp but brief pulse of heat, which a conventional aircraft structure could absorb without exceeding temperature limits. Hot areas continued to demand a titanium hot structure, which was to cover some one-eighth of the booster. However, the rest of this vehicle could make considerable use of aluminum.

How could bare aluminum, without protection, serve in a shuttle booster? It was common understanding that aluminum airframes lost strength due to aerodynamic heating at speeds beyond Mach 2, with titanium being necessary at higher speeds. However, this held true for aircraft in cruise, which faced their temperatures continually. The Boeing booster was to re-enter at Mach 7, matching the top speed of the X-15. Even so, its thermal environment resembled a fire that does not burn your hand when you whisk it through quickly. Across part of the underside, the vehicle was to protect itself by the simple method of metal with more than usual thickness to cope with the heat. Even these areas were limited in extent, with the contractors noting that "the material gauges [thicknesses] required for strength exceed the minimum heat-sink gauges over the majority of the vehicle."[53]

McDonnell Douglas went further. In mid-1971 it introduced its own external-tank orbiter that lowered the staging velocity to 6,200 feet per second. Its winged booster was 82 percent aluminum heat sink. Their selected configuration was optimized from a thermal standpoint, bringing the largest savings in the weight of thermal protection.[54] Then in March 1972 NASA selected solid-propellant rockets for the boosters. The issue of their thermal protection now went away entirely, for these big solids used steel casings that were 0.5 inch thick and that provided heat sink very effectively.[55]

Amid the design changes, NASA went over to the Air Force view and embraced the delta wing. Faget himself accepted it, making it a feature of his MSC-040 concept. Then the Office of Management and Budget asked whether NASA might return to Faget's straight wing after all, abandoning the delta and thereby saving money. Nearly a year after the Williamsburg meeting, Charles Donlan, acting director of the shuttle program office at Headquarters, ruled this out. In a memo to George Low, he wrote that high crossrange was "fundamental to the operation of the orbiter." It would enhance its maneuverability, greatly broadening the opportunities to abort a mission and perhaps save the lives of astronauts. High crossrange would also provide more frequent opportunities to return to Kennedy Space Center in the course of a normal mission.

Delta wings also held advantages that were entirely separate from crossrange. A delta orbiter would be stable in flight from hypersonic to subsonic speeds, throughout a wide range of nose-high attitudes. The aerodynamic flow over such an orbiter would be smooth and predictable, thereby permitting accurate forecasts of heating during re-entry and giving confidence in the design of the shuttle's thermal protection. In addition, the delta vehicle would experience relatively low temperatures of 600 to 800°F over its sides and upper surfaces.

By contrast, straight-wing configurations produced complicated hypersonic flow fields, with high local temperatures and severe temperature changes on the wing, body, and tail. Temperatures on the sides of the fuselage would run from 900 to 1,300°F, making the design and analysis of thermal protection more complex. During transition from supersonic to subsonic speeds, the straight-wing orbiter would experience unsteady flow and buffeting, making it harder to fly. This combination of aerodynamic and operational advantages led Donlan to favor the delta for reasons that were entirely separate from those of the Air Force.[56]

Hypersonics and the Space Shuttle

The Loss of *Columbia*

Thermal protection was delicate. The tiles lacked structural strength and were brittle. It was not possible even to bond them directly to the underlying skin, for they would fracture and break due to their inability to follow the flexing of the skin under its loads. Designers therefore placed an intermediate layer between tiles and skin that had some elasticity and could stretch in response to shuttle skin flexing without transmitting excessive strain to the tiles. It worked; there never was a serious accident due to fracturing of tiles.[57]

The same was not true of another piece of delicate work: a thermal-protection panel made of carbon that had the purpose of protecting one of the wing leading edges. It failed during re-entry in virtually the final minutes of a flight of *Columbia*, on 1 February 2003. For want of this panel, that spacecraft broke up over Texas in a shower of brilliant fireballs. All aboard were killed.

The background to this accident lay in the fact that for the nose and leading edges of the shuttle, silica RSI was not enough. These areas needed thermal protection with greater temperature resistance, and carbon was the obvious candidate. It was lighter than aluminum and could be protected against oxidation with a coating. It also had a track record, having formed the primary structure of the Dyna-Soar nose cap and the leading edge of ASSET. Graphite was the standard form, but in contrast to ablative materials, it failed to enter the aerospace mainstream. It was brittle and easily damaged and did not lend itself to use with thin-walled structures.

The development of a better carbon began in 1958, with Vought Missiles and Space Company in the forefront. The work went forward with support from the Dyna-Soar and Apollo programs and brought the advent of an all-carbon composite consisting of graphite fibers in a carbon matrix. Existing composites had names such as carbon-phenolic and graphite-epoxy; this one was carbon-carbon.

It retained the desirable properties of graphite in bulk: light weight, temperature resistance, and resistance to oxidation when coated. It had the useful property of actually gaining strength with temperature, being up to 50 percent stronger at 3,000°F than at room temperature. It had a very low coefficient of thermal expansion, which reduced thermal stress. It also had better damage tolerance than graphite.

Carbon-carbon was a composite. As with other composites, Vought engineers fabricated parts of this material by forming them as layups. Carbon cloth gave a point of departure, produced by oxygen-free pyrolysis of a woven organic fiber such as rayon. Sheets of this fabric, impregnated with phenolic resin, were stacked in a mold to form the layup and then cured in an autoclave. This produced a shape made of laminated carbon cloth phenolic. Further pyrolysis converted the resin to its basic carbon, yielding an all-carbon piece that was highly porous due to the loss of volatiles. It therefore needed densification, which was achieved through multiple cycles of reimpregnation under pressure with an alcohol, followed by further pyrolysis. These cycles continued until the part had its specified density and strength.[58]

Researchers at Vought conducted exploratory studies during the early 1960s, investigating resins, fibers, weaves, and coatings. In 1964 they fabricated a Dyna-Soar nose cap of carbon-carbon, with this exercise permitting comparison of the new nose cap with the standard versions that used graphite and zirconia tiles. In 1966 this firm crafted a heat shield for the Apollo afterbody, which lay leeward of the curved ablative front face. A year and a half later the company constructed a wind-tunnel model of a Mars probe that was designed to enter the atmosphere of that planet.[59]

These exercises did not approach the full-scale development that Dyna-Soar and ASSET had brought to hot structures. They definitely were in the realm of the preliminary. Still, as they went forward along with Lockheed's work on silica RSI and GE's studies of mullite, the work at Vought made it clear that carbon-carbon was likely to take its place amid the new generation of thermal-protection materials.

The shuttle's design specified carbon-carbon for the nose cap and leading edges, and developmental testing was conducted with care. Structural tests exercised their methods of attachment by simulating flight loads up to design limits, with design temperature gradients. Other tests, conducted within an arc-heated facility, determined the thermal responses and hot-gas leakage characteristics of interfaces between the carbon-carbon and RSI.[60]

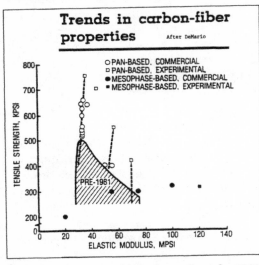

Improvements in strength of carbon-carbon after 1981. (AIAA)

Other tests used articles that represented substantial portions of the orbiter. An important test item, evaluated at NASA-Johnson, reproduced a wing leading edge and measured five by eight feet. It had two leading-edge panels of carbon-carbon set side by side, a section of wing structure that included its main spars, and aluminum skin covered with RSI. It had insulated attachments, internal insulation, and interface seals between the carbon-carbon and the RSI. It withstood simulated air loads, launch acoustics, and mission temperature-pressure environments, not once but many times.[61]

There was no doubt that left to themselves, the panels of carbon-carbon that protected the leading edges would have continued to do so. Unfortunately, they were not left to themselves. During the ascent of *Columbia*, on 16 January 2003, a

HYPERSONICS AND THE SPACE SHUTTLE

large piece of insulating foam detached itself from a strut that joined the external tank to the front of the orbiter. The vehicle at that moment was slightly more than 80 seconds into the flight, traveling at nearly Mach 2.5. This foam struck a carbon-carbon panel and delivered what proved to be a fatal wound.

Ground controllers became aware of this the following day, during a review of high-resolution film taken at the time of launch. The mission continued for two weeks, and in the words of the accident report, investigators concluded that "some localized heating damage would most likely occur during re-entry, but they could not definitively state that structural damage would result."[62]

Yet the damage was mortal. Again, in words of the accident report,

> *Columbia* re-entered Earth's atmosphere with a pre-existing breach in the leading edge of its left wing.... This breach, caused by the foam strike on ascent, was of sufficient size to allow superheated air (probably exceeding 5,000°F) to penetrate the cavity behind the RCC panel. The breach widened, destroying the insulation protecting the wing's leading edge support structure, and the superheated air eventually melted the thin aluminum wing spar. Once in the interior, the superheated air began to destroy the left wing.... Finally, over Texas,...the increasing aerodynamic forces the Orbiter experienced in the denser levels of the atmosphere overcame the catastrophically damaged left wing, causing the Orbiter to fall out of control.[63]

It was not feasible to go over to a form of thermal protection, for the wing leading edges, that would use a material other than carbon-carbon and that would be substantially more robust. Even so, three years of effort succeeded in securing the foam and the shuttle returned to flight in July 2006 with foam that stayed put.

In addition, people took advantage of the fact that most such missions had already been intended to dock with the International Space Station. It now became a rule that the shuttle could fly only if it were to go there, where it could be inspected minutely prior to re-entry and where astronauts could stay, if necessary, until a damaged shuttle was repaired or a new one brought up. In this fashion, rather than the thermal protection being shaped to fit the needs of the missions, the missions were shaped to fit the requirements of having safe thermal protection.[64]

1. "Advanced Technology Program: Technical Development Plan for Aerothermodynamic/Elastic Structural System Environmental Tests (ASSET)." Air Force Systems Command, 9 September 1963.
2. Ibid., pp. 4, 11-13; Hallion, *Hypersonic*, pp. 451, 464-65.
3. Hallion, *Hypersonic*, pp. 510-12 (quote, p. 512).
4. Ibid., pp. 512-16 (quote, p. 515); DTIC AD-357523; advertisement, *Aviation Week*, 24 May 1965, p. 62.
5. *Aviation Week*, 2 November 1964, pp. 25-26; DTIC AD-366546 (quote, p. 156); Hallion, *Hypersonic*, pp. 516-19.
6. Hallion, *Hypersonic*, pp. 524, 529, 535, 864-66.
7. NASA: SP-4220, pp. 33-39; SP-4303, pp. 148-52 (quote, p. 151).
8. NASA: SP-4220, index references; SP-4303, ch. 8; Jenkins, *Space Shuttle*, p. 34.
9. NASA SP 4220, pp. 106-09, 111-16 (quote, p. 116).
10. Miller, *X-Planes*, pp. 256-61; Martin Marietta Report ER 14465.
11. Hallion, *Hypersonic*, pp. 641, 648-49; Martin Marietta Report ER 14465, ch. VI; *Aviation Week*, 16 May 1966, pp. 64-75.
12. Martin Marietta Report ER 14465, p. I-1; Hallion, *Hypersonic*, pp. 694-702; *Aviation Week*, 10 July 1967, pp. 99-101.
13. See index references in the present volume.
14. AIAA Paper 71-443.
15. Schramm et al., "Space Shuttle" (*Lockheed Horizons*).
16. Ibid.
17. Korb et al., "Shuttle," p. 1189; Schramm, "HRSI," p. 1194.
18. NASA TM X-2273, pp. 57-58; Korb et al., "Shuttle," pp. 1189-90; CASI 72A-10764.
19. Schramm et al., "Space Shuttle," p. 11 (*Lockheed Horizons*); Schramm, "HRSI," p. 1195.
20. CASI 72A-10764; NASA TM X-2273, pp. 39-93.
21. *Aviation Week*, 17 January 1972, p. 17; North American Rockwell Report SV 72-19.
22. Schramm et al., "Space Shuttle," pp. 11-14 (*Lockheed Horizons*); *Aviation Week*, 27 March 1972, p. 48.
23. Root and Fuller, *Astronautics & Aeronautics*, January 1964, pp. 42-51; Hallion, *Hypersonic*, pp. 952-54.
24. Jenkins, *Space Shuttle*, pp. 60-62, 67; Hallion, *Hypersonic*, pp. 951, 995.
25. Hunter, "Origins"; Schnyer and Voss, "Review," pp. 17-22, 35-39; Jenkins, *Space Shuttle*, pp. 67-70.
26. NASA SP-4221, pp. 89-91; General Dynamics Reports GDC-DCB-67-031, GDC-DCB-68-017.
27. NASA SP-4221, pp. 92-94; author interview, Max Faget, 4 March 1997 (includes quote). Folder 18649, NASA Historical Reference Collection, NASA History Division, Washington, D.C. 20546.
28. Milton Silveira interview by Joe Guilmartin and John Mauer, 14 November 1984. Oral His-

torical Series, Shuttle Interviews Box 2. JSC History Collection, University of Houston-Clear Lake.
29. Faget, *Astronautics & Aeronautics*, January 1970, pp. 52-61 (quote, p. 53).
30. NASA SP-4221, pp. 218-219.
31. Author interview, Max Faget, 4 March 1997 (includes quote). Folder 18649, NASA Historical Reference Collection, NASA History Division, Washington, D.C. 20546.
32. McDonnell Douglas Report MDC E0056; Jenkins, *Space Shuttle*, pp. 84-86.
33. North American Rockwell Report SD 69-573-1; Jenkins, *Space Shuttle*, pp. 87-89.
34. Lockheed Report LMSC-A959837; Jenkins, *Space Shuttle*, pp. 90-91.
35. General Dynamics Report GDC-DCB-69-046; Jenkins, *Space Shuttle*, pp. 92-95.
36. NASA SP-4221, pp. 179-86, 219-21.
37. Ibid., pp. 215-16; Draper et al., *Astronautics & Aeronautics*, January 1971, pp. 26-35; Hallion, *Hypersonic*, pp. 459, 1032; Jenkins, *Space Shuttle*, pp. 66-68.
38. Faget, *Astronautics & Aeronautics*, January 1970, pp.52-61 (quote, p. 57).
39. AIAA Paper 70-1249; Draper et al., *Astronautics & Aeronautics*, January 1971, pp. 26-35 (quotes, p. 29).
40. Personal discussions with John Pike, Federation of American Scientists, July 1997; Pace, *Engineering*, pp. 146-49. Quote: John Mauer interview with Leroy Day, 17 October 1983, p. 41. Oral Historical Series, Shuttle Interviews Box 1, JSC History Collection, University of Houston-Clear Lake.
41. Day, *Task Group Report*, Vol. II, pp. 40-42; Faget, *Astronautics & Aeronautics*, January 1970, p. 59.
42. NASA SP-4221, p. 217; Faget and Silveira, "Fundamental."
43. Pace, *Engineering*, p. 116; NASA SP-4221, pp. 230-31, 233-34.
44. NASA SP-4221, pp. 223-25; AIAA Papers 71-804, 71-805; McDonnell Douglas Report MDC E0308.
45. North American Rockwell SV 71-28, SD 71-114-1.
46. Grumman Reports B35-43 RP-33, p. 19; North American Rockwell SV 71-50, p. 9 and SV 71-59, p. 3; McDonnell Douglas "Interim Report," p. 36.
47. NASA SP-4221, pp. 341-346; Jenkins, *Space Shuttle*, pp. 141-50.
48. Heppenheimer, *Turbulent*, pp. 209-210; author interview, Max Faget, 4 March 1997. Folder 18649, NASA Historical Reference Collection, NASA History Division, Washington, D.C. 20546.
49. CASI 81A-44344; Korb et al., "Shuttle," p. 1189.
50. John Mauer interview, Charles Donlan, 19 October 1983. Oral Historical Series, Shuttle Interviews Box 1, JSC History Collection, University of Houston-Clear Lake. Also, author interview, Max Faget, 4 March 1997. Folder 18649, NASA Historical Reference Collection, NASA History Division, Washington, D.C. 20546.
51. CASI 81A-44344.
52. Pace, *Engineering*, pp. 179-88.
53. NASA SP-4221, pp. 335-40, 348-49; Grumman Report B35-43 RP-11; *Aviation Week*, 12 July 1971, pp. 36-39.

54 McDonnell Douglas Report MDC E0376-1.
55 NASA SP-4221, pp. 420-22; Heppenheimer, *Development*, p. 188.
56 Jenkins, *Space Shuttle*, p. 147; memo, Donlan to Low, 5 December 1971 (includes quote).
57 *Space World*, June-July 1979, p. 23; CASI 77A-35304; Korb et al., "Shuttle."
58 Becker, "Leading"; Korb et al., "Shuttle."
59 Becker, "Leading"; "Technical" (Vought).
60 *Astronautics & Aeronautics*, January 1976, pp. 60, 63-64.
61 *Aviation Week*, 31 March 1975, p. 52; AIAA Paper 78-485.
62 *Columbia*, p. 38.
63 Ibid., p. 12.
64 Ibid, pp. 225-227.

7

THE FADING, THE COMEBACK

During the 1960s and 1970s, work in re-entry went from strength to strength. The same was certainly not true of scramjets, which reached a peak of activity in the Aerospaceplane era and then quickly faded. Partly it was their sheer difficulty, along with an appreciation that whatever scramjets might do tomorrow, rockets were already doing today. Yet the issues went deeper.

The 1950s saw the advent of antiaircraft missiles. Until then, the history of air power had been one of faster speeds and higher altitudes. At a stroke, though, it became clear that missiles held the advantage. A hot fighter plane, literally hot from aerodynamic heating, now was no longer a world-class dogfighter; instead it was a target for a heat-seeking missile.

When antiaircraft no longer could outrace defenders, they ceased to aim at speed records. They still needed speed but not beyond a point at which this requirement would compromise other fighting qualities. Instead, aircraft were developed with an enhanced ability to fly low, where missiles could lose themselves in ground clutter, and became stealthy. In 1952, late in the dogfight era, Clarence "Kelly" Johnson designed the F-104 as the "missile with a man in it," the ultimate interceptor. No one did this again, not after the real missiles came in.

This was bad news for ramjets. The ramjet had come to the fore around 1950, in projects such as Navaho, Bomarc, and the XF-103, because it offered Mach 3 at a time when turbojets could barely reach Mach 1. But Mach 3, when actually achieved in craft such as the XB-70 and SR-71, proved to be a highly specialized achievement that had little to do with practical air power. No one ever sent an SR-71 to conduct close air support at subsonic speed, while the XB-70 gave way to its predecessor, the B-52, because the latter could fly low whereas the XB-70 could not.

Ramjets also faltered on their merits. The ramjet was one of two new airbreathers that came forth after the war, with the other being the turbojet. Inevitably this set up a Darwinian competition in which one was likely to render the other extinct. Ramjets from the start were vulnerable, for while they had the advantage of speed, they needed an auxiliary boost from a rocket or turbojet. Nor was it small; the Navaho booster was fully as large as the winged missile itself.

The problem of compressor stall limited turbojet performance for a time. But from 1950 onward, several innovations brought means of dealing with it. They led to speedsters such as the F-104 and F-105, operational aircraft that topped Mach 2, along with the B-58 which also did this. The SR-71, in turn, exceeded Mach 3. This meant that there was no further demand for ramjets, which were not selected for new aircraft.

The ramjet thus died not only because its market was lost to advanced turbojets, but because the advent of missiles made it clear that there no longer was a demand for really fast aircraft. This, in turn, was bad news for scramjets. The scramjet was an advanced ramjet, likely to enter the aerospace mainstream only while ramjets remained there. The decline of the ramjet trade meant that there was no industry that might build scramjets, no powerful advocates that might press for them.

The scramjet still held the prospective advantage of being able to fly to orbit as a single stage. With Aerospaceplane, the Air Force took a long look as to whether this was plausible, and the answer was no, at least not soon. With this the scramjet lost both its rationale in the continuing pursuit of high speed and the prospect of an alternate mission—ascent to orbit—that might allow it to bypass this difficulty.

In its heyday the scramjet had stood on the threshold of mainstream research and development, with significant projects under way at General Electric and United Aircraft Research Laboratories, which was affiliated with Pratt & Whitney. As scramjets faded, though, even General Applied Science Laboratories (GASL), a scramjet center that had been founded by Antonio Ferri himself, had to find other activities. For a time the only complete scramjet lab in business was at NASA-Langley.

And then—lightning struck. President Ronald Reagan announced the Strategic Defense Initiative (SDI), which brought the prospect of a massive new demand for access to space. The Air Force already was turning away from the space shuttle, while General Lawrence Skantze, head of the Air Force Systems Command, was strongly interested in alternatives. He had no background in scramjets, but he embraced the concept as his own. The result was the National Aerospace Plane (NASP) effort, which aimed at airplane-like flight to orbit.

In time SDI faded as well, while lessons learned by researchers showed that NASP offered no easy path to space flight. NASP faded in turn and with it went hopes for a new day for hypersonics. Final performance estimates for the prime NASP vehicle, the X-30, were not terribly far removed from the early and optimistic estimates that had made the project appear feasible. Still, the X-30 design was so sensitive that even modest initial errors could drive its size and cost beyond what the Pentagon was willing to accept.

The Fading, the Comeback

Scramjets Pass Their Peak

From the outset, scramjets received attention for the propulsion of tactical missiles. In 1959 APL's Gordon Dugger and Frederick Billig disclosed a concept that took the name SCRAM, Supersonic Combustion Ramjet Missile. Boosted by a solid-fuel rocket, SCRAM was to cruise at Mach 8.5 and an altitude of 100,000 feet, with range of more than 400 miles. This cruise speed resulted in a temperature of 3,800°F at the nose, which was viewed as the limit attainable with coated materials.[1]

The APL researchers had a strong interest in fuels other than liquid hydrogen, which could not be stored. The standard fuel, a boron-rich blend, used ethyl decaborane. It ignited easily and gave some 25 percent more energy per pound than gasoline. Other tests used blends of pentaborane with heavy hydrocarbons, with the pentaborane promoting their ignition. The APL group went on to construct and test a complete scramjet of 10-inch diameter.[2]

Paralleling this Navy-sponsored work, the Air Force strengthened its own efforts in scramjets. In 1963 Weldon Worth, chief scientist at the Aero Propulsion Laboratory, joined with Antonio Ferri and recommended scramjets as a topic meriting attention. Worth proceeded by funding new scramjet initiatives at General Electric and Pratt & Whitney. This was significant; these firms were the nation's leading builders of turbojet and turbofan engines.

GE's complete scramjet was axisymmetric, with a movable centerbody that included the nose spike. It was water-cooled and had a diameter of nine inches, with this size being suited to the company's test facility. It burned hydrogen, which was quite energetic. Yet the engine failed to deliver net thrust, with this force being more than canceled out by drag.[3]

The Pratt & Whitney effort drew on management and facilities at nearby United Aircraft Research Laboratories. Its engine also was axisymmetric and used a long cowl that extended well to the rear, forming the outer wall of the nozzle duct. This entire cowl moved as a unit, thereby achieving variable geometry for all three major components: inlet, combustor, and nozzle. The effort culminated in fabrication of a complete water-cooled test unit of 18-inch diameter.[4]

A separate Aero Propulsion Lab initiative, the Incremental Flight Test Vehicle (IFTV), also went forward for a time. It indeed had the status of a flight vehicle, with Marquardt holding the prime contract and taking responsibility for the engine. Lockheed designed and built the vehicle and conducted wind-tunnel tests at its Rye Canyon facility, close to Marquardt's plant in Van Nuys, California.

The concept called for this craft to ride atop a solid-fuel Castor rocket, which was the second stage of the Scout launch vehicle. Castor was to accelerate the IFTV to 5,400 feet per second, with this missile then separating and entering free flight. Burning hydrogen, its engines were to operate for at least five seconds, adding an "increment" of velocity of at least 600 feet per second. Following launch over the Pacific from Vandenberg AFB, it was to telemeter its data to the ground.

This was the first attempt to develop a scramjet as the centerpiece of a flight program, and much of what could go wrong did go wrong. The vehicle grew in weight during development. It also increased its drag and found itself plagued for a time with inlets that failed to start. The scramjets themselves gave genuine net thrust but still fell short in performance.

The flight vehicle mounted four scramjets. The target thrust was 597 pounds. The best value was 477 pounds. However, the engines needed several hundred pounds of thrust merely to overcome drag on the vehicle and accelerate, and this reduction in performance meant that the vehicle could attain not quite half of the desired velocity increase of 600 feet per second.[5]

Just then, around 1967, the troubles of the IFTV were mirrored by troubles in the overall scramjet program. Scramjets had held their promise for a time, with a NASA/Air Force Ad Hoc Working Group, in a May 1965 report, calling for an expanded program that was to culminate in a piloted hypersonic airplane. The SAB had offered its own favorable words, while General Bernard Schriever, head of the Air Force Systems Command—the ARDC, its name having changed in 1961—attempted to secure $50 million in new funding.[6]

He did not get it, and the most important reason was that conventional ramjets, their predecessors, had failed to win a secure role. The ramjet-powered programs of the 1950s, including Navaho and Bomarc, now appeared as mere sidelines within a grand transformation that took the Air Force in only 15 years from piston-powered B-36 and B-50 bombers to the solid-fuel Minuteman ICBM and the powerful Titan III launch vehicle. The Air Force was happy with both and saw no reason for scramjet craft as alternatives. This was particularly true because Aerospaceplane had come up with nothing compelling.

The Aero Propulsion Laboratory had funded the IFTV and the GE and Pratt scramjets, but it had shown that it would support this engine only if it could be developed quickly and inexpensively. Neither had proved to be the case. The IFTV effort, for one, had escalated in cost from $3.5 million to $12 million, with its engine being short on power and its airframe having excessive drag and weight.[7]

After Schriever's $50-million program failed to win support, Air Force scramjet efforts withered and died. More generally, between 1966 and 1968, three actions ended Air Force involvement in broad-based hypersonic research and brought an end to a succession of halcyon years. The Vietnam War gave an important reason for these actions, for the war placed great pressure on budgets and led to cancellation of many programs that lacked urgency.

The first decision ended Air Force support for the X-15. In July 1966 the joint NASA-Air Force Aeronautics and Astronautics Coordinating Board determined that NASA was to accept all budgetary responsibility for the X-15 as of 1 January 1968. This meant that NASA was to pay for further flights—which it refused to do. This brought an end to the prospect of using this research airplane for flight testing of hypersonic engines.[8]

The second decision, in August 1967, terminated IFTV. Arthur Thomas, the Marquardt program manager, later stated that it had been a major error to embark on a flight program before ground test had established attainable performance levels. When asked why this systematic approach had not been pursued, Thomas pointed to the pressure of a fast-paced schedule that ruled out sequential development. He added that Marquardt would have been judged "nonresponsive" if its proposal had called for sequential development at the outset. In turn, this tight schedule reflected the basic attitude of the Aero Propulsion Lab: to develop a successful scramjet quickly and inexpensively, or not to develop one at all.[9]

Then in September 1968 the Navy elected to close its Ordnance Aerophysics Laboratory (OAL). This facility had stood out because it could accommodate test engines of realistic size. In turn, its demise brought a premature end to the P & W scramjet effort. That project succeeded in testing its engine at OAL at Mach 5, but only about 20 runs were conducted before OAL shut down, which was far too few for serious development. Nor could this engine readily find a new home; its 18-inch diameter had been sized to fit the capabilities of OAL. This project therefore died both from withdrawal of Air Force support and from loss of its principal test facility.[10]

As dusk fell on the Air Force hypersonics program, Antonio Ferri was among the first to face up to the consequences. After 1966 he became aware that no major new contracts would be coming from the Aero Propulsion Lab, and he decided to leave GASL, where he had been president. New York University gave him strong encouragement, offering him the endowed Astor Professorship. He took this appointment during the spring of 1967.[11]

He proceeded to build new research facilities in the Bronx, as New York University bought a parcel of land for his new lab. A landmark was a vacuum sphere for his wind tunnel, which his friend Louis Nucci called "the hallmark of hypersonic flow" as it sucks high-pressure air from a stored supply. Ferri had left a trail of such spheres at his previous appointments: NACA-Langley, Brooklyn Polytechnic, GASL. But his new facilities were far less capable than those of GASL, and his opportunities were correspondingly reduced. He set up a consulting practice within an existing firm, Advanced Technology Labs, and conducted analytical studies. Still, Nucci recalls that "Ferri's love was to do experiments. To have only [Advanced Technology Labs] was like having half a body."

GASL took a significant blow in August 1967, as the Air Force canceled IFTV. The company had been giving strong support to the developmental testing of its engine, and in Nucci's words, "we had to use our know-how in flow and combustion." Having taken over from Ferri as company president, he won a contract from the Department of Transportation to study the aerodynamics of high-speed trains running in tubes.

"We had to retread everybody," Nucci adds. Boeing held a federal contract to develop a supersonic transport; GASL studied its sonic boom. GASL also investigated the "parasol wing," a low-drag design that rode atop its fuselage at the end of a pylon. There also was work on pollution for the local utility, Long Island Lighting Company, which hoped to reduce its smog-forming emissions. The company stayed alive, but its employment dropped from 80 people in 1967 to only 45 five years later.[12]

Marquardt made its own compromises. It now was building small rocket engines, including attitude-control thrusters for Apollo and later for the space shuttle. But it too left the field of hypersonics. Arthur Thomas had managed the company's work on IFTV, and as he recalls, "I was chief engineer and assistant general manager. I got laid off. We laid off two-thirds of our people in one day." He knew that there was no scramjet group he might join, but he hoped for the next-best thing: conventional ramjets, powering high-speed missiles. "I went all over the country," he continues. "Everything in ramjet missiles had collapsed." He had to settle for a job working with turbojets, at McDonnell Douglas in St. Louis.[13]

Did these people ever doubt the value of their work? "Never," says Billig. Nucci, Ferri's old friend, gives the same answer: "Never. He always had faith." The problem they faced was not to allay any doubts of their own, but to overcome the misgivings of others and to find backers who would give them new funding. From time to time a small opportunity appeared. Then, as Billig recalls, "we were highly competitive. Who was going to get the last bits of money? As money got tighter, competition got stronger. I hope it was a friendly competition, but each of us thought he could do the job best."[14]

Amid this dark night of hypersonic research, two candles still flickered. There was APL, where a small group continued to work on missiles powered by scramjets that were to burn conventional fuels. More significantly, there was the Hypersonic Propulsion Branch at NASA-Langley, which maintained itself as the one place where important work on hydrogen-fueled scramjets still could go forward. As scramjets died within the Air Force, the Langley group went ahead, first with its Hypersonic Research Engine (HRE) and then with more advanced airframe-integrated designs.

SCRAMJETS AT NASA-LANGLEY

The road to a Langley scramjet project had its start at North American Aviation, builder of the X-15. During 1962 manager Edwin Johnston crafted a proposal to modify one of the three flight vehicles to serve as a testbed for hypersonic engines.

The Fading, the Comeback

This suggestion drew little initial interest, but in November a serious accident reopened the question. Though badly damaged, the aircraft, Tail Number 66671, proved to be repairable. It returned to flight in June 1964, with modifications that indeed gave it the option for engine testing.

The X-15 program thus had this flight-capable testbed in prospect during 1963, at a time when engines for test did not even exist on paper. It was not long, though, before NASA responded to its opportunity, as Hugh Dryden, the Agency's Deputy Administrator, joined with Robert Seamans, the Associate Administrator, in approving a new program that indeed sought to build a test engine. It took the name of Hypersonic Research Engine (HRE).

Three companies conducted initial studies: General Electric, Marquardt, and Garrett AiResearch. All eyes soon were on Garrett, as it proposed an axisymmetric configuration that was considerably shorter than the others. John Becker later wrote that it "was the smallest, simplest, easiest to cool, and had the best structural approach of the three designs." Moreover, Garrett had shown strong initiative through the leadership of its study manager, Anthony duPont.[15]

He was a member of the famous duPont family in the chemical industry. Casual and easygoing, he had already shown a keen eye for the technologies of the future. As early as 1954, as a student, he had applied for a patent on a wing made of

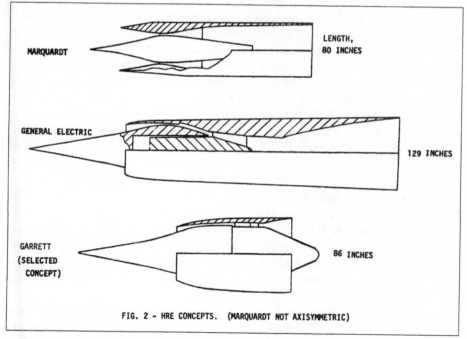

FIG. 2 - HRE CONCEPTS. (MARQUARDT NOT AXISYMMETRIC)

The HRE concepts of three competing contractors. (NASA)

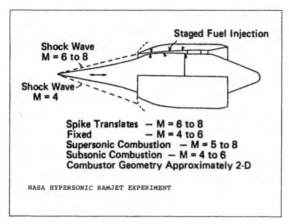

The selected HRE design. (NASA)

composite materials. He flew as a co-pilot with Pan American, commemorating those days with a framed picture of a Stratocruiser airliner in his office. He went on to Douglas Aircraft, where he managed studies of Aerospaceplane. Then Clifford Garrett, who had a strong interest in scramjets, recruited him to direct his company's efforts.[16]

NASA's managers soon offered an opportunity to the HRE competitors. The Ordnance Aerophysics Laboratory was still in business, and any of them could spend a month there testing hardware—if they could build scramjet components on short notice. Drawing on $250,000 in company funds, DuPont crafted a full-scale HRE combustor in only sixty days. At OAL, it yielded more than five hours of test data. Neither GE nor Marquardt showed similar adroitness, while DuPont's initiative suggested that the final HRE combustor would be easy to build. With this plus the advantages noted by Becker, Garrett won the contract. In July 1966 the program then moved into a phase of engine development and test.[17]

Number 66671 was flying routinely, and it proved possible to build a dummy HRE that could be mounted to the lower fin of that X-15. This led to a flight-test program that approached disaster in October 1967, when the test pilot Pete Knight flew to Mach 6.72. "We burned the engine off," Knight recalls. "I was on my way back to Edwards; my concern was to get the airplane back in one piece." He landed safely, but historian Richard Hallion writes that the airplane "resembled burnt firewood.... It was the closest any X-15 came to structural failure induced by heating."[18]

Once again it went back to the shops, marked for extensive repair. Then in mid-November another X-15 was lost outright in the accident that killed its test pilot, Mike Adams. Suddenly the X-15 was down from three flight-rated airplanes to only one, and while Number 66671 returned to the flight line the following June, it never flew again. Nor would it fly again with the HRE. This dummy engine had set up the patterns of airflow that had caused the shock-impingement heating that had nearly destroyed it.[19]

In a trice then the HRE program was completely turned on its head. It had begun with the expectation of using the X-15 for flight test of advanced engines, at a moment when no such engines existed. Now Garrett was building them—but

The Fading, the Comeback

Test pilot William "Pete" Knight initiates his record flight, which reached Mach 6.72. (NASA)

the X-15 could not be allowed to fly with them. Indeed, it soon stopped flying altogether. Thus, during 1968, it became clear that the HRE could survive only through a complete shift in focus to ground test.

Earlier plans had called for a hydrogen-cooled flightweight engine. Now the program's research objectives were to be addressed using two separate wind-tunnel versions. Each was to have a diameter of 18 inches, with a configuration and flow path matching those of the earlier flight-rated concept. The test objectives then were divided between them.

A water-cooled Aerothermodynamic Integration Model (AIM) was to serve for hot-fire testing. Lacking provision for hydrogen cooling, it stood at the technical level of the General Electric and Pratt & Whitney test scramjets. In addition, continuing interest in flightweight hydrogen-cooled engine structures brought a requirement for the Structures Assembly Model (SAM), which did not burn fuel. It operated at high temperature in Langley's eight-foot diameter High Temperature Structures Tunnel, which reached Mach 7.[20]

SAM arrived at NASA-Langley in August 1970. Under test, its inlet lip showed robustness for it stood up to the impact of small particles, some of which blocked thin hydrogen flow passages. Other impacts produced actual holes as large as 1/16 inch in diameter. The lip nevertheless rode through the subsequent shock-impingement heating without coolant starvation or damage from overheating. This represented an important advance in scramjet technology, for it demonstrated the feasibility of crafting a flightweight fuel-cooled structure that could withstand foreign object damage along with very severe heating.[21]

AIM was also on the agenda. It reached its test center at Plum Brook, Ohio, in August 1971, but the facility was not ready. It took a year before the program undertook data runs, and then most of another year before the first run that was successful. Indeed, of 63 test runs conducted across 18 months, 42 returned little or no useful data. Moreover, while scramjet advocates had hoped to achieve shock-free flow, it certainly did not do this. In addition, only about half of the injected fuel actually burned. But shocks in the subsonic-combustion zone heated the downstream flow and unexpectedly enabled the rest of the fuel to burn. In Becker's words, "without this bonanza, AIM performance would have been far below its design values."[22]

The HRE was axisymmetric. A practical engine of this type would have been mounted in a pod, like a turbojet in an airliner. An airliner's jet engines use only a small portion of the air that flows past the wings and fuselage, but scramjets have far less effectiveness. Therefore, to give enough thrust for acceleration at high Mach, they must capture and process as much as possible of the air flowing along the vehicle.

Podded engines like the HRE cannot do this. The axisymmetry of the HRE made it easy to study because it had a two-dimensional layout, but it was not suitable for an operational engine. The scramjet that indeed could capture and process most of the airflow is known as an airframe-integrated engine, in which much of the aircraft serves as part of the propulsion system. Its layout is three-dimensional and hence is more complex, but only an airframe-integrated concept has the additional power that can make it practical for propulsion.

Paper studies of airframe-integrated concepts began at Langley in 1968, breaking completely with those of HRE. These investigations considered the

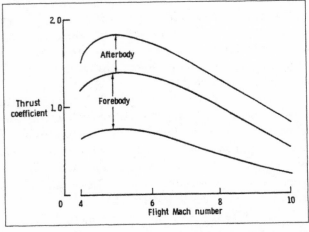

Contributions to scramjet thrust from airframe integration. (NASA)

entire undersurface of a hypersonic aircraft as an element of the propulsion system. The forebody produced a strong oblique shock that precompressed the airflow prior to its entry into the inlet. The afterbody was curved and swept upward to form a half-nozzle. This concept gave a useful shape for the airplane while retaining the advantages of airframe-integrated scramjet operation.

Within the Hypersonic Propulsion Branch, John Henry and Shimer Pinckney developed the initial concept. Their basic installation was a module, rectangular in shape, with a number of them set side by side to encircle the lower fuselage and achieve the required high capture of airflow. Their inlet had a swept opening that angled backward at 48 degrees. This provided a cutaway that readily permitted spillage of airflow, which otherwise could choke the inlet when starting.

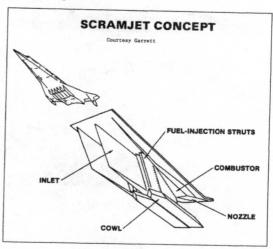

Airframe-integrated scramjet concept. (Garrett Corp.)

The bow shock gave greater compression of the flow at high Mach, thereby reducing the height of the cowl and the required size of the engine. At Mach 10 this reduction was by a factor of three. While this shock compressed the flow vertically, wedge-shaped sidewalls compressed it horizontally. This two-plane compression diminished changes in the inlet flow field with increasing Mach, making it possible to cover a broad Mach range in fixed geometry.

Like the inlet, the combustor was to cover a large Mach range in fixed geometry. This called for thermal compression, and Langley contracted with Antonio Ferri at New York University to conduct analyses. This brought Ferri back into the world of scramjets. The design called for struts as fuel injectors, swept at 48 degrees to parallel the inlet and set within the combustor flow path. They promised more effective fuel injection than the wall-mounted injectors of earlier designs.

The basic elements of the Langley concept thus included fixed geometry, airframe integration, a swept inlet, thermal compression, and use of struts for fuel injection. These elements showed strong synergism, for in addition to the aircraft undersurface contributing to the work of the inlet and nozzle, the struts also served as part of the inlet and thereby made it shorter. This happened because the flow from the inlet underwent further compression as it passed between the struts.[23]

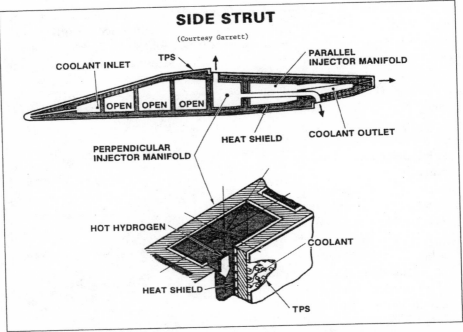

Fuel-injecting strut. Arrows show how hydrogen is injected either parallel or perpendicular to the flow. (Garrett Corporation)

Experimental work paced the Langley effort as it went forward during the 1970s and much of the 1980s. Early observations, published in 1970, showed that struts were practical for a large supersonic combustor in flight at Mach 8. This work supported the selection of strut injection as the preferred mode.[24]

Initial investigations involved inlets and combustors that were treated as separate components. These represented preludes to studies made with complete engine modules at two critical simulated flight speeds: Mach 4 and Mach 7. At Mach 4 the inlet was particularly sensitive to unstarts. The inlet alone had worked well, as had the strut, but now it was necessary to test them together and to look for unpleasant surprises. The Langley researchers therefore built a heavily instrumented engine of nickel and tested it at GASL, thereby bringing new work in hypersonics to that center as well.

Mach 7 brought a different set of problems. Unstarts now were expected to be less of a difficulty, but it was necessary to show that the fuel indeed could mix and burn within the limited length of the combustor. Mach 7 also approached the limitations of available wind tunnels. A new Langley installation, the Arc-Heated Scramjet Test Facility, reached temperatures of 3,500°F and provided the appropriate flows.

The Fading, the Comeback

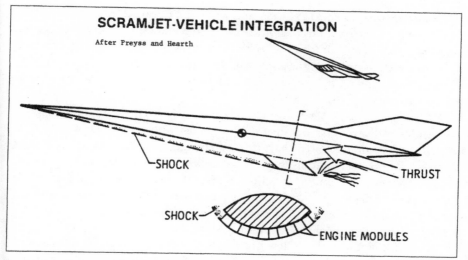

Integration of scramjets with an aircraft. (NASA)

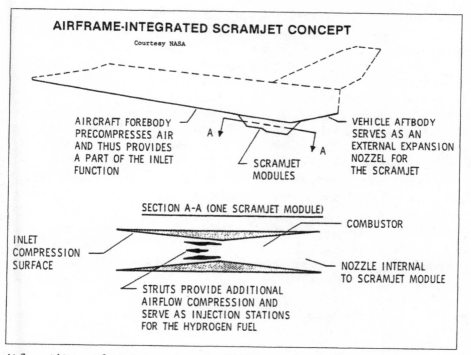

Airflow within an airframe-integrated scramjet. (NASA)

Separate engines operated at GASL and Langley. Both used heat sink, with the run times being correspondingly short. Because both engines were designed for use in research, they were built for easy substitution of components. An inlet, combustor, nozzle, or set of fuel-injecting struts could be installed without having to modify the rest of the engine. This encouraged rapid prototyping without having to construct entirely new scramjets.

More than 70 runs at Mach 4 were made during 1978, first with no fuel injection to verify earlier results from inlet tests, and then with use of hydrogen. Simple theoretical calculations showed that "thermal choking" was likely, with heat addition in the combustor limiting the achievable flow rate, and indeed it appeared. Other problems arose from fuel injection. The engine used three struts, a main one on the centerline flanked by two longer ones, and fuel from these side struts showed poor combustion when injected parallel to the flow. Some unwanted inlet-combustor interactions sharply reduced the measured thrust. These occurred because the engine ingested boundary-layer flow from the top inner surface of the wind-tunnel duct. This simulated the ingestion of an aircraft boundary layer by a flight engine.

The thermal choking and the other interactions were absent when the engine ran very fuel-lean, and the goal of the researchers was to eliminate them while burning as much fuel as possible. They eased the problem of thermal choking by returning to a fuel-injection method that had been used on the HRE, with some fuel being injected downstream as the wall. However, the combustor-inlet interactions proved to be more recalcitrant. They showed up when the struts were injecting only about half as much fuel as could burn in the available airflow, which was not the formula for a high-thrust engine.[25]

Mach 7 brought its own difficulties, as the Langley group ran off 90 tests between April 1977 and February 1979. Here too there were inlet-combustor interactions, ranging from increased inlet spillage that added drag and reduced the thrust, to complete engine unstarts. When the latter occurred, the engine would put out good thrust when running lean; when the fuel flow increased, so did the measured force. In less than a second, though, the inlet would unstart and the measured thrust would fall to zero.[26]

No simple solution appeared capable of addressing these issues. This meant that in the wake of those tests, as had been true for more than a decade, the Langley group did not have a working scramjet. Rather, they had a research problem. They addressed it after 1980 with two new engines, the Parametric Scramjet and the Step-Strut Engine. The Parametric engine lacked a strut but was built for ease of modification. In 1986 the analysts Burton Northam and Griffin Anderson wrote:

> This engine allows for easy variation of inlet contraction ratio, internal area ratio and axial fuel injection location. Sweep may be incorporated in the

inlet portion of the engine, but the remainder of the engine is unswept. In fact, the hardware is designed in sections so that inlet sweep can be changed (by substituting new inlet sidewalls) without removing the engine from the wind tunnel.

The Parametric Scramjet explored techniques for alleviating combustor-inlet interactions at Mach 4. The Step-Strut design also addressed this issue, mounting a single long internal strut fitted with fuel injectors, with a swept leading edge that resembled a staircase. Northam and Anderson wrote that it "was also tested at Mach 4 and demonstrated good performance without combustor-inlet interaction."[27]

How, specifically, did Langley develop a workable scramjet? Answers remain classified, with Northam and Anderson noting that "several of the figures have no dimension on the axes and a discussion of the figures omits much of the detail." A 1998 review was no more helpful. However, as early as 1986 the Langley researchers openly published a plot showing data taken at Mach 4 and at Mach 7. Curves showed values of thrust and showed that the scramjets of the mid-1980s

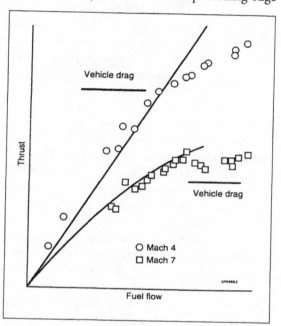

Performance of scramjets. Note that figures are missing from the axes. (NASA)

indeed could produce net thrust. Even at Mach 7, at which the thrust was less, these engines could overcome the drag of a complete vehicle and produce acceleration. In the words of Northam and Anderson, "at both Mach 4 and Mach 7 flight conditions, there is ample thrust for acceleration and cruise."[28]

THE ADVENT OF NASP

With test engines well on their way in development, there was the prospect of experimental aircraft that might exercise them in flight test. Such a vehicle might come forth as a successor to Number 66671, the X-15 that had been slated to fly the

HRE. An aircraft of this type indeed took shape before long, with the designation X-30. However, it did not originate purely as a technical exercise. Its background lay in presidential politics.

The 1980 election took place less than a year after the Soviets invaded Afghanistan. President Jimmy Carter had placed strong hope in arms control and had negotiated a major treaty with his Soviet counterpart, Leonid Brezhnev. But the incursion into Afghanistan took Carter by surprise and destroyed the climate of international trust that was essential for Senate ratification of this treaty. Reagan thus came to the White House with arms-control prospects on hold and with the Cold War once more in a deep freeze. He responded by launching an arms buildup that particularly included new missiles for Europe.[29]

Peace activist Randall Forsberg replied by taking the lead in calling for a nuclear freeze, urging the superpowers to halt the "testing, production and deployment of nuclear weapons" as an important step toward "lessening the risk of nuclear war." Her arguments touched a nerve within the general public, for within two years, support for a freeze topped 70 percent. Congressman Edward Markey introduced a nuclear-freeze resolution in the House of Representatives. It failed by a margin of only one vote, with Democratic gains in the 1982 mid-term elections making passage a near certainty. By the end of that year half the states in the Union adopted their own freeze resolutions, as did more than 800 cities, counties, and towns.[30]

To Reagan, a freeze was anathema. He declared that it "would be largely unverifiable.... It would reward the Soviets for their massive military buildup while preventing us from modernizing our aging and increasingly vulnerable forces." He asserted that Moscow held a "present margin of superiority" and that a freeze would leave America "prohibited from catching up."[31]

With the freeze ascendant, Admiral James Watkins, the Chief of Naval Operations, took a central role in seeking an approach that might counter its political appeal. Exchanges with Robert McFarlane and John Poindexter, deputies within the National Security Council, drew his thoughts toward missile defense. Then in January 1983 he learned that the Joint Chiefs were to meet with Reagan on 11 February. As preparation, he met with a group of advisors that included the physicist Edward Teller.

Trembling with passion, Teller declared that there was enormous promise in a new concept: the x-ray laser. This was a nuclear bomb that was to produce intense beams of x-rays that might be aimed to destroy enemy missiles. Watkins agreed that the broad concept of missile defense indeed was attractive. It could introduce a new prospect: that America might counter the Soviet buildup, not with a buildup of its own but by turning to its strength in advanced technology.

Watkins succeeded in winning support from his fellow Joint Chiefs, including the chairman, General John Vessey. Vessey then gave Reagan a half-hour briefing at the 11 February meeting, as he drew extensively on the views of Watkins. Reagan showed strong interest and told the Chiefs that he wanted a written proposal. Robert McFarlane, Deputy to the National Security Advisor, already had begun to explore concepts for missile defense. During the next several weeks his associates took the lead in developing plans for a program and budget.[32]

On 23 March 1983 Reagan spoke to the nation in a televised address. He dealt broadly with issues of nuclear weaponry. Toward the end of the speech, he offered new thoughts:

> "Let me share with you a vision of the future which offers hope. It is that we embark on a program to counter the awesome Soviet missile threat with measures that are defensive. Let us turn to the very strengths in technology that spawned our great industrial base and that have given us the quality of life we enjoy today.
>
> What if free people could live secure in the knowledge that their security did not rest upon the threat of instant U.S. retaliation to deter a Soviet attack, that we could intercept and destroy strategic ballistic missiles before they reached our own soil or that of our allies?...
>
> I call upon the scientific community in our country, those who gave us nuclear weapons, to turn their great talents now to the cause of mankind and world peace, to give us the means of rendering these nuclear weapons impotent and obsolete."[33]

The ensuing Strategic Defense Initiative never deployed weapons that could shoot down a missile. Yet from the outset it proved highly effective in shooting down the nuclear freeze. That movement reached its high-water mark in May 1983, as a strengthened Democratic majority in the House indeed passed Markey's resolution. But the Senate was still held by Republicans, and the freeze went no further. The SDI gave everyone something new to talk about. Reagan's speech helped him to regain the initiative, and in 1984 he swept to re-election with an overwhelming majority.[34]

The SDI brought the prospect of a major upsurge in traffic to orbit, raising the prospect of a flood of new military payloads. SDI supporters asserted that some one hundred orbiting satellites could provide an effective strategic defense, although the Union of Concerned Scientists, a center of criticism, declared that the number would be as large as 2,400. Certainly, though, an operational missile defense was likely to place new and extensive demands on means for access to space.

Within the Air Force Systems Command, there already was interest in a next-generation single-stage-to-orbit launch vehicle that was to use the existing Space Shuttle Main Engine. Lieutenant General Lawrence Skantze, Commander of the

Facing the Heat Barrier: A History of Hypersonics

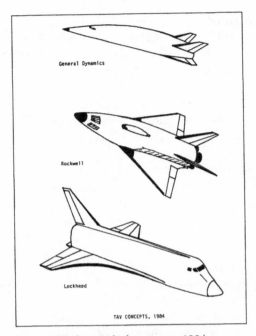

Transatmospheric Vehicle concepts, 1984. (U.S. Air Force)

Air Force Systems Command's Aeronautical Systems Division (ASD), launched work in this area early in 1982 by directing the ASD planning staff to conduct an in-house study of post-shuttle launch vehicles. It then went forward under the leadership of Stanley Tremaine, the ASD's Deputy for Development Planning, who christened these craft as Transatmospheric Vehicles. In December 1984 Tremaine set up a TAV Program Office, directed by Lieutenant Colonel Vince Rausch.[35]

Moreover, General Skantze was advancing into high-level realms of command, where he could make his voice heard. In August 1982 he went to Air Force Headquarters, where he took the post of Deputy Chief of Staff for Research, Development, and Acquisition. This gave him responsibility for all Air Force programs in these areas. In October 1983 he pinned on his fourth star as he took an appointment as Air Force Vice Chief of Staff. In August 1984 he became Commander of the Air Force Systems Command.[36]

He accepted these Washington positions amid growing military disenchantment with the space shuttle. Experience was showing that it was costly and required a long time to prepare for launch. There also was increasing concern for its safety, with a 1982 Rand Corporation study flatly predicting that as many as three shuttle orbiters would be lost to accidents during the life of the program. The Air Force was unwilling to place all its eggs in such a basket. In February 1984 Defense Secretary Caspar Weinberger approved a document stating that total reliance on the shuttle "represents an unacceptable national security risk." Air Force Secretary Edward Aldridge responded by announcing that he would remove 10 payloads from the shuttle beginning in 1988 and would fly them on expendables.[37]

Just then the Defense Advanced Research Projects Agency was coming to the forefront as an important new center for studies of TAV-like vehicles. DARPA was already reviving the field of flight research with its X-29, which featured a forward-swept wing along with an innovative array of control systems and advanced materials. Robert Cooper, DARPA's director, held a strong interest in such projects and saw them as a way to widen his agency's portfolio. He found encouragement during

214

The Fading, The Comeback

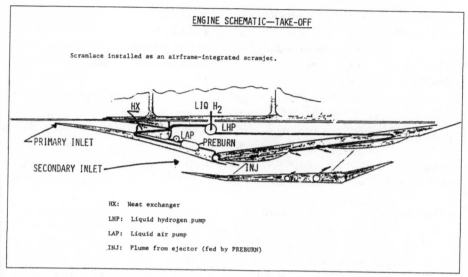

Anthony duPont's engine. (GASL)

1982 as a group of ramjet specialists met with Richard De Lauer, the Undersecretary of Defense Research and Engineering. They urged him to keep the field alive with enough new funds to prevent them from having to break up their groups. De Lauer responded with letters that he sent to the Navy, Air Force, and DARPA, asking them to help.[38]

This provided an opening for Tony duPont, who had designed the HRE. He had taken a strong interest in combined-cycle concepts and decided that the scramlace was the one he preferred. It was to eliminate the big booster that every ramjet needed, by using an ejector, but experimental versions weren't very powerful. DuPont thought he could do better by using the HRE as a point of departure, as he added an auxiliary inlet for LACE and a set of ejector nozzles upstream of the combustor. He filed for a patent on his engine in 1970 and won it two years later.[39]

In 1982 he still believed in it, and he learned that Anthony Tether was the DARPA man who had been attending TAV meetings. The two men met several times, with Tether finally sending him up to talk with Cooper. Cooper listened to duPont and sent him over to Robert Williams, one of DARPA's best aerodynamicists. Cooper declares that Williams "was the right guy; he knew the most in this area. This wasn't his specialty, but he was an imaginative fellow."[40]

Williams had come up within the Navy, working at its David Taylor research center. His specialty was helicopters; he had initiated studies of the X-wing, which was to stop its rotor in midair and fly as a fixed-wing aircraft. He also was interested in high-speed flight. He had studied a missile that was to fight what the Navy called the "outer air battle," which might use a scramjet. This had brought him into discussions with Fred Billig, who also worked for the Navy and helped him to learn his hypersonic propulsion. He came to DARPA in 1981 and joined its Tactical Technologies Office, where he became known as the man to see if anyone was interested in scramjets.[41]

Williams now phoned duPont and gave him a test: "I've got a very ambitious problem for you. If you think the airplane can do this, perhaps we can promote a program. Cooper has asked me to check you out." The problem was to achieve single-stage-to-orbit flight with a scramjet and a suite of heat-resistant materials, and duPont recalls his response: "I stayed up all night; I was more and more intrigued with this. Finally I called him back: 'Okay, Bob, it's not impossible. Now what?'"[42]

DuPont had been using a desktop computer, and Williams and Tether responded to his impromptu calculations by giving him $30,000 to prepare a report. Soon Williams was broadening his circle of scramjet specialists by talking with old-timers such as Arthur Thomas, who had been conducting similar studies a quarter-century earlier, and who quickly became skeptical. DuPont had patented his propulsion concept, but Thomas saw it differently: "I recognized it as a Marquardt engine. Tony called it the duPont cycle, which threw me off, but I recognized it as our engine. He claimed he'd improved it." In fact, "he'd made a mistake in calculating the heat capacity of air. So his engine looked so much better than ours."

Thomas nevertheless signed on to contribute to the missionary work, joining Williams and duPont in giving presentations to other conceptual-design groups. At Lockheed and Boeing, they found themselves talking to other people who knew scramjets. As Thomas recalls, "The people were amazed at the component efficiencies that had been assumed in the study. They got me aside and asked if I really believed it. Were these things achievable? Tony was optimistic everywhere: on mass fraction, on air drag of the vehicle, on inlet performance, on nozzle performance, on combustor performance. The whole thing, across the board. But what salved our conscience was that even if these weren't all achieved, we still could have something worth while. Whatever we got would still be exciting."[43]

Williams recalls that in April 1984, "I put together a presentation for Cooper called 'Resurrection of the Aerospaceplane.' He had one hour; I had 150 slides. He came in, sat down, and said Go. We blasted through those slides. Then there was silence. Cooper said, 'I want to spend a day on this.'" After hearing additional briefings, he approved a $5.5-million effort known as Copper Canyon, which brought an expanded program of studies and analyses.[44]

Copper Canyon represented an attempt to show how the SDI could achieve its access to space, and a number of high-level people responded favorably when Cooper asked to give a briefing. He and Williams made a presentation to George Keyworth, Reagan's science advisor. They then briefed the White House Science

THE FADING, THE COMEBACK

Council. Keyworth recalls that "here were people who normally would ask questions for hours. But after only about a half-hour, David Packard said, 'What's keeping us? Let's do it!'" Packard was Deputy Secretary of Defense.[45]

During 1985, as Copper Canyon neared conclusion, the question arose of expanding the effort with support from NASA and the Air Force. Cooper attended a classified review and as he recalls, "I went into that meeting with a high degree of skepticism." But technical presentations brought him around: "For each major problem, there were three or four plausible ways to deal with it. That's extraordinary. Usually it's—'Well, we don't know exactly how we'll do it, but we'll do it.' Or, 'We have *a* way to do it, which may work.' It was really a surprise to me; I couldn't pick any obvious holes in what they had done. I could find no reason why they couldn't go forward."[46]

Further briefings followed. Williams gave one to Admiral Watkins, whom Cooper describes as "very supportive, said he would commit the Navy to support of the program." Then in July, Cooper accompanied Williams as they gave a presentation to General Skantze.

They displayed their viewgraphs and in Cooper's words, "He took one look at our concept and said, 'Yeah, that's what I meant. I invented that idea.'" Not even the stars on his shoulders could give him that achievement, but his endorsement reflected the fact that he was dissatisfied with the TAV studies. He had come away appreciating that he needed something better than rocket engines—and here it was. "His enthusiasm came from the fact that this was all he had anticipated," Cooper continues. "He felt as if he owned it."

Skantze wanted more than viewgraphs. He wanted to see duPont's engine in operation. A small version was under test at GASL, without LACE but definitely with its ejector, and one technician had said, "This engine really does put out static thrust, which isn't obvious for a ramjet." Skantze saw the demonstration and came away impressed. Then, Williams adds, "the Air Force system began to move with the speed of a spaceplane. In literally a week and a half, the entire Air Force senior command was briefed."

Initial version of the duPont engine under test at GASL. (GASL)

Later that year the Secretary of Defense, Caspar Weinberger, granted a briefing. With him were members of his staff, along with senior people from NASA and the military service. After giving the presentation, Williams recalls that "there was silence in the room. The Sec-

217

retary said, 'Interesting,' and turned to his staff. Of course, all the groundwork had been laid. All of the people there had been briefed, and we could go for a yes-or-no decision. We had essentially total unanimity around the table, and he decided that the program would proceed as a major Defense Department initiative. With this, we moved immediately to issue requests for proposal to industry."[47]

In January 1986 the TAV effort was formally terminated. At Wright-Patterson AFB, the staff of its program office went over to a new Joint Program Office that now supported what was called the National Aerospace Plane. It brought together representatives from the Air Force, Navy, and NASA. Program management remained at DARPA, where Williams retained his post as the overall manager.[48]

In this fashion, NASP became a significant federal initiative. It benefited from a rare alignment of the political stars, for Reagan's SDI cried out for better launch vehicles and Skantze was ready to offer them. Nor did funding appear to be a problem, at least initially. Reagan had shown favor to aerospace through such acts as approving NASA's space station in 1984. Pentagon spending had surged, and DARPA's Cooper was asserting that an X-30 might be built for an affordable cost.

Yet NASP was a leap into the unknown. Its scramjets now were in the forefront but not because the Langley research had shown that they were ready. Instead they were a focus of hope because Reagan wanted SDI, SDI needed better access to space, and Skantze wanted something better than rockets.

The people who were making Air Force decisions, such as Skantze, did not know much about these engines. The people who did know them, such as Thomas, were well aware of duPont's optimism. There thus was abundant opportunity for high hope to give way to hard experience.

The Decline of NASP

NASP was one of Reagan's programs, and for a time it seemed likely that it would not long survive the change in administrations after he left office in 1989. That fiscal year brought a high-water mark for the program, as its budget peaked at $320 million. During the spring of that year officials prepared budgets for FY 1991, which President George H. W. Bush would send to Congress early in 1990. Military spending was already trending downward, and within the Pentagon, analyst David Chu recommended canceling all Defense Department spending for NASP. The new Secretary of Defense, Richard Cheney, accepted this proposal. With this, NASP appeared dead.

NASP had a new program manager, Robert Barthelemy, who had replaced Williams. Working through channels, he found support in the White House from Vice President Dan Quayle. Quayle chaired the National Space Council, which had been created by law in 1958 and that just then was active for the first time in a decade. He

The Fading, the Comeback

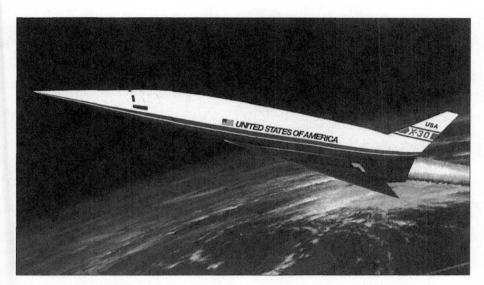

X-30 concept of 1985. (NASA)

used it to rescue NASP. He led the Space Council to recommend proceeding with the program under a reduced but stable budget, and with a schedule slip. This plan won acceptance, giving the program leeway to face a new issue: excessive technical optimism.[49]

During 1984, amid the Copper Canyon activities, Tony duPont devised a conceptual configuration that evolved into the program's baseline. It had a gross weight of 52,650 pounds, which included a 2,500-pound payload that it was to carry to polar orbit. Its weight of fuel was 28,450 pounds. The propellant mass fraction, the ratio of these quantities, then was 0.54.[50]

The fuel had low density and was bulky, demanding high weight for the tankage and airframe. To save weight, duPont's concept had no landing gear. It lacked reserves of fuel; it was to reach orbit by burning its last drops. Once there it could not execute a controlled deorbit, for it lacked maneuvering rockets as well as fuel and oxidizer for them. DuPont also made no provision for a reserve of weight to accommodate normal increases during development.[51]

Williams's colleagues addressed these deficiencies, although they continued to accept duPont's optimism in the areas of vehicle drag and engine performance. The new concept had a gross weight of 80,000 pounds. Its engines gave a specific impulse of 1,400 seconds, averaged over the trajectory, which corresponded to a mean exhaust velocity of 45,000 feet per second. (That of the SSME was 453.5 seconds in vacuum, or 14,590 feet per second.) The effective velocity increase for the X-30 was calculated at 47,000 feet per second, with orbital velocity being 25,000 feet per second; the difference represented loss due to drag. This version of the X-30 was designated the "government baseline" and went to the contractors for further study.[52]

The initial round of contract awards was announced in April 1986. Five airframe firms developed new conceptual designs, introducing their own estimates of drag and engine performance along with their own choices of materials. They gave the following weight estimates for the X-30:

Rockwell International	175,000 pounds
McDonnell Douglas	245,000
General Dynamics	280,000
Boeing	340,000
Lockheed	375,000

A subsequent downselection, in October 1987, eliminated the two heaviest concepts while retaining Rockwell, McDonnell Douglas, and General Dynamics for further work.[53]

What brought these weight increases? Much of the reason lay in a falloff in estimated engine performance, which fell as low as 1,070 seconds of averaged specific impulse. New estimates of drag pushed the required effective velocity increase during ascent to as much as 52,000 feet per second.

A 1989 technical review, sponsored by the National Research Council, showed what this meant. The chairman, Jack Kerrebrock, was an experienced propulsion specialist from MIT. His panel included other men of similar background: Seymour Bogdonoff of Princeton, Artur Mager of Marquardt, Frank Marble from Caltech. Their report stated that for the X-30 to reach orbit as a single stage, "a fuel fraction of approximately 0.75 is required."[54]

One gains insight by considering three hydrogen-fueled

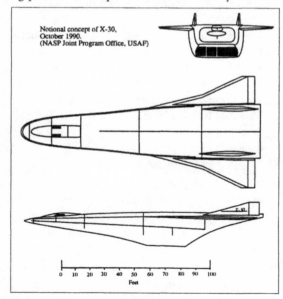

X-30 concept of 1990, which had grown considerably. (U.S. Air Force)

The Fading, the Comeback

rocket stages of NASA and calculating their values of propellant mass fraction if both their hydrogen and oxygen tanks were filled with NASP fuel. This was slush hydrogen, a slurry of the solid and liquid. The stages are the S-II and S-IVB of Apollo and the space shuttle's external tank. Liquid hydrogen has 1/16 the density of liquid oxygen. With NASP slush having 1.16 times the density of liquid hydrogen,[55] the propellant mass fractions are as follows:[56]

S-IVB, third stage of the Saturn V	0.722
S-II, second stage of the Saturn V	0.753
External Tank	0.868

The S-II, which comes close to Kerrebrock's value of 0.75, was an insulated shell that mounted five rocket engines. It withstood compressive loads along its length that resulted from the weight of the S-IVB and the Apollo moonship but did not require reinforcement to cope with major bending loads. It was constructed of aluminum alloy and lacked landing gear, thermal protection, wings, and a flight deck.

How then did NASP offer an X-30 concept that constituted a true hypersonic airplane rather than a mere rocket stage? The answer lay in adding weight to the fuel, which boosted the propellant mass fraction. The vehicle was not to reach orbit entirely on slush-fueled scramjets but was to use a rocket for final ascent. It used tanked oxygen—with nearly 14 times the density of slush hydrogen. In addition, design requirements specified a tripropellant system that was to burn liquid methane during the early part of the flight. This fuel had less energy than hydrogen, but it too added weight because it was relatively dense. The recommended mix called for 69 percent hydrogen, 20 percent oxygen, and 11 percent methane.[57]

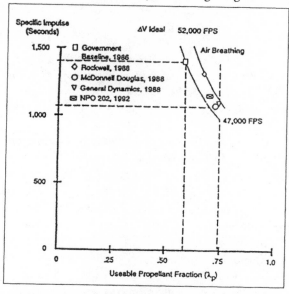

Evolution of the X-30. The government baseline of 1986 had Isp of 1,400 seconds, delta-V to reach orbit of 47,000 feet per second, and propellant mass fraction of 0.54. Its 1992 counterpart had less Isp, more drag, propellant mass fraction of 0.75, and could not reach orbit. (NASP National Program Office)

221

In 1984, with optimism at its height, Cooper had asserted that the X-30 would be the size of an SR-71 and could be ready in three years. DuPont argued that his concept could lead to a "5-5-50" program by building a 50,000-pound vehicle in five years for $5 billion.[58] Eight years later, in October 1990, the program had a new chosen configuration. It was rectangular in cross section, with flat sides. Three scramjet engines were to provide propulsion. Two small vertical stabilizers were at the rear, giving better stability than a single large one. A single rocket engine of approximately 60,000 pounds of thrust, integrated into the airframe, completed the layout. Other decisions selected the hot structure as the basic approach to thermal protection. The primary structure was to be of titanium-matrix composite, with insulated panels of carbon to radiate away the heat.[59]

This 1990 baseline design showed little resemblance to its 1984 ancestor. As revised in 1992, it no longer was to fly to a polar orbit but would take off on a due-east launch from Kennedy Space Center, thereby gaining some 1,340 feet per second of launch velocity. Its gross weight was quoted at 400,000 pounds, some 40 percent heavier than the General Dynamics weight that had been the heaviest acceptable in the 1987 downselect. Yet even then the 1992 concept was expected to fall short of orbit by some 3,000 feet per second. An uprated version, with a gross weight of at least 450,000 pounds, appeared necessary to reach orbital velocity. The prospective program budget came to $15 billion or more, with the time to first flight being eight to ten years.[60]

During 1992 both the Defense Science Board (DSB) and Congress's General Accounting Office (GAO) conducted major program reviews. The immediate issue was whether to proceed as planned by making a commitment that would actually build and fly the X-30. Such a decision would take the program from its ongoing phase of research and study into a new phase of mainstream engineering development.

Both reviews focused on technology, but international issues were in the background, for the Cold War had just ended. The Soviet Union had collapsed in 1991, with communists falling from power while that nation dissolved into 15 constituent states. Germany had already reunified; the Berlin Wall had fallen, and the whole of Eastern Europe had won independence from Moscow. The western border of Russia now approximated that of 1648, at the end of the Thirty Years' War. Two complete tiers of nominally independent nations now stood between Russia and the West.

These developments greatly diminished the military urgency of NASP, while the reviews' conclusions gave further reason to reduce its priority. The GAO noted that program managers had established 38 technical milestones that were to be satisfied before proceeding to mainstream development. These covered the specific topics of X-30 design, propulsion, structures and materials, and use of slush hydrogen as a fuel. According to the contractors themselves, only 17 of those milestones—fewer

The Fading, the Comeback

than half—were to be achieved by September 1993. The situation was particularly worrisome in the critical area of structures and materials, for which only six of 19 milestones were slated for completion. The GAO therefore recommended delaying a commitment to mainstream development "until critical technologies are developed and demonstrated."[61]

The DSB concurred, highlighting specific technical deficiencies. The most important involved the prediction of scramjet performance and of boundary-layer transition. In the latter, an initially laminar or smoothly flowing boundary layer becomes turbulent. This brings large increases in heat transfer and skin friction, a major source of drag. The locations of transition thus had to be known.

The scramjet-performance problem arose because of basic limitations in the capabilities of ground-test facilities. The best of them could accommodate a complete engine, with inlet, combustor, and nozzle, but could conduct tests only below Mach 8. "Even at Mach 8," the DSB declared, "the scramjet cycle is just beginning to be established and consequently, there is uncertainty associated with extrapolating the results into the higher Mach regime. At speeds above Mach 8, only small components of the scramjet can be tested." This brought further uncertainty when predicting the performance of complete engines.

Boundary-layer transition to turbulence also demanded attention: "It is essential to understand the boundary-layer behavior at hypersonic speeds in order to ensure thermal survival of the airplane structure as designed, as well as to accurately predict the propulsion system performance and airplane drag. Excessive conservatism in boundary-layer predictions will lead to an overweight design incapable of achieving [single stage to orbit], while excessive optimism will lead to an airplane unable to survive in the hypersonic flight environment."

The DSB also showed strong concern over issues of control in flight of the X-30 and its engines. These were not simple matters of using ailerons or pushing throttles. The report stated that "controllability issues for NASP are so complex, so widely ranging in dynamics and frequency, and so interactive between technical disciplines as to have no parallels in aeronautical history...the most fundamental initial requirements for elementary aircraft control are not yet fully comprehended." An onboard computer was to manage the vehicle and its engines in flight, but an understanding of the pertinent forces and moments "is still in an embryonic state." Active cooling of the vehicle demanded a close understanding of boundary-layer transition. Active cooling of the engine called for resolution of "major uncertainties...connected with supersonic burning." In approaching these issues, "very great uncertainties exist at a fundamental level."

The DSB echoed the GAO in calling for extensive additional research before proceeding into mainstream development of the X-30:

> We have concluded [that] fundamental uncertainties will continue to exist in at least four critical areas: boundary-layer transition; stability and controllability; propulsion performance; and structural and subsystem weight. Boundary-layer transition and scramjet performance cannot be validated in existing ground-test facilities, and the weight estimates have insufficient reserves for the inevitable growth attendant to material allowables, fastening and joining, and detailed configuration issues.... Using optimistic assumptions on transition and scramjet performance, and the present weight estimates on material performance and active cooling, the vehicle design does not yet close; the velocity achieved is short of orbital requirements.[62]

Faced with the prospect that the flight trajectory of the X-30 would merely amount to a parabola, budget makers turned the curve of program funding into a parabola as well. The total budget had held at close to $250 million during FY 1990 and 1991, falling to $205 million in 1992. But in 1993 it took a sharp dip to $140 million. The NASP National Program Office tried to rescue the situation by proposing a six-year program with a budget of $2 billion, called Hyflite, that was to conduct a series of unmanned flight tests. The Air Force responded with a new technical group, the Independent Review Team, that turned thumbs down on Hyflite and called instead for a "minimum" flight test program. Such an effort was to address the key problem of reducing uncertainties in scramjet performance at high Mach.

The National Program Office came back with a proposal for a new program called HySTP. Its budget request came to $400 million over five years, which would have continued the NASP effort at a level only slightly higher than its allocation of $60 million for FY 1994. Yet even this minimal program budget proved to be unavailable. In January 1995 the Air Force declined to approve the HySTP budget and initiated the formal termination of the NASP program.[63]

In this fashion, NASP lived and died. Like SDI and the space station, one could view it as another in a series of exercises in Reaganesque optimism that fell short. Yet from the outset, supporters of NASP had emphasized that it was to make important contributions in such areas as propulsion, hypersonic aerodynamics, computational fluid dynamics, and materials. The program indeed did these things and thereby laid groundwork for further developments.

1. AIAA Paper 93-2329.
2. *Johns Hopkins APL Technical Digest*, Vol. 13, No. 1 (1992), pp. 63-65.
3. Hallion, *Hypersonic*, pp. 754-55; Harshman, "Design and Test."
4. Waltrup et al., "Supersonic," pp. 42-18 to 42-19; DTIC AD-386653.
5. Hallion, *Hypersonic*, pp. VI-xvii to VI-xx; "Scramjet Flight Test Program" (Marquardt brochure, September 1965); DTIC AD-388239.
6. Hallion, *Hypersonic*, pp. VI-xiv, 780; "Report of the USAF Scientific Advisory Board Aerospace Vehicles Panel," February 1966.
7. Hallion, *Hypersonic*, p. 780.
8. NASA SP-2000-4518, p. 63; NASA SP-4303, pp. 125-26.
9. AIAA Paper 93-2328; interoffice memo (GASL), E. Sanlorenzo to L. Nucci, 24 October 1967.
10. Mackley, "Historical"; DTIC AD-393374.
11. *Journal of Aircraft*, January-February 1968, p. 3.
12. Author interviews, Louis Nucci, 13 November 1987 and 24 June 1988 (includes quotes). Folder 18649, NASA Historical Reference Collection, NASA History Division, Washington, D.C. 20546.
13. Author interviews, Arthur Thomas, 24 September 1987 and 24 June 1988 (includes quotes). Folder 18649, NASA Historical Reference Collection, NASA History Division, Washington, D.C. 20546.
14. Author interviews: Fred Billig, 27 June 1987 and Louis Nucci, 24 June 1988. Folder 18649, NASA Historical Reference Collection, NASA History Division, Washington, D.C. 20546.
15. Hallion, *Hypersonic*, pp. 756-78; Miller, *X-Planes*, pp. 189-90.
16. Author interview, Anthony duPont, 23 November 1987. Folder 18649, NASA Historical Reference Collection, NASA History Division, Washington, D.C. 20546.
17. Hallion, *Hypersonic*, pp. 774-78; "Handbook of Texas Online: Ordnance Aerophysics Laboratory," internet website.
18. Hallion, *Hypersonic*, pp. 142-45; Miller, *X-Planes*, pp. 190-91, 194-95; NASA SP-4303, pp. 121-22; author interview, William "Pete" Knight, 24 September 1987. Folder 18649, NASA Historical Reference Collection, NASA History Division, Washington, D.C. 20546.
19. NASA SP-2000-4518, pp. 59-60.
20. Hallion, *Hypersonic*, pp. 792-96; AIAA Paper 93-2323; NASA TM X-2572.
21. Hallion, *Hypersonic*, pp. 798-802; AIAA Paper 93-2323; NASA TM X-2572.
22. Hallion, *Hypersonic*, pp. 802-22 (quote, p. 818); Mackley, "Historical."
23. Waltrup et al., "Supersonic," pp. 42-13 to 42-14; NASA SP-292, pp. 157-77; NASA TM X-2895; *Astronautics & Aeronautics*, February 1978, pp. 38-48; AIAA Paper 86-0159.
24. AIAA Papers 70-715, 75-1212.
25. AIAA Papers 79-7045, 98-2506.
26. AIAA Paper 79-7045.
27. AIAA Paper 86-0159 (quotes, p. 7).

28 Ibid. (quotes, pp. 1, 7); AIAA Paper 98-2506.
29 Malia, *Soviet*, pp. 378-80; Walker, *Cold War*, pp. 250-57.
30 Fitzgerald, *Way Out*, pp. 179-82, 190-91 (quotes, p. 180).
31 "Address by the President to the Nation." Washington, DC: The White House, 23 March 1983.
32 Baucom, *Origins*, pp. 184-92; Broad, *Teller's War*, pp. 121-24; Fitzgerald, *Way Out*, pp. 195-98, 200-02.
33 "Address by the President to the Nation." Washington, DC: The White House, 23 March 1983.
34 Fitzgerald, *Way Out*, p. 225.
35 Schweikart, *Quest*, pp. 20-22; Hallion, *Hypersonic*, pp. 1336-46; AIAA Paper 84-2414; *Air Power History*, Spring 1994, 39.
36 "Biography: General Lawrence A. Skantze." Washington, DC: Office of Public Affairs, Secretary of the Air Force, August 1984.
37 *Science*, 29 June 1984, pp. 1407-09 (includes quote).
38 Miller, *X-Planes*, ch. 33; author interview, Robert Cooper, 12 May 1986. Folder 18649, NASA Historical Reference Collection, NASA History Division, Washington, D.C. 20546.
39 U.S. Patent 3,690,102 (duPont). Also author interview, Anthony duPont, 23 November 1987. Folder 18649, NASA Historical Collection, NASA History Division, Washington, D.C. 20546.
40 Author interviews, Robert Cooper, 12 May 1986 (includes quote); Anthony duPont, 23 November 1987. Folder 18649, NASA Historical Reference Collection, NASA History Division, Washington, D.C. 20546.
41 Schweikart, *Quest*, p. 20; author interview, Robert Williams, 1 May 1986 and 23 November 1987. Folder 18649, NASA Historical Reference Collection, NASA History Division, Washington, D.C. 20546.
42 Author interviews, Robert Williams, 1 May 1986 and 23 November 1987. Folder 18649, NASA Historical Reference Collection, NASA History Division, Washington, D.C. 20546.
43 Author interviews, Arthur Thomas, 24 September 1987 and 22 July 1988. Folder 18649, NASA Historical Reference Collection, NASA History Divison, Washington, D.C. 20546.
44 Author interview, Robert Williams, 1 May 1986. Folder 18649, NASA Historical Reference Collection, NASA History Division, Washington, D.C. 20546.
45 Author interview, George Keyworth, 23 May 1986. Folder 18469, NASA Historical Reference Collection, NASA History Division, Washington, D.C. 20546.
46 Author interview, Robert Cooper, 12 May 1986. Folder 18469, NASA Historical Reference Collection, NASA History Division, Washington, D.C. 20546.
47 Ibid.; author interview, Robert Williams, 1 May 1986. Folder 18469, NASA Historical Reference Collection, NASA History Division, Washington, D.C. 20546.
48 Schweikart, *Quest*, pp. 47-51; Hallion, *Hypersonic*, pp. 1345-46, 1351, 1364-65.
49 Schweikart, *Quest*, pp. 161-62.
50 AIAA Paper 95-6031.
51 Schweikart, *Quest*, pp. 29, 199.
52 AIAA Paper 95-6031. SSME: Rocketdyne Report RI/RD87-142.

THE FADING, THE COMEBACK

53 Schweikart, *Quest*, pp. 51, 199-200; Miller, *X-Planes*, p. 310.
54 Air Force Studies Board, *Hypersonic* (quote, p. 5).
55 Sutton, *Rocket*, pp. 170-71; AIAA Paper 89-5014.
56 NASA SP-2000-4029, p. 284; Jenkins, *Space Shuttle*, pp. 421, 424.
57 DTIC ADB-197189, Item 063.
58 *Aerospace America*, June 1984, p. 1.
59 *Aviation Week*: 29 October 1990, pp. 36-37, 46; 1 April 1991, p. 80.
60 AIAA Paper 95-6031.
61 General Accounting Office Report NSIAD-93-71.
62 DTIC ADA-274530.
63 Schweikart, *Quest*, table opposite p. 182; AIAA Paper 95-6031.

8

Why NASP Fell Short

NASP was founded on optimism, but it involved a good deal more than blind faith. Key technical areas had not been properly explored and offered significant prospects of advance. These included new forms of titanium, along with the use of an ejector to eliminate the need for an auxiliary engine as a separate installation, for initial boost of a scramjet. There also was the highly promising field of computational fluid dynamics (CFD), which held the prospect of supplementing flight test and work in wind tunnels with sophisticated mathematical simulation.

Still NASP fell short, and there were reasons. CFD proved not to be an exact science, particularly at high Mach. Investigators worked with the complete equations of fluid mechanics, which were exact, but were unable to give precise treatments in such crucial areas as transition to turbulence and the simulation or modeling of turbulence. Their discussions introduced approximations that took away the accuracy and left NASP with more drag and less engine performance than people had sought.

In the field of propulsion, ejectors had not been well studied and stood as a topic that was ripe for deeper investigation. Even so, the ejectors offered poor performance at the outset, and subsequent studies did not bring substantial improvements. This was unfortunate, for use of a highly capable ejector was a key feature of Anthony duPont's patented engine cycle, which had provided technical basis for NASP.

With drag increasing and engine performance falling off, metallurgists might have saved the day by offering new materials. They indeed introduced Beta-21S titanium, which approached the heat resistance of Rene 41, the primary structural material of Dyna-Soar, but had only half the density. Yet even this achievement was not enough. Structural designers needed still more weight saving, and while they experimented with new types of beryllium and carbon-carbon, they came up with no significant contributions to the state of the art.

Aerodynamics

In March 1984, with the Copper Canyon studies showing promise, a classified program review was held near San Diego. In the words of George Baum, a close associate of Robert Williams, "We had to put together all the technology pieces to make it credible to the DARPA management, to get them to come out to a meeting in La Jolla and be willing to sit down for three full days. It wasn't hard to get people out to the West Coast in March; the problem was to get them off the beach."

One of the attendees, Robert Whitehead of the Office of Naval Research, gave a talk on CFD. Was the mathematics ready; were computers at hand? Williams recalls that "he explained, in about 15 minutes, the equations of fluid mechanics, in a memorable way. With a few simple slides, he could describe their nature in almost an offhand manner, laying out these equations so the computer could solve them, then showing that the computer technology was also there. We realized that we could compute our way to Mach 25, with high confidence. That was a high point of the presentations."[1]

Whitehead's point of departure lay in the fundamental equations of fluid flow: the Navier-Stokes equations, named for the nineteenth-century physicists Claude-Louis-Marie Navier and Sir George Stokes. They form a set of nonlinear partial differential equations that contain 60 partial derivative terms. Their physical content is simple, comprising the basic laws of conservation of mass, momentum, and energy, along with an equation of state. Yet their solutions, when available, cover the entire realm of fluid mechanics.[2]

An example of an important development, contemporaneous with Whitehead's presentation, was a 1985 treatment of flow over a complete X-24C vehicle at Mach 5.95. The authors, Joseph Shang and S. J. Scheer, were at the Air Force's Wright Aeronautical Laboratories. They used a Cray X-MP supercomputer and gave lift and drag coefficients:[3]

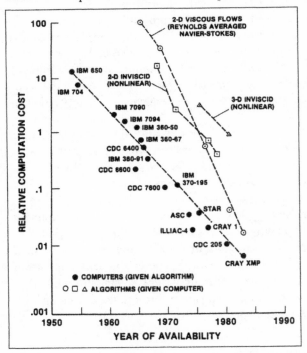

Development of CFD prior to NASP. In addition to vast improvement in computers, there also was similar advance in the performance of codes. (NASA)

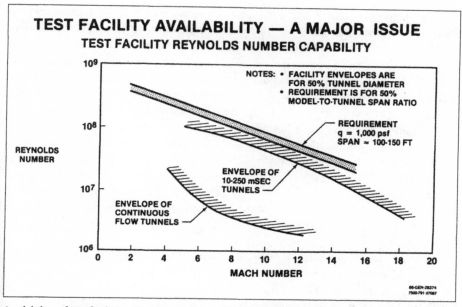

Availability of test facilities. Continuous-flow wind tunnels are far below the requirements of realistic simulation of full-size aircraft in flight. Impulse facilities, such as shock tunnels, come close to the requirements but are limited by their very short run times. (NASA)

	C_D	C_L	L/D
Experimental data	0.03676	0.03173	1.158
Numerical results	0.03503	0.02960	1.183
Percent error	4.71	6.71	2.16

(Source: AIAA Paper 85-1509)

    In that year the state of the art permitted extensive treatments of scramjets. Complete three-dimensional simulations of inlets were available, along with two-dimensional discussions of scramjet flow fields that covered the inlet, combustor, and nozzle. In 1984 Fred Billig noted that simulation of flow through an inlet using complete Navier-Stokes equations typically demanded a grid of 80,000 points and up to 12,000 time steps, with each run demanding four hours on a Control Data Cyber 203 supercomputer. A code adapted for supersonic flow was up to a hundred times faster. This made it useful for rapid surveys of a number of candidate inlets, with full Navier-Stokes treatments being reserved for a few selected choices.[4]

CFD held particular promise because it had the potential of overcoming the limitations of available facilities. These limits remained in place all through the NASP era. A 1993 review found "adequate" test capability only for classical aerodynamic experiments in a perfect gas, namely helium, which could support such work to Mach 20. Between Mach 13 and 17 there was "limited" ability to conduct tests that exhibited real-gas effects, such as molecular excitation and dissociation. Still, available facilities were too small to capture effects associated with vehicle size, such as determining the location of boundary-layer transition to turbulence.

For scramjet studies, the situation was even worse. There was "limited" ability to test combustors out to Mach 7, but at higher Mach the capabilities were "inadequate." Shock tunnels supported studies of flows in rarefied air from Mach 16 upward, but the whole of the nation's capacity for such tests was "inadequate." Some facilities existed that could study complete engines, either by themselves or in airframe-integrated configurations, but again the whole of this capability was "inadequate."[5]

Yet it was an exaggeration in 1984, and remains one to this day, to propose that CFD could remedy these deficiencies by computing one's way to orbital speeds "with high confidence." Experience has shown that CFD falls short in two areas: prediction of transition to turbulence, which sharply increases drag due to skin friction, and in the simulation of turbulence itself.

For NASP, it was vital not only to predict transition but to understand the properties of turbulence after it appeared. One could see this by noting that hypersonic propulsion differs substantially from propulsion of supersonic aircraft. In the latter, the art of engine design allows engineers to ensure that there is enough margin of thrust over drag to permit the vehicle to accelerate. A typical concept for a Mach 3 supersonic airliner, for instance, calls for gross thrust from the engines of 123,000 pounds, with ram drag at the inlets of 54,500. The difference, nearly 80,000 pounds of thrust, is available to overcome skin-friction drag during cruise, or to accelerate.

At Mach 6, a representative hypersonic-transport design shows gross thrust of 330,000 pounds and ram drag of 220,000. Again there is plenty of margin for what, after all, is to be a cruise vehicle. But in hypersonic cruise at Mach 12, the numbers typically are 2.1 million pounds for gross thrust—and 1.95 million for ram drag! Here the margin comes to only 150,000 pounds of thrust, which is narrow indeed. It could vanish if skin-friction drag proves to be higher than estimated, perhaps because of a poor forecast of the location of transition. The margin also could vanish if the thrust is low, due to the use of optimistic turbulence models.[6]

Any high-Mach scramjet-powered craft must not only cruise but accelerate. In turn, the thrust driving this acceleration appears as a small difference between two quantities: total drag and net thrust, the latter being net of losses within the engines. Accordingly, valid predictions concerning transition and turbulence are matters of the first importance.

WHY NASP FELL SHORT

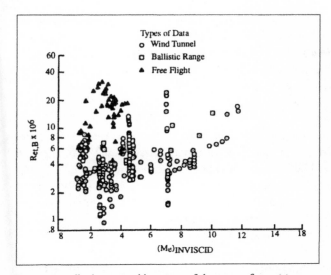

Experimentally determined locations of the onset of transition to turbulent flow. The strong scatter of the data points defeats attempts to find a predictive rule. (NASA)

NASP-era analysts fell back on the "e^N method," which gave a greatly simplified summary of the pertinent physics but still gave results that were often viewed as useful. It used the Navier-Stokes equations to solve for the overall flow in the laminary boundary layer, upstream of transition. This method then introduced new and simple equations derived from the original Navier-Stokes. These were linear and traced the growth of a small disturbance as one followed the flow downstream. When it had grown by a factor of 22,000—e^{10}, with N = 10—the analyst accepted that transition to turbulence had occurred.[7]

One can obtain a solution in this fashion, but transition results from local roughnesses along a surface, and these can lead to results that vary dramatically. Thus, the repeated re-entries of the space shuttle, during dozens of missions, might have given numerous nearly identical data sets. In fact, transition has occurred at Mach numbers from 6 to 19! A 1990 summary presented data from wind tunnels, ballistic ranges, and tests of re-entry vehicles in free flight. There was a spread of as much as 30 to one in the measured locations of transition, with the free-flight data showing transition positions that typically were five times farther back from a nose or leading edge than positions observed using other methods. At Mach 7, observed locations covered a range of 20 to one.[8]

One may ask whether transition can be predicted accurately even in principle because it involves minute surface roughnesses whose details are not known a priori and may even change in the course of a re-entry. More broadly, the state of transition was summarized in a 1987 review of problems in NASP hypersonics that was written by three NASA leaders in CFD:

> Almost nothing is known about the effects of heat transfer, pressure gradient, three-dimensionality, chemical reactions, shock waves, and other

233

influences on hypersonic transition. This is caused by the difficulty of conducting meaningful hypersonic transition experiments in noisy ground-based facilities and the expense and difficulty of carrying out detailed and carefully controlled experiments in flight where it is quiet. Without an adequate, detailed database, development of effective transition models will be impossible.[9]

Matters did not improve in subsequent years. In 1990 Mujeeb Malik, a leader in studies of transition, noted "the long-held view that conventional, noisy ground facilities are simply not suitable for simulation of flight transition behavior." A subsequent critique added that "we easily recognize that there is today no reasonably reliable predictive capability for engineering applications" and commented that "the reader...is left with some feeling of helplessness and discouragement."[10] A contemporary review from the Defense Science Board pulled no punches: "Boundary layer transition...cannot be validated in existing ground test facilities."[11]

There was more. If transition could not be predicted, it also was not generally possible to obtain a valid simulation, from first principles, of a flow that was known to be turbulent. The Navier-Stokes equations carried the physics of turbulence at all scales. The problem was that in flows of practical interest, the largest turbulent eddies were up to 100,000 times bigger than the smallest ones of concern. This meant that complete numerical simulations were out of the question.

Late in the nineteenth century the physicist Osborne Reynolds tried to bypass this difficulty by rederiving these equations in averaged form. He considered the flow velocity at any point as comprising two elements: a steady-flow part and a turbulent part that contained all the motion due to the eddies. Using the Navier-Stokes equations, he obtained equations for averaged quantities, with these quantities being based on the turbulent velocities.

He found, though, that the new equations introduced additional unknowns. Other investigators, pursuing this approach, succeeded in deriving additional equations for these extra unknowns—only to find that these introduced still more unknowns. Reynolds's averaging procedure thus led to an infinite regress, in which at every stage there were more unknown variables describing the turbulence than there were equations with which to solve for them. This contrasted with the Navier-Stokes equations themselves, which in principle could be solved because the number of these equations and the number of their variables was equal.

This infinite regress demonstrated that it was not sufficient to work from the Navier-Stokes equations alone—something more was needed. This situation arose because the averaging process did not preserve the complete physical content of the Navier-Stokes formulation. Information had been lost in the averaging. The problem of turbulence thus called for additional physics that could replace the lost information, end the regress, and give a set of equations for turbulent flow in which the number of equations again would match the number of unknowns.[12]

The standard means to address this issue has been a turbulence model. This takes the form of one or more auxiliary equations, either algebraic or partial-differential, which are solved simultaneously with the Navier-Stokes equations in Reynolds-averaged form. In turn, the turbulence model attempts to derive one or more quantities that describe the turbulence and to do so in a way that ends the regress.

Viscosity, a physical property of every liquid and gas, provides a widely used point of departure. It arises at the molecular level, and the physics of its origin is well understood. In a turbulent flow, one may speak of an "eddy viscosity" that arises by analogy, with the turbulent eddies playing the role of molecules. This quantity describes how rapidly an ink drop will mix into a stream—or a parcel of hydrogen into the turbulent flow of a scramjet combustor.[13]

Like the e^N method in studies of transition, eddy viscosity presents a view of turbulence that is useful and can often be made to work, at least in well-studied cases. The widely used Baldwin-Lomax model is of this type, and it uses constants derived from experiment. Antony Jameson of Princeton University, a leading writer of flow codes, described it in 1990 as "the most popular turbulence model in the industry, primarily because it's easy to program."[14]

This approach indeed gives a set of equations that are solvable and avoid the regress, but the analyst pays a price: Eddy viscosity lacks standing as a concept supported by fundamental physics. Peter Bradshaw of Stanford University virtually rejects it out of hand, declaring, "Eddy viscosity does not even deserve to be described as a 'theory' of turbulence!" He adds more broadly, "The present state is that even the most sophisticated turbulence models are based on brutal simplification of the N-S equations and hence cannot be relied on to predict a large range of flows with a fixed set of empirical coefficients."[15]

Other specialists gave similar comments throughout the NASP era. Thomas Coakley of NASA-Ames wrote in 1983 that "turbulence models that are now used for complex, compressible flows are not well advanced, being essentially the same models that were developed for incompressible attached boundary layers and shear flows. As a consequence, when applied to compressible flows they yield results that vary widely in terms of their agreement with experimental measurements."[16]

A detailed critique of existing models, given in 1985 by Budugur Lakshminarayana of Pennsylvania State University, gave pointed comments on algebraic models, which included Baldwin-Lomax. This approach "provides poor predictions" for flows with "memory effects," in which the physical character of the turbulence does not respond instantly to a change in flow conditions but continues to show the influence of upstream effects. Such a turbulence model "is not suitable for flows with curvature, rotation, and separation. The model is of little value in three-dimensional complex flows and in situations where turbulence transport effects are important."

"Two-equation models," which used two partial differential equations to give more detail, had their own faults. In the view of Lakshminarayana, they "fail to capture many of the features associated with complex flows." This class of models "fails for flows with rotation, curvature, strong swirling flows, three-dimensional flows, shock-induced separation, etc."[17]

Rather than work with eddy viscosity, some investigators used "Reynolds stress" models. Reynolds stresses were not true stresses, which contributed to drag. Rather, they were terms that appeared in the Reynolds-averaged Navier-Stokes equations alongside other terms that indeed represented stress. Models of this type offered greater physical realism, but again this came at the price of severe computational difficulty.[18]

A group at NASA-Langley, headed by Thomas Gatski, offered words of caution in 1990: "…even in the low-speed incompressible regime, it has not been possible to construct a turbulence closure model which can be applied over a wide class of flows…. In general, Reynolds stress closure models have not been very successful in handling the effects of rotation or three-dimensionality even in the incompressible regime; therefore, it is not likely that these effects can be treated successfully in the compressible regime with existing models."[19]

Anatol Roshko of Caltech, widely viewed as a dean of aeronautics, has his own view: "History proves that each time you get into a new area, the existing models are found to be inadequate." Such inadequacies have been seen even in simple flows, such as flow over a flat plate. The resulting skin friction is known to an accuracy of around one percent. Yet values calculated from turbulence models can be in error by up to 10 percent. "You can always take one of these models and fix it so it gives the right answer for a particular case," says Bradshaw. "Most of us choose the flat plate. So if you can't get the flat plate right, your case is indeed piteous."[20]

Another simple case is flow within a channel that suddenly widens. Downstream of the point of widening, the flow shows a zone of strongly whirling circulation. It narrows until the main flow reattaches, flowing in a single zone all the way to the now wider wall. Can one predict the location of this reattachment point? "This is a very severe test," says John Lumley of Cornell University. "Most of the simple models have trouble getting reattachment within a factor of two." So-called "k-epsilon models," he says, are off by that much. Even so, NASA's Tom Coakley describes them as "the most popular two-equation model," whereas Princeton University's Jameson speaks of them as "probably the best engineering choice around" for such problems as…flow within a channel.[21]

Turbulence models have a strongly empirical character and therefore often fail to predict the existence of new physics within a flow. This has been seen to cause difficulties even in the elementary case of steady flow past a cylinder at rest, a case so simple that it is presented in undergraduate courses. Nor do turbulence models cope with another feature of some flows: their strong sensitivity to slight changes in conditions. A simple example is the growth of a mixing layer.

In this scenario, two flows that have different velocities proceed along opposite sides of a thin plate, which terminates within a channel. The mixing layer then forms and grows at the interface between these streams. In Roshko's words, "a one-percent periodic disturbance in the free stream completely changes the mixing layer growth." This has been seen in experiments and in highly detailed solutions of the Navier-Stokes equations that solve the complete equations using a very fine grid. It has not been seen in solutions of Reynolds-averaged equations that use turbulence models.[22]

And if simple flows of this type bring such difficulties, what can be said of hypersonics? Even in the free stream that lies at some distance from a vehicle, one finds strong aerodynamic heating along with shock waves and the dissociation, recombination, and chemical reaction of air molecules. Flow along the aircraft surface adds a viscous boundary layer that undergoes shock impingement, while flow within the engine adds the mixing and combustion of fuel.

As William Dannevik of Lawrence Livermore National Laboratory describes it, "There's a fully nonlinear interaction among several fields: an entropy field, an acoustic field, a vortical field." By contrast, in low-speed aerodynamics, "you can often reduce it down to one field interacting with itself." Hypersonic turbulence also brings several channels for the flow and exchange of energy: internal energy, density, and vorticity. The experimental difficulties can be correspondingly severe.[23]

Roshko sees some similarity between turbulence modeling and the astronomy of Ptolemy, who flourished when the Roman Empire was at its height. Ptolemy represented the motions of the planets using epicycles and deferents in a purely empirical fashion and with no basis in physical theory. "Many of us have used that example," Roshko declares. "It's a good analogy. People were able to continually keep on fixing up their epicyclic theory, to keep on accounting for new observations, and they were completely wrong in knowing what was going on. I don't think we're that badly off, but it's illustrative of another thing that bothers some people. Every time some new thing comes around, you've got to scurry and try to figure out how you're going to incorporate it."[24]

A 1987 review concluded, "In general, the state of turbulence modeling for supersonic, and by extension, hypersonic, flows involving complex physics is poor." Five years later, late in the NASP era, little had changed, for a Defense Science Board program review pointed to scramjet development as the single most important issue that lay beyond the state of the art.[25]

Within NASP, these difficulties meant that there was no prospect of computing one's way in orbit, or of using CFD to make valid forecasts of high-Mach engine performance. In turn, these deficiencies forced the program to fall back on its test facilities, which had their own limitations.

Propulsion

In the spring of 1992 the NASP Joint Program Office presented a final engine design called the E22A. It had a length of 60 feet and included an inlet ramp, cowled inlet, combustor, and nozzle. An isolator, located between the inlet and combustor, sought to prevent unstarts by processing flow from the inlet through a series of oblique shocks, which increased the backpressure from the combustor.

Program officials then constructed two accurately scaled test models. The Subscale Parametric Engine (SXPE) was built to one-eighth scale and had a length of eight feet. It was tested from April 1993 to March 1994. The Concept Demonstrator Engine (CDE), which followed, was built to a scale of 30 percent. Its length topped 16 feet, and it was described as "the largest airframe-integrated scramjet engine ever tested."[26]

In working with the SXPE, researchers had an important goal in achieving combustion of hydrogen within its limited length. To promote rapid ignition, the engine used a continuous flow of a silane-hydrogen mixture as a pilot, with the silane igniting spontaneously on exposure to air. In addition, to promote mixing, the model incorporated an accurate replication of the spacing between the fuel-injecting struts and ramps, with this spacing being preserved at the model's one-eighth scale. The combustor length required to achieve the desired level of mixing then scaled in this fashion as well.

The larger CDE was tested within the Eight-Foot High-Temperature Tunnel, which was Langley's biggest hypersonic facility. The tests mapped the flowfield entering the engine, determined the performance of the inlet, and explored the potential performance of the design. Investigators varied the fuel flow rate, using the combustors to vary its distribution within the engine.

Boundary-layer effects are important in scramjets, and the tests might have replicated the boundary layers of a full-scale engine by operating at correspondingly higher flow densities. For the CDE, at 30 percent scale, the appropriate density would have been 1/0.3 or 3.3 times that of the atmospheric density at flight altitude. For the SXPE, at one-eighth scale, the test density would have shown an eightfold increase over atmospheric. However, the SXPE used an arc-heated test facility that was limited in the power that drove its arc, and it provided its engine with air at only one-fiftieth of that density. The High Temperature Tunnel faced limits on its flow rate and delivered its test gas at only one-sixth of the appropriate density.

Engineers sought to compensate by using analytical methods to determine the drag in a full-scale engine. Still, this inability to replicate boundary-layer effects meant that the wind-tunnel tests gave poor simulations of internal drag within the test engines. This could have led to erroneous estimates of true thrust, net of drag. In turn, this showed that even when working with large test models and with test facilities of impressive size, true simulations of the boundary layer were ruled out from the start.[27]

For takeoff from a runway, the X-30 was to use a Low-Speed System (LSS). It comprised two principal elements: the Special System, an ejector ramjet; and the Low Speed Oxidizer System, which used LACE.[28] The two were highly synergistic. The ejector used a rocket, which might have been suitable for the final ascent to orbit, with ejector action increasing its thrust during takeoff and acceleration. By giving an exhaust velocity that was closer to the vehicle velocity, the ejector also increased the fuel economy.

The LACE faced the standard problem of requiring far more hydrogen than could be burned in the air it liquefied. The ejector accomplished some derichening by providing a substantial flow of entrained air that burned some of the excess. Additional hydrogen, warmed in the LACE heat exchanger, went into the fuel tanks, which were full of slush hydrogen. By melting the slush into conventional liquid hydrogen (LH_2), some LACE coolant was recycled to stretch the vehicle's fuel supply.[29]

There was good news in at least one area of LACE research: deicing. LACE systems have long been notorious for their tendency to clog with frozen moisture within the air that they liquefy. "The largest LACE ever built made around half a pound per second of liquid air," Paul Czysz of McDonnell Douglas stated in 1986. "It froze up at six percent relative humidity in the Arizona desert, in 38 seconds." Investigators went on to invent more than a dozen methods for water alleviation. The most feasible approach called for injecting antifreeze into the system, to enable the moisture to condense out as liquid water without freezing. A rotary separator eliminated the water, with the dehumidified air being so cold as to contain very little residual water vapor.[30]

The NASP program was not run by shrinking violets, and its managers stated that its LACE was not merely to operate during hot days in the desert near Phoenix. It was to function even on rainy days, for the X-30 was to be capable of flight from anywhere in the world. At NASA-Lewis, James Van Fossen built a water-alleviation system that used ethylene glycol as the antifreeze, spraying it directly onto the cold tubes of a heat exchanger. Water, condensing on those tubes, dissolved some of the glycol and remained liquid as it swept downstream with the flow. He reported that this arrangement protected the system against freezing at temperatures as low as −55°F, with the moisture content of the chilled air being reduced to 0.00018 pounds in each pound of this air. This represented removal of at least 99 percent of the humidity initially present in the airflow.[31]

Pratt & Whitney conducted tests of a LACE precooler that used this arrangement. A company propulsion manager, Walt Lambdin, addressed a NASP technical review meeting in 1991 and reported that it completely eliminated problems of reduced performance of the precooler due to formation of ice. With this, the problem of ice in a LACE system appeared amenable to control.[32]

It was also possible to gain insight into the LACE state of the art by considering contemporary work that was under way in Japan. The point of departure in that country was the H-2 launch vehicle, which first flew to orbit in February 1994. It was a two-stage expendable rocket, with a liquid-fueled core flanked by two solid boosters. LACE was pertinent because a long-range plan called for upgrades that could replace the solid strap-ons with new versions using LACE engines.[33]

Mitsubishi Heavy Industries was developing the H-2's second-stage engine, designated LE-5. It burned hydrogen and oxygen to produce 22,000 pounds of thrust. As an initial step toward LACE, this company built heat exchangers to liquefy air for this engine. In tests conducted during 1987 and 1988, the Mitsubishi heat exchanger demonstrated liquefaction of more than three pounds of air for every pound of LH_2. This was close to four to one, the theoretical limit based on the thermal properties of LH_2 and of air. Still, it takes 34.6 pounds of air to burn a pound of hydrogen, and an all-LACE LE-5 was to run so fuel-rich that its thrust was to be only 6,000 pounds.

But the Mitsubishi group found their own path to prevention of ice buildup. They used a freeze-thaw process, melting ice by switching periodically to the use of ambient air within the cooler after its tubes had become clogged with ice from LH_2. The design also provided spaces between the tubes and allowed a high-speed airflow to blow ice from them.[34]

LACE nevertheless remained controversial, and even with the moisture problem solved, there remained the problem of weight. Czysz noted that an engine with 100,000 pounds of thrust would need 600 pounds per second of liquid air: "The largest liquid-air plant in the world today is the AiResearch plant in Los Angeles, at 150 pounds per second. It covers seven acres. It contains 288,000 tubes welded to headers and 59 miles of 3/32-inch tubing."[35]

Still, no law required the use of so much tubing, and advocates of LACE have long been inventive. A 1963 Marquardt concept called for an engine with 10,000 pounds of thrust, which might have been further increased by using an ejector. This appeared feasible because LACE used LH_2 as the refrigerant. This gave far greater effectiveness than the AiResearch plant, which produced its refrigerant on the spot by chilling air through successive stages.[36]

For LACE heat exchangers, thin-walled tubing was essential. The Japanese model, which was sized to accommodate the liquid-hydrogen flow rate of the LE-5, used 5,400 tubes and weighed 304 pounds, which is certainly noticeable when the engine is to put out no more than 6,000 pounds of thrust. During the mid-1960s investigators at Marquardt and AiResearch fabricated tubes with wall thicknesses as low as 0.001 inch, or one mil. Such tubes had not been used in any heat exchanger subassemblies, but 2-mil tubes of stainless steel had been crafted into a heat exchanger core module with a length of 18 inches.[37]

Even so, this remained beyond the state of the art for NASP, a quarter-century later. Weight estimates for the X-30 LACE heat exchanger were based on the assumed use of 3-mil Weldalite tubing, but a 1992 Lockheed review stated, "At present, only small quantities of suitable, leak free, 3-mil tubing have been fabricated." The plans of that year called for construction of test prototypes using 6-mil Weldalite tubing, for which "suppliers have been able to provide significant quantities." Still, a doubled thickness of the tubing wall was not the way to achieve low weight.[38]

Other weight problems arose in seeking to apply an ingenious technique for derichening the product stream by increasing the heat capacity of the LH_2 coolant. Molecular hydrogen, H_2, has two atoms in its molecule and exists in two forms: para and ortho, which differ in the orientation of the spins of their electrons. The ortho form has parallel spin vectors, while the para form has spin vectors that are oppositely aligned. The ortho molecule amounts to a higher-energy form and loses energy as heat when it transforms into the para state. The reaction therefore is exothermic.

The two forms exist in different equilibrium concentrations, depending on the temperature of the bulk hydrogen. At room temperature the gas is about 25 percent para and 75 percent ortho. When liquefied, the equilibrium state is 100 percent para. Hence it is not feasible to prepare LH_2 simply by liquefying the room-temperature gas. The large component of ortho will relax to para over several hours, producing heat and causing the liquid to boil away. The gas thus must be exposed to a catalyst to convert it to the para form before it is liquefied.

These aspects of fundamental chemistry also open the door to a molecular shift that is endothermic and that absorbs heat. One achieves this again by using a catalyst to convert the LH_2 from para to ortho. This reaction requires heat, which is obtained from the liquefying airflow within the LACE. As a consequence, the air chills more readily when using a given flow of hydrogen refrigerant. This effect is sufficiently strong to increase the heat-sink capacity of the hydrogen by as much as 25 percent.[39]

This concept also dates to the 1960s. Experiments showed that ruthenium metal deposited on aluminum oxide provided a suitable catalyst. For 90 percent para-to-ortho conversion, the LACE required a "beta," a ratio of mass to flow rate, of five to seven pounds of this material for each pound per second of hydrogen flow. Data published in 1988 showed that a beta of five pounds could achieve 85 percent conversion, with this value showing improvement during 1992. However, X-30 weight estimates assumed a beta of two pounds, and this performance remained out of reach.[40]

During takeoff, the X-30 was to be capable of operating from existing runways and of becoming airborne at speeds similar to those of existing aircraft. The low-speed system, along with its accompanying LACE and ejector systems, therefore needed substantial levels of thrust. The ejector, again, called for a rocket exhaust to serve as a primary flow within a duct, entraining an airstream as the secondary flow. Ejectors gave good performance across a broad range of flight speeds, showing an effectiveness that increased with Mach. In the SR-71 at Mach 2.2, they accounted for 14 percent of the thrust in afterburner; at Mach 3.2 this was 28.4 percent. Nor did the SR-71 ejectors burn fuel. They functioned entirely as aerodynamic devices.[41]

It was easy to argue during the 1980s that their usefulness might be increased still further. The most important unclassified data had been published during the 1950s. A good engine needed a high pressure increase, but during the mid-1960s studies at Marquardt recommended a pressure rise by a factor of only about 1.5, when turbojets were showing increases that were an order of magnitude higher.[42] The best theoretical treatment of ejector action dated to 1974. Its author, NASA's B. H. Anderson, also wrote a computer program called REJECT that predicted the performance of supersonic ejectors. However, he had done this in 1974, long before the tools of CFD were in hand. A 1989 review noted that since then "little attention has been directed toward a better understanding of the details of the flow mechanism and behavior."[43]

Within the NASP program, then, the ejector ramjet stood as a classic example of a problem that was well suited to new research. Ejectors were known to have good effectiveness, which might be increased still further and which stood as a good topic for current research techniques. CFD offered an obvious approach, and NASP activities supplemented computational work with an extensive program of experiment.[44]

The effort began at GASL, where Tony duPont's ejector ramjet went on a static test stand during 1985 and impressed General Skantze. DuPont's engine design soon took the title of the Government Baseline Engine and remained a topic of active experimentation during 1986 and 1987. Some work went forward at NASA-Langley, where the Combustion Heated Scramjet Test Facility exercised ejectors over the range of Mach 1.2 to 3.5. NASA-Lewis hosted further tests, at Mach 0.06 and from Mach 2 to 3.5 within its 10 by 10 foot Supersonic Wind Tunnel.

The Lewis engine was built to accommodate growth of boundary layers and placed a 17-degree wedge ramp upstream of the inlet. Three flowpaths were mounted side by side, but only the center duct was fueled; the others were "dummies" that gave data on unfueled operation for comparison. The primary flow had a pressure of 1,000 pounds per square inch and a temperature of 1,340°F, which simulated a fuel-rich rocket exhaust. The experiments studied the impact of fuel-to-air ratio on performance, although the emphasis was on development of controls.

Even so, the performance left much to be desired. Values of fuel-to-air ratio greater than 0.52, with unity representing complete combustion, at times brought "buzz" or unwanted vibration of the inlet structure. Even with no primary flow, the inlet failed to start. The main burner never achieved thermal choking, where the flow rate would rise to the maximum permitted by heat from burning fuel. Ingestion of the boundary layer significantly degraded engine performance. Thrust measurements were described as "no good" due to nonuniform thermal expansion across a break between zones of measurement. As a contrast to this litany of woe, operation of the primary gave a welcome improvement in the isolation of the inlet from the combustor.

Also at GASL, again during 1987, an ejector from Boeing underwent static test. It used a markedly different configuration that featured an axisymmetric duct and a fuel-air mixer. The primary flow was fuel-rich, with temperatures and pressures similar to those of NASA-Lewis. On the whole, the results of the Boeing tests were encouraging. Combustion efficiencies appeared to exceed 95 percent, while measured values of thrust, entrained airflow, and pressures were consistent with company predictions. However, the mixer performance was no more than marginal, and its length merited an increase for better performance.[45]

In 1989 Pratt & Whitney emerged as a major player, beginning with a subscale ejector that used a flow of helium as the primary. It underwent tests at company facilities within the United Technologies Research Center. These tests addressed the basic issue of attempting to increase the entrainment of secondary flow, for which non-combustible helium was useful. Then, between 1990 and 1992, Pratt built three versions of its Low Speed Component Integration Rig (LSCIR), testing them all within facilities of Marquardt.

LSCIR-1 used a design that included a half-scale X-30 flowpath. It included an inlet, front and main combustors, and nozzle, with the inlet cowl featuring fixed geometry. The tests operated using ambient air as well as heated air, with and without fuel in the main combustor, while the engine operated as a pure ramjet for several runs. Thermal choking was achieved, with measured combustion efficiencies lying within 2 percent of values suitable for the X-30. But the inlet was unstarted for nearly all the runs, which showed that it needed variable geometry. This refinement was added to LSCIR-2, which was put through its paces in July 1991, at Mach 2.7. The test sequence would have lasted longer but was terminated prematurely due to a burnthrough of the front combustor, which had been operating at 1,740°F. Thrust measurements showed only limited accuracy due to flow separation in the nozzle.

LSCIR-3 followed within months. The front combustor was rebuilt with a larger throat area to accommodate increased flow and received a new ignition system that used silane. This gas ignited spontaneously on contact with air. In tests, leaks developed between the main combustor, which was actively cooled, and the uncooled nozzle. A redesigned seal eliminated the leakage. The work also validated a method for calculating heat flux to the wall due to impingement of flow from primaries.

Other results were less successful. Ignition proceeded well enough using pure silane, but a mix of silane and hydrogen failed as an ignitant. Problems continued to recur due to inlet unstarts and nozzle flow separation. The system produced 10,000 pounds of thrust at Mach 0.8 and 47,000 pounds at Mach 2.7, but this performance still was rated as low.

Within the overall LSS program, a Modified Government Baseline Engine went under test at NASA-Lewis during 1990, at Mach 3.5. The system now included hydraulically-operated cowl and nozzle flaps that provided variable geometry, along with an isolator with flow channels that amounted to a bypass around the combustor. This helped to prevent inlet unstarts.

Once more the emphasis was on development of controls, with many tests operating the system as a pure ramjet. Only limited data were taken with the primaries on. Ingestion of the boundary layer gave significant degradation in engine performance, but in other respects most of the work went well. The ramjet operations were successful. The use of variable geometry provided reliable starting of the inlet, while operation in the ejector mode, with primaries on, again improved the inlet isolation by diminishing the effect of disturbances propagating upstream from the combustor.[46]

Despite these achievements, a 1993 review at Rocketdyne gave a blunt conclusion: "The demonstrated performance of the X-30 special system is lower than the performance level used in the cycle deck...the performance shortfall is primarily associated with restrictions on the amount of secondary flow." (Secondary flow is entrained by the ejector's main flow.) The experimental program had taught much concerning the prevention of inlet unstarts and the enhancement of inlet-combustor isolation, but the main goal—enhanced performance of the ejector ramjet—still lay out of reach.

Simple enlargement of a basic design offered little promise; Pratt & Whitney had tried that, in LSCIR-3, and had found that this brought inlet flow separation along with reduced inlet efficiency. Then in March 1993, further work on the LSS was canceled due to budget cuts. NASP program managers took the view that they could accelerate an X-30 using rockets for takeoff, as an interim measure, with the LSS being added at a later date. Thus, although the LSS was initially the critical item in duPont's design, in time it was put on hold and held off for another day.[47]

Materials

No aircraft has ever cruised at Mach 5, and an important reason involves structures and materials. "If I cruise in the atmosphere for two hours," says Paul Czysz of McDonnell Douglas, "I have a thousand times the heat load into the vehicle that the shuttle gets on its quick transit of the atmosphere." The thermal environment of

Why NASP Fell Short

the X-30 was defined by aerodynamic heating and by the separate issue of flutter.[48]

A single concern dominated issues of structural design: The vehicle was to fly as low as possible in the atmosphere during ascent to orbit. Re-entry called for flight at higher altitudes, and the loads during ascent therefore were higher than those of re-entry. Ascent at lower altitude—200,000 feet, for instance, rather than 250,000—increased the drag on the X-30. But it also increased the thrust, giving a greater margin between thrust and drag that led to increased acceleration. Considerations of ascent, not re-entry, therefore shaped the selection of temperature-resistant materials.

Yet the aircraft could not fly too low, or it would face limits set by aerodynamic flutter. This resulted from forces on the vehicle that were not steady but oscillated, at frequencies of oscillation that changed as the vehicle accelerated and lost weight. The wings tended to vibrate at characteristic frequencies, as when bent upward and released to flex up and down. If the frequency of an aerodynamic oscillation matched that at which the wings were prone to flex, the aerodynamic forces could tear the wings off. Stiffness in materials, not strength, was what resisted flutter, and the vehicle was to fly a "flutter-limited trajectory," staying high enough to avoid the problem.

The mechanical properties of metals depend on their fine-grained structure. An ingot of metal consists of a mass of interlaced grains or crystals, and small grains give higher strength. Quenching, plunging hot metal into water, yields small grains but often makes the metal brittle or hard to form. Alloying a metal, as by adding small quantities of carbon to make steel, is another traditional practice. However, some additives refuse to dissolve or separate out from the parent metal as it cools.

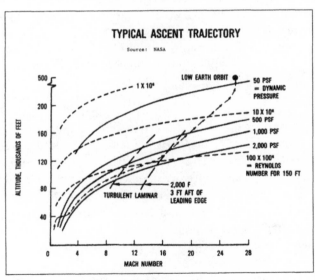

Ascent trajectory of an airbreather. (NASA)

To overcome such restrictions, techniques of powder metallurgy were in the forefront. These methods gave direct control of the microstructure of metals by forming

245

them from powder, with the grains of powder sintering or welding together by being pressed in a mold at high temperature. A manufacturer could control the grain size independently of any heat-treating process. Powder metallurgy also overcame restrictions on alloying by mixing in the desired additives as powdered ingredients.

Several techniques existed to produce the powders. Grinding a metal slab to sawdust was the simplest, yielding relatively coarse grains. "Splat-cooling" gave better control. It extruded molten metal onto the chilled rim of a rotating wheel, which cooled it instantly into a thin ribbon. This represented a quenching process that produced a fine-grained microstructure in the metal. The ribbon then was chemically treated with hydrogen, which made it brittle, so that it could be ground into a fine powder. Heating the powder then drove off the hydrogen.

The Plasma Rotating Electrode Process, developed by the firm of Nuclear Metals, showed particular promise. The parent metal was shaped into a cylinder that rotated at up to 30,000 revolutions per minute and served as an electrode. An electric arc melted the spinning metal, which threw off droplets within an atmosphere of cool inert helium. The droplets plummeted in temperature by thousands of degrees within milliseconds, and their microstructures were so fine as to approach an amorphous state. Their molecules did not form crystals, even tiny ones, but arranged themselves in formless patterns. This process, called "rapid solidification," promised particular gains in high-temperature strength.

Standard titanium alloys, for instance, lost strength at temperatures above 700 to 900°F. By using rapid solidification, McDonnell Douglas raised this limit to 1,100°F prior to 1986. Philip Parrish, the manager of powder metallurgy at DARPA, noted that his agency had spent some $30 million on rapid-solidification technology since 1975. In 1986 he described it as "an established technology. This technology now can stand along such traditional methods as ingot casting or drop forging."[49]

Nevertheless 1,100°F was not enough, for it appeared that the X-30 needed a material that was rated at 1,700°F. This stemmed from the fact that for several years, NASP design and trajectory studies indicated that a flight vehicle indeed would face such temperatures on its fuselage. But after 1990 the development of new baseline configurations led to an appreciation that the pertinent areas of the vehicle would face temperatures no higher than 1,500°F. At that temperature, advanced titanium alloys could serve in "metal matrix composites," with thin-gauge metals being reinforced with fibers.

The new composition came from the firm of Titanium Metals and was designated Beta-21S. That company developed it specifically for the X-30 and patented it in 1989. It consisted of titanium along with 15 percent molybdenum, 2.8 percent columbium, 3 percent aluminum, and 0.2 percent silicon. Resistance to oxidation proved to be its strong suit, with this alloy showing resistance that was two orders of magnitude greater than that of conventional aircraft titanium. Tests showed that it

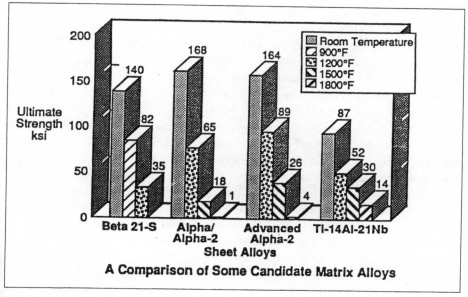

Comparison of some matrix alloys. (NASA)

also could be exposed repeatedly to leaks of gaseous hydrogen without being subject to embrittlement. Moreover, it lent itself readily to being rolled to foil-gauge thicknesses of 4 to 5 mil when metal matrix composites were fabricated.[50]

Such titanium-matrix composites were used in representative X-30 structures. The Non-Integral Fuselage Tank Article (NIFTA) represented a section of X-30 fuselage at one-fourth scale. It was oblong in shape, eight feet long and measuring four by seven feet in cross section, and it contained a splice. Its skin thickness was 0.040 inches, about the same as for the X-30. It held an insulated tank that could hold either liquid nitrogen or LH_2 in tests, which stood as a substantial engineering item in its own right.

The tank had a capacity of 940 gallons and was fabricated of graphite-epoxy composite. No liner protected the tankage on the inside, for graphite-epoxy was impervious to damage by LH_2. However, the exterior was insulated with two half-inch thicknesses of Q-felt, a quartz-fiber batting with density of only 3.5 pounds per cubic foot. A thin layer of Astroquartz high-temperature cloth covered the Q-felt. This insulation filled space between the tank wall and the surrounding wall of the main structure, with both this space and the Q-felt being purged with helium.[51]

The test sequence for NIFTA duplicated the most severe temperatures and stresses of an ascent to orbit. These stresses began on the ground, with the vehicle being heavy with fuel and subject to a substantial bending load. There was also a large shear load, with portions of the vehicle being pulled transversely in opposite directions. This happened because the landing gear pushed upward to support the entire weight of the craft, while the weight of the hydrogen tank pushed downward only a few feet away. Other major bending and shear loads arose during subsonic climbout, with the X-30 executing a pullup maneuver.

Significant stresses arose near Mach 6 and resulted from temperature differences across the thickness of the stiffened skin. Its outer temperature was to be 800°F, but the tops of the stiffeners, a few inches away, were to be 350°F. These stiffeners were spot-welded to the skin panels, which raised the issue of whether the welds would hold amid the different thermal expansions. Then between Mach 10 and 16, the vehicle was to reach peak temperatures of 1,300°F. The temperature differences between the top and bottom of the vehicle also would be at their maximum.

The tests combined both thermal and mechanical loads and were conducted within a vacuum chamber at Wyle Laboratories during 1991. Banks of quartz lamps applied up to 1.5 megawatts of heat, while jacks imposed bending or shear forces that reached 100 percent of the design limits. Most tests placed nonflammable liquid nitrogen in the tank for safety, but the last of them indeed used LH_2. With this supercold fuel at -423°F, the lamps raised the exterior temperature of NIFTA to the full 1,300°F, while the jacks applied the full bending load. A 1993 paper noted "100% successful completion of these tests," including the one with LH_2 that had been particularly demanding.[52]

NIFTA, again, was at one-fourth scale. In a project that ran from 1991 through the summer of 1994, McDonnell Douglas engineers designed and fabricated the substantially larger Full Scale Assembly. Described as "the largest and most representative NASP fuselage structure built," it took shape as a component measuring 10 by 12 feet. It simulated a section of the upper mid-fuselage, just aft of the crew compartment.

A 1994 review declared that it "was developed to demonstrate manufacturing and assembly of a full scale fuselage panel incorporating all the essential structural details of a flight vehicle fuselage assembly." Crafted in flightweight, it used individual panels of titanium-matrix composite that were as large as four by eight feet. These were stiffened with longitudinal members of the same material and were joined to circumferential frames and fittings of Ti-1100, a titanium alloy that used no fiber reinforcement. The complete assembly posed manufacturing challenges because the panels were of minimum thickness, having thinner gauges than had been used previously. The finished article was completed just as NASP was reaching its end, but it showed that the thin panels did not introduce significant problems.[53]

The firm of Textron manufactured the fibers, designated SCS-6 and -9, that reinforced the composites. As a final touch, in 1992 this company opened the world's first manufacturing plant dedicated to the production of titanium-matrix

Why NASP Fell Short

composites. "We could get the cost down below a thousand dollars a pound if we had enough volume," Bill Grant, a company manager, told *Aerospace America*. His colleague Jim Henshaw added, "We think SCS/titanium composites are fully developed for structural applications."[54]

Such materials served to 1,500°F, but on the X-30 substantial areas were to withstand temperatures approaching 3,000°F, which is hotter than molten iron. If a steelworker were to plunge a hand into a ladle of this metal, the hand would explode from the sudden boiling of water in its tissues. In such areas, carbon-carbon was necessary. It had not been available for use in Dyna-Soar, but the Pentagon spent $200 million to fund its development between 1970 and 1985.[55]

Much of this supported the space shuttle, on which carbon-carbon protected such hot areas as the nose cap and wing leading edges. For the X-30, these areas expanded to cover the entire nose and much of the vehicle undersurface, along with the rudders and both the top and bottom surfaces of the wings. The X-30 was to execute 150 test flights, exposing its heat shield to prolonged thermal soaks while still in the atmosphere. This raised the problem of protection against oxidation.[56]

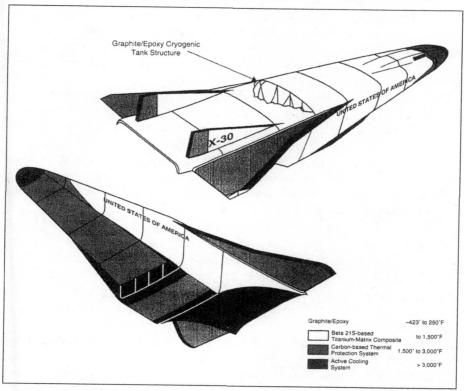

Selection of NASP materials based on temperature. (General Accounting Office)

Standard approaches called for mixing oxidation inhibitors into the carbon matrix and covering the surface with a coating of silicon carbide. However, there was a mismatch between the thermal expansions of the coating and the carbon-carbon substrate, which led to cracks. An interlayer of glass-forming sealant, placed between them, produced an impervious barrier that softened at high temperatures to fill the cracks. But these glasses did not flow readily at temperatures below 1,500°F. This meant that air could penetrate the coating and reach the carbon through open cracks to cause loss by oxidation.[57]

The goal was to protect carbon-carbon against oxidation for all 150 of those test flights, or 250 hours. These missions included 75 to orbit and 75 in hypersonic cruise. The work proceeded initially by evaluating several dozen test samples that were provided by commercial vendors. Most of these materials proved to resist oxidation for only 10 to 20 hours, but one specimen from the firm of Hitco reached 70 hours. Its surface had been grooved to promote adherence of the coating, and it gave hope that long operational life might be achieved.[58]

Complementing the study of vendors' samples, researchers ordered new types of carbon-carbon and conducted additional tests. The most durable came from the firm of Rohr, with a coating by Science Applications International. It easily withstood 2,000°F for 200 hours and was still going strong at 2,500 °F when the tests stopped after 150 hours. This excellent performance stemmed from its use of large quantities of oxidation inhibitors, which promoted long life, and of multiple glass layers in the coating.

But even the best of these carbon-carbons showed far poorer performance when tested in arcjets at 2,500°F. The high-speed airflows forced oxygen into cracks and pores within the material, while promoting evaporation of the glass sealants. Powerful roars within the arcjets imposed acoustic loads that contributed to cracking, with other cracks arising from thermal shock as test specimens were suddenly plunged into a hot flow stream. The best results indicated lifetimes of less than two hours.

Fortunately, actual X-30 missions were to impose 2,500°F temperatures for only a few minutes during each launch and reentry. Even a single hour of lifetime therefore could permit panels of carbon-carbon to serve for a number of flights. A 1992 review concluded that "maximum service temperatures should be limited to 2,800°F; above this temperature the silicon-based coating systems afford little practical durability," due to active oxidation. In addition, "periodic replacement of parts may be inevitable."[59]

New work on carbon-carbon, reported in 1993, gave greater encouragement as it raised the prospect of longer lifetimes. The effort evaluated small samples rather than fabricated panels and again used the arcjet installations of NASA-Johnson and Ames. Once again there was an orders-of-magnitude difference in the observed lifetimes of the carbon-carbon, but now the measured lifetimes extended into the hundreds of minutes. A formulation from the firm of Carbon-Carbon Advanced

Why NASP Fell Short

Technologies gave the best results, suggesting 25 reuses for orbital missions of the X-30 and 50 reuses for the less-demanding missions of hypersonic cruise.[60]

There also was interest in using carbon-carbon for primary structure. Here the property that counted was not its heat resistance but its light weight. In an important experiment, the firm of LTV fabricated half of an entire wing box of this material. An airplane's wing box is a major element of aircraft structure that joins the wings and provides a solid base for attachment of the fuselage fore and aft. Indeed, one could compare it with the keel of a ship. It extends to left and right of the aircraft centerline, and LTV's box constituted the portion to the left of this line. Built at full scale, it represented a hot-structure wing proposed by General Dynamics. It measured five by eight feet with a maximum thickness of 16 inches. Three spars ran along its length; five ribs were mounted transversely, and the complete assembly weighed 802 pounds.

The test plan called for it to be pulled upward at the tip to reproduce the bending loads of a wing in flight. Torsion or twisting was to be applied by pulling more strongly on the front or rear spar. The maximum load corresponded to having the X-30 execute a pullup maneuver at Mach 2.2, with the wing box at room temperature. With the ascent continuing and the vehicle undergoing aerodynamic heating, the next key event brought the maximum difference in the temperatures of the top and bottom of the wing box, with the former being 994°F and the latter at 1,671°F. At that moment the load on the wing box corresponded to 34 percent of the Mach 2.2 maximum. Farther along, the wing box was to reach its peak temperature, 1,925°F, on the lower surface. These three points were to be reproduced through mechanical forces applied at the ends of the spars and through the use of graphite heaters.

But several key parts delaminated during their fabrication, seriously compromising the ability of the wing box to bear its specified load. Plans to impose the peak or Mach 2.2 load were abandoned, with the maximum planned load being reduced to the 34 percent associated with the maximum temperature difference. For the same reason, the application of torsion was deleted from the test program. Amid these reductions in the scope of the structural tests, two exercises went forward during December 1991. The first took place at room temperature and successfully reached the mark of 34 percent, without causing further damage to the wing box.

The second test, a week later, reproduced the condition of peak temperature difference while briefly applying the calculated load of 34 percent. The plan then called for further heating to the peak temperature of 1,925°F. As the wing box approached this value, a problem arose due to the use of metal fasteners in its assembly. Some were made from coated columbium and were rated for 2,300°F, but most were of a nickel alloy that had a permissible temperature of 2,000°F. However, an instrumented nickel-alloy fastener overheated and reached 2,147°F. The wing box showed a maximum temperature of 1,917°F at that moment, and the test was terminated because the strength of the fasteners now was in question. This test nevertheless counted as a success because it had come within 8°F of the specified temperature.[61]

Both tests thus were marked as having achieved their goals, but their merits were largely in the mind of the beholder. The entire project would have been far more impressive if it had avoided delamination, successfully achieved the Mach 2.2 peak load, incorporated torsion, and subjected the wing box to repeated cycles of bending, torsion, and heating. This effort stood as a bold leap toward a future in which carbon-carbon might take its place as a mainstream material, suitable for a hot primary structure, but it was clear that this future would not arrive during the NASP program.

Then there was beryllium. It had only two-thirds the density of aluminum and possessed good strength, but its temperature range was limited. The conventional metal had a limit of some 850°F, but an alloy from Lockheed called Lockalloy, which contained 38 percent aluminum, was rated only for 600°F. It had never become a mainstream engineering material like titanium, but for NASP it offered the advantage of high thermal conductivity. Work with titanium had greatly increased its temperatures of use, and there was hope of achieving similar results with beryllium.

Initial efforts used rapid-solidification techniques and sought temperature limits as high as 1,500°F. These attempts bore no fruit, and from 1988 onward the temperature goal fell lower and lower. In May 1990 a program review shifted the emphasis away from high-temperature formulations toward the development of beryllium as a material suitable for use at cryogenic temperatures. Standard forms of this metal became unacceptably brittle when only slightly colder than -100°F, but cryo-beryllium proved to be out of reach as well. By 1992 investigators were working with ductile alloys of beryllium and were sacrificing all prospect of use at temperatures beyond a few hundred degrees but were winning only modest improvements in low-temperature capability. Terence Ronald, the NASP materials director, wrote in 1995 of rapid-solidification versions with temperature limits as low as 500°F, which was not what the X-30 needed to reach orbit.[62]

In sum, the NASP materials effort scored a major advance with Beta-21S, but the genuinely radical possibilities failed to emerge. These included carbon-carbon as primary structure, along with alloys of beryllium that were rated for temperatures well above 1,000°F. The latter, if available, might have led to a primary structure with the strength and temperature resistance of Beta-21S but with less than half the weight. Indeed, such weight savings would have ramified through the entire design, leading to a configuration that would have been smaller and lighter overall.

Overall, work with materials fell well short of its goals. In dealing with structures and materials, the contractors and the National Program Office established 19 program milestones that were to be accomplished by September 1993. A General Accounting Office program review, issued in December 1992, noted that only six of them would indeed be completed.[63] This slow progress encouraged conservatism in drawing up the bill of materials, but this conservatism carried a penalty.

Why NASP Fell Short

When the scramjets faltered in their calculated performance and the X-30 gained weight while falling short of orbit, designers lacked recourse to new and very light materials—structural carbon-carbon, high-temperature beryllium—that might have saved the situation. With this, NASP spiraled to its end. It also left its supporters with renewed appreciation for rockets as launch vehicles, which had been flying to orbit for decades.

1. Author interviews, George Baum, 15 May 1986; Robert Williams, 1 May 1986. Folder 18649, NASA Historical Reference Collection, NASA History Division, Washington, D.C. 20546.
2. Jameson, *Science*, 28 July 1989, pp. 361-71.
3. AIAA Paper 85-1509.
4. AIAA Paper 84-0387.
5. AIAA Paper 93-5101.
6. *Aerospace America*, September 1987, p. 37.
7. AIAA Paper 90-5232.
8. AIAA Papers 90-5231, 90-5233; *Journal of Spacecraft and Rockets*, January-February 1999, p. 13.
9. Dwoyer et al., *Aerospace America*, October 1987, pp. 32-35 (quote, p. 35).
10. AIAA Papers 90-5232 (quote, p. 12), 92-5049 (quote, p. 2).
11. DTIC ADA-274530 (quote, p. i).
12. Heppenheimer, "Tractable"; Lin, *Statistical*, pp. 15-17, 43-45.
13. Lee et al., *Statistical*, pp. 78-83; Kuethe and Chow, *Foundations*, pp. 393-94; AIAA Paper 83-1693.
14. AIAA Paper 78-257. Quote: Author interview, Anthony Jameson, 4 June 1990. See *Mosaic*, Spring 1991, p. 30.
15. Quote: AIAA Paper 90-1480, p. 3.
16. Quote: AIAA Paper 83-1693, p. 1.
17. AIAA Paper 85-1652 (quotes, pp. 9, 12).
18. Shapiro, *Compressible*, p. 1085.
19. Quotes: AIAA Paper 90-5247, p. 1.
20. Author interviews: Anatol Roshko, 8 June 1990; Peter Bradshaw, 11 June 1990. See *Mosaic*, Spring 1991, p. 38.
21. Quotes: AIAA Paper 93-0200, p. 2. Author interviews: John Lumley, 5 June 1990; Anthony Jameson, 6 June 1990. See *Mosaic*, Spring 1991, pp. 32, 38.
22. Author interview, Anatol Roshko, 8 June 1990, as noted.
23. Author interview, William Dannevik, 11 June 1990. See *Mosaic*, Spring 1991, p. 38.
24. Author interview, Anatol Roshko, 8 June 1990, as noted.
25. DTIC ADA-274530. Quote: *Aerospace America*, October 1987, 35.
26. AIAA Paper 95-6055, p. 3 (includes quote).
27. Ibid.
28. 1989 NASP Technical Maturation Plan, task 2.1.02; *Air Power History*, Fall 2000, diagram, p. 26; DTIC ADB-191715, p. ES-4; ADB-968993, p. 24.
29. AIAA Paper 86-1680.
30. AIAA Paper 91-1944; author interview, Paul Czysz, 13 March 1986. Folder 18649, NASA Historical Reference Collection, NASA History Division, Washington, D.C. 20546.
31. DTIC ADB-191464, pp. 1, 10.
32. DTIC ADB-192311, task 1561.

33 *Aviation Week*, 3 April 1989, p. 58; Thompson, *Space Log*, Vol. 30 (1994), pp. 5-6.
34 *Aviation Week*, 11 May 1992, p. 54; AIAA Paper 92-5023; General Accounting Office Report NSIAD-92-5.
35 Author interview, Paul Czysz, 13 March 1986. Folder 18649, NASA Historical Reference Collection, NASA History Division, Washington, D.C. 20546.
36 DTIC AD-350952; also available as CASI 64X-17126.
37 AIAA Paper 92-3499; *Aviation Week*, 11 May 1992, p. 54.
38 DTIC ADB-191715 (quote, p. 3-22).
39 Escher, "Cryogenic," pp. 5-6; DTIC ADB-191715, p. 3-20.
40 AIAA Paper 91-1944; DTIC ADB-191715, p. 3-20.
41 Crickmore, *SR-71*, p. 95.
42 Ejector ramjets: AIAA Paper 86-1680, Fig. 3. Turbojets: *Aviation Week*: 3 May 1954, p. 44; 2 May 1960, p. 145.
43 AIAA Paper 89-0008 (quote, p. 1).
44 See also 1989 NASP Technology Maturation Plan, task 2.1.02.
45 DTIC ADB-197041, Attachment D.
46 Ibid.
47 Schweikart, *Quest*, p. 300; quote: DTIC ADB-197041, p. 2.
48 Author interview, Paul Czysz, 13 March 1986. Notes available, Folder 18649, NASA Historical Reference Collection, NASA History Division, Washington, D.C. 20546.
49 Heppenheimer, *High Technology*, September 1986, pp. 54-55; author interview, Philip Parrish, 21 March 1986. Folder 18649, NASA Historical Reference Collection, NASA History Division, Washington, D.C. 20546.
50 AIAA Paper 90-5207; DTIC ADB-192559; *Aviation Week*, 3 May 1993, p. 36.
51 AIAA Papers 91-5008, 93-2564.
52 AIAA Paper 93-2564 (quote, p. 8); *Aviation Week*, 3 February 1992, p. 52.
53 DTIC ADB-196085.
54 Brown, *Aerospace America*, August 1992, pp. 66-67.
55 AIAA Paper 90-5206; Preston, *American Steel*, pp. 8-9.
56 General Accounting Office Report NSIAD-93-71; DTIC ADB-192559.
57 AIAA Paper 89-5010.
58 DTIC ADB-191699, ADB-192559; ADB-190937, pp. 3-5.
59 DTIC ADB-191699 (quotes, p. 9).
60 DTIC ADB-191402, pp. 117-49.
61 DTIC ADB-191627.
62 DTIC ADB-191898; AIAA Paper 95-6131.
63 General Accounting Office Report NSIAD-93-71, table, p. 20.

9

Hypersonics After NASP

On 7 December 1995 the entry probe of the Galileo spacecraft plunged into the atmosphere of Jupiter. It did not plummet directly downward but sliced into that planet's hydrogen-rich envelope at a gentle angle as it followed a trajectory that took it close to Jupiter's edge. The probe entered at Mach 50, with its speed of 29.5 miles per second being four times that of a return to Earth from the Moon. Peak heating came to 11,800 BTU per square foot-second, corresponding to a radiative equilibrium temperature of 12,000°F. The heat load totaled 141,800 BTU per square foot, enough to boil 150 pounds of water for each square foot of heatshield surface.[1] The deceleration peaked at 228 g, which was tantamount to slamming from a speed of 5,000 miles per hour to a standstill in a single second. Yet the probe survived. It deployed a parachute and transmitted data from every one of its onboard instruments for nearly an hour, until it was overwhelmed within the depths of the atmosphere.[2]

It used an ablative heatshield, and as an exercise in re-entry technology, the design was straightforward. The nose cap was of chopped and molded carbon phenolic composite; the rest of the main heatshield was tape-wrapped carbon phenolic. The maximum thickness was 5.75 inches. The probe also mounted an aft heatshield, which was of phenolic nylon. The value of these simple materials, under the extreme conditions of Jupiter atmosphere entry, showed beyond doubt that the problem of re-entry was well in hand.[3]

Other activities have done less well. The X-33 and X-34 projects, which sought to build next-generation shuttles using NASP materials, failed utterly. Test scramjets have lately taken to flight but only infrequently. Still, work in CFD continues to flourish. Today's best supercomputers offer a million times more power than the ancestral Illiac 4, the top computer of the mid-1970s. This ushers in the important new topic of Large Eddy Simulation (LES). It may enable us to learn, via computation, just how good scramjets may become.

The DC-X, which flew, and the X-33, which did not. (NASA)

THE X-33 AND X-34

During the early 1990s, as NASP passed its peak of funding and began to falter, two new initiatives showed that there still was much continuing promise in rockets. The startup firm of Orbital Sciences Corporation had set out to become the first company to develop a launch vehicle as a commercial venture, and this rocket, called Pegasus, gained success on its first attempt. This occurred in April 1990, as NASA's B-52 took off from Edwards AFB and dropped it into flight. Its first stage mounted wings and tail surfaces. Its third stage carried a small satellite and placed it in orbit.[4]

In a separate effort, the Strategic Defense Initiative Office funded the DC-X project of McDonnell Douglas. This single-stage vehicle weighed some 40,000 pounds when fueled and flew with four RL10 rocket engines from Pratt & Whitney. It took off and landed vertically, like Flash Gordon's rocket ship, using rocket thrust during the descent and avoiding the need for a parachute. It went forward as an exercise in rapid prototyping, with the contract being awarded in August 1991 and the DC-X being rolled out in April 1993. It demonstrated both reusability and low cost, flying with a ground crew of only 15 people along with three more in its control center. It flew no higher than a few thousand feet, but it became the first rocket in history to abort a flight and execute a normal landing.[5]

The Clinton Administration came to Washington in January 1993. Dan Goldin, the NASA Administrator, soon chartered a major new study of launch options called Access to Space. Arnold Aldrich, Associate Administrator for Space Systems Development, served as its director. With NASP virtually on its deathbed, the work comprised three specific investigations. Each addressed a particular path toward a new generation of launch vehicles, which could include a new shuttle.

Managers at NASA Headquarters and at NASA-Johnson considered how upgrades to current expendables, and to the existing shuttle, might maintain them in service through the year 2030. At NASA-Marshall, a second group looked at prospects for new expendables that could replace existing rockets, including the shuttle, beginning in 2005. A collaboration between Headquarters and Marshall also considered a third approach: development of an entirely new reusable launch vehicle, to replace the shuttle and current expendables beginning in 2008.[6]

Engineers in industry were ready with ideas of their own. At Lockheed's famous Skunk Works, manager David Urie already had a concept for a fully-reusable single-stage vehicle that was to fly to orbit. It used a lifting-body configuration that drew on an in-house study of a vehicle to rescue crews from the space station. Urie's design was to be built as a hot structure with metal external panels for thermal protection and was to use high-performing rocket engines from Rocketdyne that would burn liquid hydrogen and liquid oxygen. This concept led to the X-33.[7]

Orbital Sciences was also stirring the pot. During the spring of 1993, this company conducted an internal study that examined prospects for a Pegasus follow-on. Pegasus used solid propellant in all three of its stages, but the new effort specifically considered the use of liquid propellants for higher performance. Its concept took shape as an air-launched two-stage vehicle, with the first stage being winged and fully reusable while the second stage, carried internally, was to fly to orbit without being recovered. Later that year executives of Orbital Sciences approached officials of NASA-Marshall to ask whether they might be interested, for this concept might complement that of Lockheed by lifting payloads of much lesser weight. This initiative led in time to the X-34.[8]

NASA's Access to Space report was in print in January 1994. Managers of the three option investigations had sought to make as persuasive a case as possible for their respective alternatives, and the view prevailed that technology soon would be in hand to adopt Lockheed's approach. In the words of the report summary,

> The study concluded that the most beneficial option is to develop and deploy a fully reusable single-stage-to-orbit (SSTO) pure-rocket launch

vehicle fleet incorporating advanced technologies, and to phase out current systems beginning in the 2008 time period....

The study determined that while the goal of achieving SSTO fully reusable rocket launch vehicles had existed for a long time, recent advances in technology made such a vehicle feasible and practical in the near term provided that necessary technologies were matured and demonstrated prior to start of vehicle development.[9]

Within weeks NASA followed with a new effort, the Advanced Launch Technology Program. It sought to lay technical groundwork for a next-generation shuttle, as it solicited initiatives from industry that were to pursue advances in structures, thermal protection, and propulsion.[10]

The Air Force had its own needs for access to space and had generally been more conservative than NASA. During the late 1970s, while that agency had been building the shuttle, the Air Force had pursued the Titan 34D as a new version of its Titan 3. More recently that service had gone forward with its upgraded Titan 4.[11] In May 1994 Lieutenant General Thomas Moorman, Vice Commander of the Air Force's Space Command, released his own study that was known as the Space Launch Modernization Plan. It considered a range of options that paralleled NASA's, including development of "a new reusable launch system." However, whereas NASA had embraced SSTO as its preferred direction, the Air Force study did not even mention this as a serious prospect. Nor did it recommend a selected choice of launch system. In a cover letter to the Deputy Secretary of Defense, John Deutch, Moorman wrote that "this study does not recommend a specific program approach" but was intended to "provide the Department of Defense a range of choices." Still, the report made a number of recommendations, one of which proved to carry particular weight: "Assign DOD the lead role in expendable launch vehicles and NASA the lead in reusables."[12]

The NASA and Air Force studies both went to the White House, where in August the Office of Science and Technology Policy issued a new National Space Transportation Policy. It divided the responsibilities for new launch systems in the manner that the Air Force had recommended and gave NASA the opportunity to pursue its own wishes as well:

> The Department of Defense (DoD) will be the lead agency for improvement and evolution of the current U.S. expendable launch vehicle (ELV) fleet, including appropriate technology development.
>
> The National Aeronautics and Space Administration (NASA) will provide for the improvement of the Space Shuttle system, focusing on reliability, safety, and cost-effectiveness.

> The National Aeronautics and Space Administration will be the lead agency for technology development and demonstration for next generation reusable space transportation systems, such as the single-stage-to-orbit concept.[13]

The Pentagon's assignment led to the Evolved Expendable Launch Vehicle Program, which brought development of the Delta 4 family and of new versions of the Atlas.[14]

The new policy broke with past procurement practices, whereby NASA had paid the full cost of the necessary research and development and had purchased flight vehicles under contract. Instead, the White House took the view that the private sector could cover these costs, developing the next space shuttle as if it were a new commercial airliner. NASA's role still was critical, but this was to be the longstanding role of building experimental flight craft to demonstrate pertinent technologies. The policy document made this clear:

> The objective of NASA's technology development and demonstration effort is to support government and private sector decisions by the end of this decade on development of an operational next generation reusable launch system.
>
> Research shall be focused on technologies to support a decision no later than December 1996 to proceed with a sub-scale flight demonstration which would prove the concept of single-stage-to-orbit....
>
> It is envisioned that the private sector could have a significant role in managing the development and operation of a new reusable space transportation system. In anticipation of this role, NASA shall actively involve the private sector in planning and evaluating its launch technology activities.[15]

This flight demonstrator became the X-33, with the smaller X-34 being part of the program as well. In mid-October NASA issued Cooperative Agreement Notices, which resembled requests for proposals, for the two projects. At a briefing to industry representatives held at NASA-Marshall on 19 October 1994, agency officials presented year-by-year projections of their spending plans. The X-33 was to receive $660 million in federal funds—later raised to $941 million—while the X-34 was slated for $70 million. Contractors were to add substantial amounts of their own and to cover the cost of overruns. Orbital Sciences was a potential bidder and held no contract, but its president, David Thompson, was well aware that he needed deeper pockets. He turned to Rockwell International and set up a partnership.[16]

The X-34 was the first to go to contract, as NASA selected the Orbital Sciences proposal in March 1995. Matching NASA's $70 million, this company and Rockwell each agreed to put up $60 million, which meant that the two corporations together were to provide more than 60 percent of the funding. Their partnership, called American Space Lines, anticipated developing an operational vehicle, the X-34B, that would carry 2,500 pounds to orbit. Weighing 108,500 pounds when fully fueled, it was to fly from NASA's Boeing 747 that served as the shuttle's carrier aircraft. Its length of 88 feet compared with 122 feet for the space shuttle orbiter.[17]

Very quickly an imbroglio developed over the choice of rocket engine for NASA's test craft. The contract called for use of a Russian engine, the Energomash RD-120 that was being marketed by Pratt & Whitney. Rockwell, which owned Rocketdyne, soon began demanding that its less powerful RS-27 engine be used instead. "The bottom line is Rockwell came in two weeks ago and said 'Use our engine or we'll walk,'" a knowledgeable industry observer told *Aviation Week*.[18]

As the issue remained unresolved, Orbital Sciences missed program milestone dates for airframe design and for selecting between configurations. Early in November NASA responded by handing Orbital a 14-day suspension notice. This led to further discussions, but even the personal involvement of Dan Goldin failed to resolve the matter. In addition, the X-34B concept had grown to as much as 140,000 pounds. Within the program, strong private-sector involvement meant that private-sector criteria of profitability were important, and Orbital determined that the new and heavy configuration carried substantial risk of financial loss. Early in 1996 company officials called for a complete redesign of NASA's X-34 that would substantially reduce its size. The agency responded by issuing a stop-work order. Rockwell then made its move by bailing out as well. With this, the X-34 appeared dead.

But it soon returned to life, as NASA prepared to launch it anew. It now was necessary to go back to square one and again ask for bids and proposals, and again Orbital Sciences was in the running, this time without a partner. The old X-34 had amounted to a prototype of the operational X-34B, approaching it in size and weight while also calling for use of NASA's Boeing 747. The company's new concept was only 58 feet long compared with 83; its gross weight was to be 45,000 pounds rather than 120,000. It was not to launch payloads into orbit but was to serve as a technology demonstrator for an eventual (and larger) first stage by flying to Mach 8. In June 1996 NASA selected Orbital again as the winner, choosing its proposal over competing concepts from such major players as McDonnell Douglas, Northrop Grumman, Rockwell, and the Lockheed Martin Skunk Works.[19]

Preparations for the X-33 had meanwhile been going forward as well. Design studies had been under way, with Lockheed Martin, Rockwell, and McDonnell Douglas as the competitors. In July 1996 Vice President Albert Gore announced that Lockheed had won the prize. This company envisioned a commercial SSTO craft named VentureStar as its eventual goal. It was to carry a payload of 59,000 pounds to low Earth orbit, topping the 51,000 pounds of the shuttle. Lockheed's X-33 amounted to a version of this vehicle built at 53 percent scale. It was to fly to

Mach 15, well short of orbital velocity, but would subject its thermal protection to a demanding test.[20]

No rocket craft of any type had ever flown to orbit as a single stage. NASA hoped that vehicles such as VentureStar not only would do this but would achieve low cost, cutting the cost of a pound in orbit from the $10,000 of the space shuttle to as little as $1,000.[21] The X-33 was to demonstrate the pertinent technology, which was being pursued under NASA's Advanced Launch Technology Program of 1994. Developments based on this program were to support the X-34 as well.

Lightweight structures were essential, particularly for the X-33. Accordingly, there was strong interest in graphite-composite tanks and primary structure. This represented a continuation of NASP activity, which had anticipated a main hydrogen tank of graphite-epoxy. The DC-X supported the new work, as NASA took it over and renamed it the DC-XA. Its oxygen tank had been aluminum; a new one, built in Russia, used an aluminum-lithium alloy. Its hydrogen tank, also of aluminum, gave way to one of graphite-epoxy with lightweight foam for internal insulation. This material also served for an intertank structure and a feedline and valve assembly.[22]

Rapid turnaround offered a particularly promising road to low launch costs, and the revamped DC-XA gave support in this area as well. Two launches, conducted in June 1996, demonstrated turnaround and reflight in only 26 hours, again with its ground crew of only 15.[23]

Thermal protection raised additional issues. The X-34 was to fly only to Mach 8 and drew on space shuttle technology. Its surface was to be protected with insulation blankets that resembled those in use on the shuttle orbiter. These included the High Heat Blanket for the X-34 undersurface, rated for 2,000°F, with a Nextel 440 fabric and Saffil batting. The nose cap as well as the wing and rudder leading edges were protected with Fibrous Refractory Composite Insulation, which formed the black silica tiles of the shuttle orbiter. For the X-34, these tiles were to be impregnated with silicone to make them water resistant, impermeable to flows of hot gas, and easier to repair.[24]

VentureStar faced the demands of entry from orbit, but its re-entry environment was to be more benign than that of the shuttle. The shuttle orbiter was compact in size and relatively heavy and lost little of its orbital energy until well into the atmosphere. By contrast, VentureStar would resemble a big lightweight balloon when it re-entered after expending its propellants. The VentureStar thermal protection system was to be tested in flight on the X-33. It had the form of a hot structure, with radiative surface panels of carbon-carbon, Inconel 617 nickel alloy, and titanium, depending on the temperature.[25]

In an effort separate from that of the X-33, elements of this thermal protection were given a workout by being mounted to the space shuttle *Endeavour* and tested during re-entry. Thoughts of such tests dated to 1981 and finally were realized during Mission STS-77 in May 1996. Panels of Inconel 617 and of Ti-1100 titanium, measuring 7 by 10 inches, were mounted in recessed areas of the fuselage that lay near the vertical tail and which were heated only to approximately 1,000°F during re-entry. Both materials were rated for considerably higher temperatures, but this successful demonstration put one more arrow in NASA's quiver.[26]

For both VentureStar and its supporting X-33, light weight was critical. The X-30 of NASP had been designed for SSTO operation, with a structural mass fraction—the ratio of unfueled weight to fully fueled weight—of 25 percent.[27] This requirement was difficult to achieve because most of the fuel was slush hydrogen, which has a very low density. This ballooned the size of the X-30 and increased the surface area that needed structural support and thermal protection. VentureStar was to use rockets, which had less performance than scramjets. It therefore needed more fuel, and its structural mass fraction, including payload, engines, and thermal protection, was less than 12 percent. However, this fuel included a great deal of liquid oxygen, which was denser than water and drove up the weight of the propellant. This low structural mass fraction therefore appeared within reach, and for the X-33, the required value was considerably less stringent. Its design called for an empty weight of 63,000 pounds and a loaded weight of 273,000, for a structural mass fraction of 23 percent.[28]

Even this design goal imposed demands, for while liquid oxygen was dense and compact, liquid hydrogen still was bulky and again enlarged the surface area. Designers thus made extensive use of lightweight composites, specifying graphite-epoxy for the hydrogen tanks. A similar material, graphite-bismaleimide, was to serve for load-bearing trusses as well as for the outer shell that was to support the thermal protection. This represented the X-30's road not taken, for the NASP thermal environment during ascent had been so severe that its design had demanded a primary structure of titanium-matrix composite, which was heavier. The lessened requirements of VentureStar's thermal protection meant that Lockheed could propose to reach orbit using materials that were considerably less heavy—that indeed were lighter than aluminum. The X-33 design saved additional weight because it was to be unpiloted, needing no flight deck and no life-support system for a crew.[29]

But aircraft often gain weight during development, and the X-33 was no exception. Starting in mid-1996 with a dry weight of 63,000 pounds, it was at 80,000 a year later, although a weight-reduction exercise trimmed this to 73,000.[30] Managers responded by cutting the planned top speed from Mach 15 or more to Mach 13.8. Jerry Rising, vice president at the Skunk Works that was the X-33's home, explained that such a top speed still would permit validation of the thermal protection in flight test. The craft would lift off from Edwards AFB and follow a boost-glide trajectory, reaching a peak altitude of 300,000 feet. The vehicle then would be lower in the atmosphere than previously planned, and the heating rate would consequently be higher to properly exercise the thermal protection. The X-33 then was to glide onward to a landing at Malmstrom AFB in northern Montana, 950 miles from Edwards.[31]

The original program plan called for rollout of a complete flight vehicle on 1 November 1998. When that date arrived, though, the effort faced a five-month schedule slip. This resulted from difficulties with the rocket engines.[32] Then in December, two days before Christmas, the program received a highly unwelcome present. A hydrogen fuel tank, under construction at a Lockheed Martin facility in Sunnyvale, California, sustained major damage within an autoclave. An inner wall of the tank showed delamination over 90 percent of its area, while another wall sprang loose from its frame. The tank had been inspected using ultrasound, but this failed to disclose the incipient problem, which raised questions as to the adequacy of inspection procedures as well as of the tank design itself. Another delay was at hand of up to seven months.

By May 1999 the weight at main engine cutoff was up to 83,000 pounds, including unburned residual propellant. Cleon Lacefield, the Lockheed Martin program manager, continued to insist bravely that the vehicle would reach at least Mach 13, but working engineers told *Aviation Week* that the top speed had been Mach 10 for quite some time and that "the only way it's getting to Malmstrom is on the back of a truck."[33] The commercial VentureStar concept threatened to be far more demanding, and during that month Peter Teets, president and CEO of Lockheed Martin, told the U.S. Senate Commerce and Science Committee that he could not expect to attract the necessary private-sector financing. "Wall Street has spoken," he declared. "They have picked the status quo; they will finance systems with existing technology. They will not finance VentureStar."[34]

By then the VentureStar design had gone over to aluminum tanks. These were heavier than tanks of graphite-epoxy, but the latter brought unacceptable technical risks because no autoclave existed that was big enough to fabricate such tankage. Lockheed Martin designers reshaped VentureStar and accepted a weight increase from 2.6 million pounds to 3.3 million. (It had been 2.2 million in 1996.) The use of graphite-epoxy in the X-33 tank now no longer was relevant to VentureStar, but this was what the program held in hand, and a change to aluminum would have added still more weight to the X-33.

During 1999 a second graphite-epoxy hydrogen tank was successfully assembled at Lockheed Martin and then was shipped to NASA-Marshall for structural tests. Early in November it experienced its own failure, showing delamination and a ripped outer skin along with several fractures or breaks in the skin. Engineers had been concerned for months about structural weakness, with one knowledgeable specialist telling *Aviation Week*, "That tank belonged in a junkyard, not a test stand." The program now was well on its way to becoming an orphan. It was not beloved by NASA, which refused to increase its share of funding above $941 million, while the in-house cost at Lockheed Martin was mounting steadily.[35]

The X-33 effort nevertheless lingered through the year 2000. This was an election year, not a good time to cancel a billion-dollar federal program, and Al Gore was running for president. He had announced the contract award in 1996, and in the words of a congressional staffer, "I think NASA will have a hard time walking away from the X-33 until after the election. For better or worse, Al Gore now has ownership of it. They can't admit it's a failure."[36]

The X-34 was still in the picture, as a substantial effort in its own right. Its loaded weight of 47,000 pounds approached the 56,000 of the X-15 with external tanks, built more than 30 years earlier.[37] Yet despite this reduced weight, the X-34 was to reach Mach 8, substantially exceeding the Mach 6.7 of the X-15. This reflected the use of advanced materials, for whereas the X-15 had been built of heavy Inconel X, the X-34 design specified lightweight composites for the primary structure and fuel tank, along with aluminum for the liquid-oxygen tank.[38]

Its construction went forward without major mishaps because it was much smaller than the X-33. The first of them reached completion in February 1999, but during the next two years it never came close to powered flight. The reason was that the X-34 program called for use of an entirely new engine, the 60,000-pound-thrust Fastrak of NASA-Marshall that burned liquid oxygen and kerosene. This engine encountered development problems, and because it was not ready, the X-34 could not fly under power.[39]

Early in March 2001, with George W. Bush in the White House, NASA pulled the plug. Arthur Stephenson, director of NASA-Marshall, canceled the X-34. This reflected the influence of the Strategic Defense Initiative Office, which had maintained a continuing interest in low-cost access to orbit and had determined that the X-34's costs outweighed the benefits. Stephenson also announced that the cooperative agreement between NASA and Lockheed Martin, which had supported the X-33, would expire at the end of the month. He then pronounced an epitaph on both programs: "One of the things we have learned is that our technology has not yet advanced to the point that we can successfully develop a new reusable launch vehicle that substantially improves safety, reliability, and affordability."[40]

One could say that the X-30 effort went farther than the X-33, for the former successfully exercised a complete hydrogen tank within its NIFTA project, whereas the latter did not. But the NIFTA tank was subscale, whereas those of the X-33 were full-size units intended for flight. The reason that NIFTA appears to have done better is that NASP never got far enough to build and test a full-size tank for its hydrogen slush. Because that tank also was to have been of graphite-epoxy, as with the X-33, it is highly plausible that the X-30 would have run aground on the same shoal of composite-tank structural failure that sank Lockheed Martin's rocket craft.[41]

Hypersonics After NASP

Scramjets Take Flight

On 28 November 1991 a Soviet engine flew atop an SA-5 surface-to-air missile in an attempt to demonstrate supersonic combustion. The flight was launched from the Baikonur center in Kazakhstan and proceeded ballistically, covering some 112 miles. The engine did not produce propulsive thrust but rode the missile while mounted to its nose. The design had an axisymmetric configuration, resembling that of NASA's Hypersonic Research Engine, and the hardware had been built at Moscow's Central Institute of Aviation Motors (CIAM).

As described by Donat Ogorodnikov, the center director, the engine performed two preprogrammed burns during the flight. The first sought to demonstrate the important function of transition from subsonic to supersonic combustion. It was initiated at 59,000 feet and Mach 3.5, as the rocket continued to accelerate. Ogorodnikov asserted that after fifteen seconds, near Mach 5, the engine went over to supersonic combustion and operated in this mode for five seconds, while the rocket accelerated to Mach 6 at 92,000 feet. Within the combustor, internal flow reached a measured speed of Mach 3. Pressures within the combustor were one to two atmospheres.

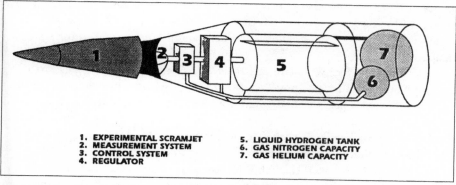

Russian flight-test scramjet. (Aviation Week and Space Technology)

The second engine burn lasted ten seconds. This one had the purpose of verifying the design of the engine's ignition system. It took place on the downward leg of the trajectory, as the vehicle descended from 72,000 feet and Mach 4.5 to 59,000 feet and Mach 3.5. This burn involved only subsonic combustion. Vyacheslav Vinogradov, chief of engine gasdynamics at CIAM, described the engine as mounting three rows of fuel injectors. Choice of an injector row, out of the three available, was to help in changing the combustion mode.

The engine diameter at the inlet was 9.1 inches; its length was 4.2 feet. The spike, inlet, and combustor were of stainless steel, with the spike tip and cowl leading edge being fabricated using powder metallurgy. The fuel was liquid hydrogen, and the system used no turbopump. Pressure, within a fuel tank that also was stainless steel, forced the hydrogen to flow. The combustor was regeneratively cooled; this vaporized the hydrogen, which flowed through a regulator at rates that varied from 0.33 pounds per second in low-Mach flight to 0.11 at high Mach.[42]

The Russians made these extensive disclosures because they hoped for financial support from the West. They obtained initial assistance from France and conducted a second flight test a year later. The engine was slightly smaller and the trajectory was flatter, reaching 85,000 feet. It ignited near Mach 3.5 and sustained subsonic combustion for several seconds while the rocket accelerated to Mach 5. The engine then transitioned to supersonic combustion and remained in this mode for some fifteen seconds, while acceleration continued to Mach 5.5. Burning then terminated due to exhaustion of the fuel.[43]

On its face, this program had built a flightworthy scramjet, had achieved a supersonic internal airflow, and had burned hydrogen within this flow. Even so, this was not necessarily the same as accomplishing supersonic combustion. The alleged transition occurred near Mach 5, which definitely was at the low end for a scramjet.[44] In addition, there are a number of ways whereby pockets of subsonic flow might have existed within an internal airstream that was supersonic overall. These could have served as flameholders, localized regions where conditions for combustion were particularly favorable.[45]

In 1994 CIAM received a contract from NASA, with NASA-Langley providing technical support. The goal now was Mach 6.5, at which supersonic combustion appeared to hold a particularly strong prospect. The original Russian designs had been rated for Mach 6 and were modified to accommodate the higher heat loads at this higher speed. The flight took place in February 1998 and reached Mach 6.4 at 70,000 feet, with the engine operating for 77 seconds.[46]

It began operation near Mach 3.5. Almost immediately the inlet unstarted due to excessive fuel injection. An onboard control system detected the unstart and reduced the fuel flow, which enabled the inlet to start and to remain started. However, the onboard control failed to detect this restart and failed to permit fuel to flow through the first of the three rows of fuel injectors. Moreover, the inlet performance fell short of predictions due to problems in fabrication.

At Mach 5.5 and higher, airflow entered the fuel-air mixing zone within the combustor at speeds near Mach 2. However, only the two rear rows of injectors were active, and burning of their fuel forced the internal Mach number to subsonic values. The flow reaccelerated to sonic velocity at the combustor exit. The combination of degraded inlet performance and use of only the rear fuel injectors ensured that even at the highest flight speeds, the engine operated primarily in a subsonic-combustion mode and showed little if any supersonic combustion.[47]

Hypersonics After NASP

It nevertheless was clear that with better quality control in manufacturing and with better fault tolerance in the onboard control laws, full success might readily be achieved. However, the CIAM design was axisymmetric and hence was of a type that NASA had abandoned during the early 1970s. Such scramjets had played no role in NASP, which from the start had focused on airframe-integrated configurations. The CIAM project had represented an existing effort that was in a position to benefit from even the most modest of allocations; the 1992 flight, for instance, received as little as $200,000 from France.[48] But NASA had its eye on a completely American scramjet project that could build on the work of NASP. It took the name Hyper-X and later X-43A.

Its background lay in a 1995 study conducted by McDonnell Douglas, with Pratt & Whitney providing concepts for propulsion. This effort, the Dual-Fuel Airbreathing Hypersonic Vehicle Study, gave conceptual designs for vehicles that could perform two significant missions: weapons delivery and reconnaissance, and operation as the airbreathing first stage of a two-stage-to-orbit launch system. This work drew interest at NASA Headquarters and led the Hypersonic Vehicles Office at NASA-Langley to commission the conceptual design of an experimental airplane that could demonstrate critical technologies required for the mission vehicles.

The Hyper-X design grew out of a concept for a Mach 10 cruise aircraft with length of 200 feet and range of 8,500 nautical miles. It broke with the NASP approach of seeking a highly integrated propulsion package that used an ejector ramLACE as a low-speed system. Instead it returned to the more conservative path of installing separate types of engine. Hydrocarbon-fueled turboramjets were to serve for takeoff, acceleration to Mach 4, and subsonic cruise and landing. Hydrogen-burning scramjets were to take the vehicle to Mach 10. The shape of this vehicle defined that of Hyper-X, which was designed as a detailed scale model that was 12 feet long rather than 200.[49]

Like the Russian engines, Hyper-X was to fly to its test Mach using a rocket booster. But Hyper-X was to advance beyond the Russian accomplishments by separating from this booster to execute free flight. This separation maneuver proved to be trickier than it looked. Subsonic bombers had been dropping rocket planes into flight since the heyday of Chuck Yeager, and rocket stages had separated in near-vacuum at the high velocities of a lunar mission. However, Hyper-X was to separate at speeds as high as Mach 10 and at 100,000 feet, which imposed strong forces from the airflow. As the project manager David Reubush wrote in 1999, "To the program's knowledge there has never been a successful separation of two vehicles (let alone a separation of two non-axisymmetric vehicles) at these conditions. Therefore, it soon became obvious that the greatest challenge for the Hyper-X program was, not the design of an efficient scramjet engine, but the development of a separation scenario and the mechanism to achieve it."[50]

Engineers at Sandia National Laboratory addressed this issue. They initially envisioned that the rocket might boost Hyper-X to high altitude, with the separation taking place in near-vacuum. The vehicle then could re-enter and light its scramjet. This approach fell by the wayside when the heat load at Mach 10 proved to exceed the capabilities of the thermal protection system. The next concept called for Hyper-X to ride the underside of its rocket and to be ejected downward as if it were a bomb. But this vehicle then would pass through the bow shock of the rocket and would face destabilizing forces that its control system could not counter.

Sandia's third suggestion called for holding the vehicle at the front of the rocket using a hinged adapter resembling a clamshell or a pair of alligator jaws. Pyrotechnics would blow the jaws open, releasing the craft into flight. The open jaws then were to serve as drag brakes, slowing the empty rocket casing while the flight vehicle sailed onward. The main problem was that if the vehicle rolled during separation, one of its wings might strike this adapter as it opened. Designers then turned to an adapter that would swing down as a single piece. This came to be known as the "drop-jaw," and it served as the baseline approach for a time.[51]

NASA announced the Hyper-X Program in October 1996, citing a budget of $170 million. In February 1997 Orbital Sciences won a contract to provide the rocket, which again was to be a Pegasus. A month later the firm of Micro Craft Inc. won the contract for the Hyper-X vehicle, with GASL building the engine. Work at GASL went forward rapidly, with that company delivering a scramjet to NASA-Langley in August 1998. NASA officials marked the occasion by changing the name of the flight aircraft to X-43A.[52]

The issue of separation in flight proved not to be settled, however, and developments early in 1999 led to abandonment of the drop-jaw. This adapter extended forward of the end of the vehicle, and there was concern that while opening it would form shock waves that would produce increased pressures on the rear underside of the flight craft, which again could overtax its control system. Wind-tunnel tests showed that this indeed was the case, and a new separation mechanism again was necessary. This arrangement called for holding the X-43A in position with explosive bolts. When they were fired, separate pyrotechnics were to actuate pistons that would push this craft forward, giving it a relative speed of at least 13 feet per second. Further studies and experiments showed that this concept indeed was suitable.[53]

The minimal size of the X-43A meant that there was little need to keep its weight down, and it came in at 2,800 pounds. This included 900 pounds of tungsten at the nose to provide ballast for stability in flight while also serving as a heat sink. High stiffness of the vehicle was essential to prevent oscillations of the structure that could interfere with the Pegasus flight control system. The X-43A thus was built with steel longerons and with steel skins having thickness of one-fourth inch. The wings were stubby and resembled horizontal stabilizers; they did not mount ailerons but moved as a whole to provide sufficient control authority. The wings and tail surfaces were constructed of temperature-resistant Haynes 230 alloy. Leading edges of the nose, vertical fins, and wings used carbon-carbon. For thermal protection, the vehicle was covered with Alumina Enhanced Thermal Barrier tiles, which resembled the tiles of the space shuttle.[54]

Additional weight came from the scramjet. It was fabricated of a copper alloy called Glidcop, which was strengthened with very fine particles of aluminum oxide dispersed within. This increased its strength at high temperatures, while retaining the excellent thermal conductivity of copper. This alloy formed the external surface, sidewalls, cowl, and fuel injectors. Some internal surfaces were coated with zirconia to form a thermal barrier that protected the Glidcop in areas of high heating. The engine did not use its hydrogen fuel as a coolant but relied on water cooling for the sidewalls and cowl leading edge. Internal engine seals used braided ceramic rope.[55]

Because the X-43A was small, its engine tests were particularly realistic. This vehicle amounted to a scale model of a much larger operational craft of the future, but the engine testing involved ground-test models that were full size for the X-43A. Most of the testing took place at NASA-Langley, where the two initial series were conducted at the Arc-Heated Scramjet Test Facility. This wind tunnel was described in 1998 as "the primary Mach 7 scramjet test facility at Langley."[56]

Development tests began at the very outset of the Hyper-X Program. The first test article was the Dual-Fuel Experiment (DFX), with a name that reflected links to the original McDonnell Douglas study. The DFX was built in 1996 by modifying existing NASP engine hardware. It provided a test scramjet that could be modified rapidly and inexpensively for evaluation of changes to the flowpath. It was fabricated primarily of copper and used no active cooling, relying on heat sink. This ruled out tests at the full air density of a flight at Mach 7, which would have overheated this engine too quickly for it to give useful data. Even so, tests at reduced air densities gave valuable guidance in designing the flight engine.

The DFX reproduced the full-scale height and length of the Hyper-X engine, correctly replicating details of the forebody, cowl, and sidewall leading edge. The forebody and afterbody were truncated, and the engine width was reduced to 44 percent of the true value so that this test engine could fit with adequate clearances in the test facility. This effort conducted more than 250 tests of the DFX, in four different configurations. They verified predicted engine forces and moments as well as inlet and combustor component performances. Other results gave data on ignition requirements, flameholding, and combustor-inlet interactions.

Within that same facility, subsequent tests used the Hyper-X Engine Module (HXEM). It resembled the DFX, including the truncations fore and aft, and it too was of reduced width. But it replicated the design of the flight engine, thereby overcoming limitations of the DFX. The HXEM incorporated the active cooling of the flight version, which opened the door to tests at Mach 7 and at full air density. These took place within the large Eight-Foot High Temperature Tunnel (HTT).

The HTT had a test section that was long enough to accommodate the full 12-foot length of the X-43A underside, which provided major elements of the inlet and nozzle with its airframe-integrated forebody and afterbody. This replica of the underside initially was tested with the HXEM, thereby giving insight into the aerodynamic effects of the truncations. Subsequent work continued to use the HTT and replaced the HXEM with the full-width Hyper-X Flight Engine (HXFE). This was a flight-spare Mach 7 scramjet that had been assigned for use in ground testing.

Mounted on its undersurface, this configuration gave a geometrically accurate nose-to-tail X-43A propulsion flowpath at full scale. NASA-Langley had conducted previous tests of airframe-integrated scramjets, but this was the first to replicate the size and specific details of the propulsion system of a flight vehicle. The HTT heated its air by burning methane, which added large quantities of carbon dioxide and water vapor to the test gas. But it reproduced the Mach, air density, pressure, and temperature of flight at altitude, while gaseous oxygen, added to the airflow, enabled the engine to burn hydrogen fuel. Never before had so realistic a test series been accomplished.[57]

The thrust of the engine was classified, but as early as 1997 Vince Rausch, the Hyper-X manager at NASA-Langley, declared that it was the best-performing scramjet that had been tested at his center. Its design called for use of a cowl door that was to protect the engine by remaining closed during the rocket-powered ascent, with this door opening to start the inlet. The high fidelity of the HXFE, and of the test conditions, gave confidence that its mechanism would work in flight. The tests in the HTT included 14 unfueled runs and 40 with fuel. This cowl door was actuated 52 times under the Mach 7 test conditions, and it worked successfully every time.[58]

Aerodynamic wind-tunnel investigations complemented the propulsion tests and addressed a number of issues. The overall program covered all phases of the flight trajectory, using 15 models in nine wind tunnels. Configuration development alone demanded more than 5,800 wind-tunnel runs. The Pegasus rocket called for evaluation of its own aerodynamic characteristics when mated with the X-43A, and these had to be assessed from the moment of being dropped from the B-52 to separation of the flight vehicle. These used the Lockheed Martin Vought High Speed Wind Tunnel in Grand Prairie, Texas, along with facilities at NASA-Langley that operated at transonic as well as hypersonic speeds.[59]

Much work involved evaluating stability, control, and performance characteristics of the basic X-43A airframe. This effort used wind tunnels of McDonnell Douglas and Rockwell, with the latter being subsonic. At NASA-Langley, activity focused on that center's 20-inch Mach 6 and 31-inch Mach 10 facilities. The test models were only one foot in length, but they incorporated movable rudders and wings. Eighteen-inch models followed, which were as large as these tunnels could accommodate, and gave finer increments of the control-surface deflections. Thirty-inch models brought additional realism and underwent supersonic and transonic tests in the Unitary Plan Wind Tunnel and the 16-Foot Transonic Tunnel.[60]

Similar studies evaluated the methods proposed for separation of the X-43A from its Pegasus booster. Initial tests used Langley's Mach 6 and Mach 10 tunnels. These were blowdown facilities that did not give long run times, while their test sections were too small to permit complete representations of vehicle maneuvers during separation. But after the drop-jaw concept had been selected, testing moved to tunnel B of the Von Karman Facility at the Arnold Engineering Development Center. This wind tunnel operated with continuous flow, in contrast to the blowdown installations of Langley, and provided a 50-inch-diameter test section for use at Mach 6. It was costly to test in that tunnel but highly productive, and it accommodated models that demonstrated a full range of relative orientations of Pegasus and the X-43A during separation.[61]

This wind-tunnel work also contributed to inlet development. To enhance overall engine performance, it was necessary for the boundary layer upstream of this inlet to be turbulent. Natural transition to turbulence could not be counted on, which meant that an aerodynamic device of some type was needed to trip the boundary layer into turbulence. The resulting investigations ran from 1997 into 1999 and used both the Mach 6 and Mach 10 Langley wind tunnels, executing more than 300 runs. Hypulse, a shock tunnel at GASL, conducted more than two dozen additional tests.[62]

Computational fluid dynamics was used extensively. The wind-tunnel tests that supported studies of X-43A separation all were steady-flow experiments, which failed to address issues such as unsteady flow in the gap between the two vehicles as they moved apart. CFD dealt with this topic. Other CFD analyses examined relative orientations of the separating vehicles that were not studied at AEDC. To scale wind-tunnel results for use with flight vehicles, CFD solutions were generated both for the small models under wind-tunnel conditions and for full-size vehicles in flight.[63]

Flight testing was to be conducted at NASA-Dryden. The first X-43A flight vehicle arrived there in October 1999, with its Pegasus booster following in December. Tests of this Pegasus were completed in May 2000, with the flight being attempted a year later. The plan called for acceleration to Mach 7 at 95,000 feet, followed by 10 seconds of powered scramjet operation. This brief time reflected the fact that the engine was uncooled and relied on copper heat sink, but it was long enough to take data and transmit them to the ground. In the words of NASA manager Lawrence Huebner, "we have ground data, we have ground CFD, we have flight CFD—all we need is the flight data."[64]

Launch finally occurred in June 2001. Ordinarily, when flying to orbit, Pegasus was air-dropped at 38,000 feet, and its first stage flew to 207,000 feet prior to second-stage ignition. It used solid propellant and its performance could not readily be altered; therefore, to reduce its peak altitude to the 95,000 feet of the X-43A, it was to be air-dropped at 24,000 feet, even though this lower altitude imposed greater loads.

The B-52 took off from Edwards AFB and headed over the Pacific. The Pegasus fell away; its first stage ignited five seconds later and it flew normally for some eight seconds that followed. During those seconds, it initiated a pullout to begin its climb. Then one of its elevons came off, followed almost immediately by another. As additional parts fell away, this booster went out of control. It fell tumbling toward the ocean, its rocket motor still firing, and a safety officer sent a destruct signal. The X-43A never had a chance to fly, for it never came close to launch conditions.[65]

A year later, while NASA was trying to recoup, a small group in Australia beat the Yankees to the punch by becoming the first in the world to fly a scramjet and achieve supersonic combustion. Their project, called HyShot, cost under $2 million, compared with $185 million for the X-43A program. Yet it had plenty of technical sophistication, including tests in a shock tunnel and CFD simulations using a supercomputer.

Allan Paull, a University of Queensland researcher, was the man who put it together. He took a graduate degree in applied mathematics in 1985 and began working at that university with Ray Stalker, an engineer who had won a global reputation by building a succession of shock tunnels. A few years later Stalker suffered a stroke, and Paull found himself in charge of the program. Then opportunity came knocking, in the form of a Florida-based company called Astrotech Space Operations. That firm was building sounding rockets and wanted to expand its activities into the Asia and Pacific regions.

In 1998 the two parties signed an agreement. Astrotech would provide two Terrier-Orion sounding rockets; Paull and his colleagues would construct experimental scramjets that would ride those rockets. The eventual scramjet design was not airframe-integrated, like that of the X-43A. It was a podded axisymmetric configuration. But it was built in two halves, with one part being fueled with hydrogen while the other part ran unfueled for comparison.[66]

Paull put together a team of four people—and found that the worst of his problems was what he called an "amazing legal nightmare" that ate up half his time. In the words of the magazine *Air & Space*, "the team had to secure authorizations from various state government agencies, coordinate with aviation bodies and insurance companies in both Australia and the United States (because of the involvement of U.S. funding), perform environmental assessments, and ensure their launch debris would steer clear of land claimed by Aboriginal tribes.... All told, the preparations took three and a half years."[67]

HYPERSONICS AFTER NASP

The flight plan called for each Terrier-Orion to accelerate its scramjet onto a ballistic trajectory that was to reach an altitude exceeding 300 kilometers. Near the peak of this flight path, an attitude-control system was to point the rocket downward. Once it re-entered the atmosphere, below 40 kilometers, its speed would fall off and the scramjet would ignite. This engine was to operate while continuing to plunge downward, covering distance into an increasingly dense atmosphere, until it lost speed in the lower atmosphere and crashed into the outback.

The flights took place at Woomera Instrumented Range, north of Adelaide. The first launch attempt came at the end of October 2001. It flopped; the first stage performed well, but the second stage went off course. But nine months later, on 30 July 2002, the second shot gained full success. The rocket was canted slightly away from the vertical as it leaped into the air, accelerating at 22 g as it reached Mach 3.6 in only six seconds.

This left it still at low altitude while topping the speed of the SR-71, so after the second stage with payload separated, it coasted for 16 seconds while continuing to ascend. The second stage then ignited, and this time its course was true. It reached a peak speed of Mach 7.7. The scramjet went over the top; it pointed its nose downward, and at an altitude of 36 kilometers with its speed approaching Mach 7.8, gaseous hydrogen caused it to begin producing thrust. This continued until HyShot reached 25 kilometers, when it shut down.

It fired for only five seconds. But it returned data over 40 channels, most of which gave pressure readings. NASA itself provided support, with Lawrence Huebner, the X-43A manager, declaring, "We're very hungry for flight data." For the moment, at least, the Aussies were in the lead.[68]

But the firm of Micro Craft had built two more X-43As, and the second flight took place in March 2004. This time the Pegasus first stage had been modified by having part of its propellant removed, to reduce its performance, and the drop altitude was considerably higher.[69] In the words of *Aviation Week*,

> The B-52B released the 37,500-lb. stack at 40,000 ft. and the Pegasus booster ignited 5 sec. later.... After a few seconds it pulled up and reached a maximum dynamic pressure of 1,650 psf. at Mach 3.5 climbing through 47,000 ft. Above 65,000 ft. it started to push over to a negative angle of attack to kill the climb rate and gain more speed. Burnout was 84 sec. after drop, and at 95 sec. a pair of pistons pushed the X-43A away from the booster at a target condition of Mach 7 and 95,000 ft. and a dynamic pressure of 1,060 psf. in a slight climb before the top of a ballistic arc.
>
> After a brief period of stabilization, the X-43A inlet door was opened to let air in through the engine.... The X-43A stabilized again because the engine airflow changed the trim.... Then silane, a chemical that burns upon contact with air, was injected for 3 sec. to establish flame to ignite the

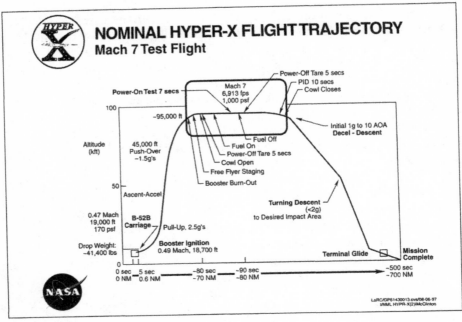

X-43A mission to Mach 7. (NASA)

hydrogen. Injection of the gaseous hydrogen fuel ramped up as the silane ramped down, lasting 8 sec. The hydrogen flow rate increased through and beyond a stoichiometric mixture ratio, and then ramped down to a very lean ratio that continued to burn until the fuel was shut off.... The hydrogen was stored in 8,000-psi bottles.

Accelerometers showed the X-43A gained speed while fuel was on.... Data was gathered all the way to the splashdown 450 naut. mi. offshore at about 11 min. after drop.

Aviation Week added that the vehicle accelerated "while in a slight climb at Mach 7 and 100,000 ft. altitude. The scramjet field is sufficiently challenging that producing thrust greater than drag on an integrated airframe/engine is considered a major accomplishment."[70]

In this fashion, NASA executed its first successful flight of a scramjet. The overall accomplishment was not nearly as ambitious as that planned for the Incremental Flight Test Vehicle of the 1960s, for which the velocity increase was to have been much greater. Nor did NASA have a follow-on program in view that could draw on the results of the X-43A. Still, the agency now could add the scramjet to its list of flight engines that had been successfully demonstrated.

HYPERSONICS AFTER NASP

The program still had one unexpended X-43A vehicle that was ready to fly, and it flew successfully as well, in November. The goal now was Mach 10. This called for beefing up the thermal structure by adding leading edges of solid carbon-carbon to the vertical tails along with a coating of hafnium carbide and by making the nose blunter to increase the detachment of the bow shock. These changes indeed were necessary. Nose temperatures reached 3,600°F, compared with 2,600°F on the Mach 7 flight, and heating rates were twice as high.

The Pegasus rocket, with the X-43A at its front, fell away from its B-52 carrier aircraft at 40,000 feet. Its solid rocket took the combination to Mach 10 at 110,000 feet. Several seconds after burnout, pistons pushed the X-43A away at Mach 9.8. Then, 2.5 seconds after separation, the engine inlet door opened and the engine began firing at Mach 9.65. It ran initially with silane to ensure ignition; then the engine continued to operate with silane off, for comparison. It fired for a total of 10 to 12 seconds and then continued to operate with the fuel off. Twenty-one seconds after separation, the inlet door closed and the vehicle entered a hypersonic glide. This continued for 14 minutes, with the craft returning data by telemetry until it struck the Pacific Ocean and sank.

This flight gave a rare look at data taken under conditions that could not be duplicated on the ground using continuous-flow wind tunnels. The X-43A had indeed been studied in 0.005-second runs within shock tunnels, and *Aviation Week* noted that Robert Bakos, vice president of GASL, described such tests as having done "a very good job of predicting the flight." Dynamic pressure during the flight was 1,050 pounds per square foot, and the thrust approximately equaled the drag. In addition, the engine achieved true supersonic combustion, without internal pockets of subsonic flow. This meant that the observations could be scaled to still higher Mach values.[71]

RECENT ADVANCES IN FLUID MECHANICS

The methods of this field include ground test, flight test, and CFD. Ground-test facilities continue to show their limitations, with no improvements presently in view that would advance the realism of tests beyond Mach 10. A recently announced Air Force project, Mariah, merely underscores this point. This installation, to be built at AEDC, is to produce flows up to Mach 15 that are to run for as long as 10 seconds, in contrast to the milliseconds of shock tunnels. Mariah calls for a powerful electron beam to create an electrically charged airflow that can be accelerated with magnets. But this installation will require an e-beam of 200 megawatts. This is well beyond the state of the art, and even with support from a planned research program, Mariah is not expected to enter service until 2015.[72]

Similar slow progress is evident in CFD, for which the flow codes of recent projects have amounted merely to updates of those used in NASP. In designing the X-43A, the most important such code was the General Aerodynamic Simulation Program (GASP). NASP had used version 2.0; the X-43A used 3.0. The latter continued to incorporate turbulence models. Results from the codes often showed good agreement with test, but this was because the codes had been benchmarked extensively with wind-tunnel data. It did not reflect reliance on first principles at higher Mach.

Engine studies for the X-43A used their own codes, which again amounted to those of NASP. GASP 3.0 had the relatively recent date of 1996, but other pertinent literature showed nothing more recent than 1993, with some papers dating to the 1970s.[73]

The 2002 design of ISTAR, a rocket-based combined-cycle engine, showed that specialists were using codes that were considerably more current. Studies of the forebody and inlet used OVERFLOW, from 1999, while analysis of the combustor used VULCAN version 4.3, with a users' manual published in March 2002. OVERFLOW used equilibrium chemistry while VULCAN included finite-rate chemistry, but both solved the Navier-Stokes equations by using a two-equation turbulence model. This was no more than had been done during NASP, more than a decade earlier.[74]

The reason for this lack of progress can be understood with reference to Karl Marx, who wrote that people's thoughts are constrained by their tools of production. The tools of CFD have been supercomputers, and during the NASP era the best of them had been rated in gigaflops, billions of floating-point operations per second.[75] Such computations required the use of turbulence models. But recent years have seen the advent of teraflop machines. A list of the world's 500 most powerful is available on the Internet, with the accompanying table giving specifics for the top 10 of November 2004, along with number 500.

One should not view this list as having any staying power. Rather, it gives a snapshot of a technology that is advancing with extraordinary rapidity. Thus, in 1980 NASA was hoping to build the Numerical Aerodynamic Simulator, and to have it online in 1986. It was to be the world's fastest supercomputer, with a speed of one gigaflop (0.001 teraflop), but it would have fallen below number 500 as early as 1994. Number 500 of 2004, rated at 850 gigaflops, would have been number one as recently as 1996. In 2002 Japan's Earth Simulator was five times faster than its nearest rivals. In 2004 it had fallen to third place.[76]

Today's advances in speed are being accomplished both by increasing the number of processors and by multiplying the speed of each such unit. The ancestral Illiac-4, for instance, had 64 processors and was rated at 35 megaflops.[77] In 2004 IBM's BlueGene was two million times more powerful. This happened both because it had 512 times more processors—32,768 rather than 64—and because each individual processor had 4,000 times more power. Put another way, a single BlueGene processor could do the work of two Numerical Aerodynamic Simulator concepts of 1980.

Analysts are using this power. The NASA-Ames aerodynamicist Christian Stemmer, who has worked with a four-teraflop machine, notes that it achieved this speed by using vectors, strings of 256 numbers, but that much of its capability went unused when his vector held only five numbers, representing five chemical species. The computation also slowed when finding the value of a single constant or when taking square roots, which is essential when calculating the speed of sound. Still, he adds, "people are happy if they get 50 percent" of a computer's rated performance. "I do get 50 percent, so I'm happy."[78]

THE WORLD'S FASTEST SUPERCOMPUTERS (Nov. 2004; updated annually)

	Name	Manufacturer	Location	Year	Rated speed teraflops	Number of processors
1	BlueGene	IBM	Rochester, NY	2004	70,720	32,768
2	Numerical Aerodynamic Simulator	Silicon Graphics	NASA-Ames	2004	51,870	10,160
3	Earth Simulator	Nippon Electric	Yokohama, Japan	2002	35,860	5,120
4	Mare Nostrum	IBM	Barcelona, Spain	2004	20,530	3,564
5	Thunder	California Digital Corporation	Lawrence Livermore National Laboratory	2004	19,940	4,096
6	ASCI Q	Hewlett-Packard	Los Alamos National Laboratory	2002	13,880	8,192
7	System X	Self-made	Virginia Tech	2004	12,250	2,200
8	BlueGene (prototype)	IBM, Livermore	Rochester, NY	2004	11,680	8,192
9	eServer p Series 655	IBM	Naval Oceanographic Office	2004	10,310	2,944
10	Tungsten	Dell	National Center for Supercomputer Applications	2003	9,819	2,500
500	Superdome 875	Hewlett-Packard	SBC Service, Inc.	2004	850.6	416

Source: http://www.top500.org/list/2004/11

Teraflop ratings, representing a thousand-fold advance over the gigaflops of NASP and subsequent projects, are required because the most demanding problems in CFD are four-dimensional, including three physical dimensions as well as time. William Cabot, who uses the big Livermore machines, notes that "to get an increase in resolution by a factor of two, you need 16" as the increase in computational speed because the time step must also be reduced. "When someone says, 'I have a new computer that's an order of magnitude better,'" Cabot continues, "that's about a factor of 1.8. That doesn't impress people who do turbulence."[79]

But the new teraflop machines increase the resolution by a factor of 10. This opens the door to two new topics in CFD: Large-Eddy Simulation (LES) and Direct Numerical Simulation (DNS).

One approaches the pertinent issues by examining the structure of turbulence within a flow. The overall flowfield has a mean velocity at every point. Within it, there are turbulent eddies that span a very broad range of stress. The largest carry most of the turbulent energy and accomplish most of the turbulent mixing, as in a combustor. The smaller eddies form a cascade, in which those of different sizes are intermingled. Energy flows down this cascade, from the larger to the smaller ones, and while turbulence is often treated as a phenomenon that involves viscosity, the transfer of energy along the cascade takes place through inviscid processes. However, viscosity becomes important at the level of the smallest eddies, which were studied by Andrei Kolmogorov in the Soviet Union and hence define what is called the Kolmogorov scale of turbulence. At this scale, viscosity, which is an intermolecular effect, dissipates the energy from the cascade into heat. The British meteorologist Lewis Richardson, who introduced the concept of the cascade in 1922, summarized the matter in a memorable sendup of a poem by England's Jonathan Swift:

> Big whorls have little whorls
> Which feed on their velocity;
> And little whorls have lesser whorls,
> And so on to viscosity.[80]

In studying a turbulent flow, DNS computes activity at the Kolmogorov scale and may proceed into the lower levels of the cascade. It cannot go far because the sizes of the turbulent eddies span several orders of magnitude, which cannot be captured using computational grids of realistic size. Still, DNS is the method of choice for studies of transition to turbulence, which may predict its onset. Such simulations directly reproduce the small disturbances within a laminar flow that grow to produce turbulence. They do this when they first appear, making it possible to observe their growth. DNS is very computationally intensive and remains far from ready for use with engineering problems. Even so, it stands today as an active topic for research.

LES is farther along in development. It directly simulates the large energy-bearing eddies and goes onward into the upper levels of the cascade. Because its computations do not capture the complete physics of turbulence, LES continues to rely on turbulence models to treat the energy flow in the cascade along with the Kolmogorov-scale dissipation. But in contrast to the turbulence models of present-day codes, those of LES have a simple character that applies widely across a broad range of flows. In addition, their errors have limited consequence for a flow as a whole, in an inlet or combustor under study, because LES accurately captures the physics of the large eddies and therefore removes errors in their modeling at the outset.[81]

The first LES computations were published in 1970 by James Deardorff of the National Center for Atmospheric Research.[82] Dean Chapman, Director of Astronautics at NASA-Ames, gave a detailed review of CFD in the 1979 AIAA Dryden Lectureship in Research, taking note of the accomplishments and prospects of LES.[83] However, the limits of computers restricted the development of this field. More than a decade later Luigi Martinelli of Princeton University, a colleague of Antony Jameson who had established himself as a leading writer of flow codes, declared that "it would be very nice if we could run a large-eddy simulation on a full three-dimensional configuration, even a wing." Large eddies were being simulated only for simple cases such as flow in channels and over flat plates, and even then the computations were taking as long as 100 hours on a Cray supercomputer.[84]

Since 1995, however, the Center for Turbulence Research has come to the forefront as a major institution where LES is being developed for use as an engineering tool. It is part of Stanford University and maintains close ties both with NASA-Ames and with Lawrence Livermore National Laboratory. At this center, Kenneth Jansen published LES studies of flow over a wing in 1995 and 1996, treating a NACA 4412 airfoil at maximum lift.[85] More recent work has used LES in studies of reacting flows within a combustor of an existing jet engine of Pratt & Whitney's PW6000 series. The LES computation found a mean pressure drop across the injector of 4,588 pascals, which differs by only two percent from the observed value of 4,500 pascals. This compares with a value of 5,660 pascals calculated using a Reynolds-averaged Navier-Stokes code, which thus showed an error of 26 percent, an order of magnitude higher.[86]

Because LES computes turbulence from first principles, by solving the Navier-Stokes equations on a very fine computational grid, it holds high promise as a means for overcoming the limits of ground testing in shock tunnels at high Mach. The advent of LES suggests that it indeed may become possible to compute one's way to orbit, obtaining accurate results even for such demanding problems as flow in a scramjet that is flying at Mach 17.

Parviz Moin, director of the Stanford center, cautions that such flows introduce shock waves, which do not appear in subsonic engines such as the PW6000 series, and are difficult to treat using currently available methods of LES. But his colleague

Heinz Pitsch anticipates rapid progress. He predicted in 2003 that LES will first be applied to scramjets in university research, perhaps as early as 2005. He adds that by 2010 "LES will become the state of the art and will become the method of choice" for engineering problems, as it emerges from universities and begins to enter the mainstream of CFD.[87]

HYPERSONICS AND THE AVIATION FRONTIER

Aviation has grown through reliance upon engines, and three types have been important: the piston motor, turbojet, and rocket. Hypersonic technologies have made their largest contributions, not by adding the scramjet to this list, but by enhancing the value and usefulness of rockets. This happened when these technologies solved the re-entry problem.

This problem addressed critical issues of the national interest, for it was essential to the success of Corona and of the return of film-carrying capsules from orbit. It also was a vital aspect of the development of strategic missiles. Still, if such weapons had proven to be technically infeasible, the superpowers would have fallen back on their long-range bombers. No such backup was available within the Corona program. During the mid-1960s the Lunar Orbiter Program used a high-resolution system for scanning photographic film, with the data being returned using telemetry.[88] But this arrangement had a rather slow data rate and was unsuitable for the demands of strategic reconnaissance.

Success in re-entry also undergirded the piloted space program. In 40 years of effort, this program has failed to find a role in the mainstream of technical activity akin to the importance of automated satellites in telecommunications. Still, piloted flight brought the unforgettable achievements of Apollo, which grow warmer in memory as the decades pass.

In a related area, the advent of thermal-protection methods led to the development of aircraft that burst all bounds on speed and altitude. These took form as the X-15 and the space shuttle. On the whole, though, this work has led to disappointment. The Air Force had anticipated that airbreathing counterparts of the X-15, powered perhaps by ramjets, would come along in the relatively near future. This did not happen; the X-15 remains sui generis, a thing unto itself. In turn, the shuttle failed to compete effectively with expendable launch vehicles.

This conclusion remains valid in the wake of the highly publicized flights of SpaceShipOne, built by the independent inventor Burt Rutan. Rutan showed an uncanny talent for innovation in 1986, when his Voyager aircraft, piloted by his brother Dick and by Dick's former girlfriend Jeana Yeager, circled the world on a single load of fuel. This achievement had not even been imagined, for no science-fiction writer had envisioned such a nonstop flight around the world. What made it possible was the use of composites in construction. Indeed, Voyager was built at

Rutan's firm of Scaled Composites.[89] Such lightweight materials also found use in the construction of SpaceShipOne, which was assembled within that plant.

SpaceShipOne brought the prospect of routine commercial flights having the performance of the X-15. Built entirely as a privately funded venture, it used a simple rocket engine that burned rubber, with nitrous oxide as the oxidizer, and reached altitudes as high as 70 miles. A movable set of wings and tail booms, rotating upward, provided stability in attitude during re-entry and kept the craft's nose pointing upward as well. The craft then glided to a landing.

There was no commercial follow-on to Voyager, but today there is serious interest in building commercial versions of SpaceShipOne that will take tourists on brief hops into space—and enable them to win astronauts' wings in the process. Richard Branson, founder of Virgin Airways, is currently sponsoring a new enterprise, Virgin Galactic, that aims to do just that. He has formed a partnership with Scaled, has sold more than 100 tickets at $200,000 each, and hopes for his first flight late in 2008.

And yet…. The top speed of SpaceShipOne was only 2,200 miles per hour, or Mach 3.3. Rutan's vehicle thus stands today as a brilliant exercise in rocketry and the design of reusable piloted spacecraft. But it is too slow to qualify as a project in hypersonics.[90]

Is that it, then? Following more than half a century of effort, does the re-entry problem stand as the single unambiguous contribution of hypersonics? Air Force historian Richard Hallion has written of a "hypersonic revolution," but from this perspective, one may regard hypersonics less as an extension of aeronautics than as a branch of materials science, akin to metallurgy. Specialists in that field introduced superalloys that extended the temperature limits of jet engines, thereby enhancing their range and fuel economy. Similarly, the hypersonics community developed lightweight thermal-protection systems that have found use even in exploring the planet Jupiter. Yet one does not speak of a "superalloy revolution," and hypersonics has had similarly limited application.

There remains the issue of the continuing effort to develop the scramjet. This work has gone forward as part of an ongoing hope that better methods might be devised for ascent to orbit, corresponding perhaps to the jet airliners that drove their piston-driven counterparts to the boneyard. Access to space holds undeniable importance, and one may speak without challenge of a "satellite revolution" when we consider the vital role of such craft in a host of areas: weather forecasting, navigation, tactical warfare, reconnaissance, as well as telecommunications. Yet low-cost access remains out of reach and hence continues to justify work on advanced technologies, including scramjets.

Still, despite 40 years of effort, the scramjet continues to stand at two removes from importance. The first goal is simply to make it work, by demonstrating flight to orbit in a vehicle that uses such engines for propulsion. The X-30 was to fly in this fashion, although present-day thinking leans more toward using it merely in an airbreathing first stage. But at least within the next decade the most that anyone hopes for is to accelerate a small test vehicle of the X-43 class.[91]

Yet even if a large launch vehicle indeed should fly using scramjets, it then will face a subsequent test, for it will have to win success in the face of competition from existing launchers. The history of aerospace shows several types of craft that indeed flew well but that failed in the market. The classic example was the dirigible, which was abandoned because it could not be made safe.[92]

The world still remembers the *Hindenburg*, but the problems ran deeper than the use of hydrogen. Even with nonflammable helium, such airships proved to be structurally weak. The U.S. Navy built three large ones—the *Shenandoah*, *Akron*, and *Macon*—and quickly lost them all in storms and severe weather. Nor has this problem been solved. Dirigibles might be attractive today as aerial cruise ships, offering unparalleled views of Caribbean islands, but the safety problem persists.

More recently the Concorde supersonic airliner flew with great style and panache but faltered due to its high costs. The Saturn V Moon rocket proved to be too large to justify continued production; it lacked payloads that demanded its heft. Piloted space flight raises its own questions. It too is very costly, and in the light of experience with the shuttle, perhaps it too cannot be made completely safe.

Yet though scramjets face obstacles both in technology and in the market, they will continue to tantalize. Hallion writes that faith in a future for hypersonics "is akin to belief in the Second Coming: one knows and trusts that it *will* occur, but one can't be certain *when*." Scramjet advocates will continue to echo the defiant words of Eugen Sänger: "Nevertheless, my silver birds will fly!"[93]

1. Milos et al., *Journal of Spacecraft and Rockets*, May-June 1999, pp. 298-99.
2. Young et al., *Science*, 10 May 1996, pp. 837-38 (228 g: Table 3).
3. Milos et al., *Journal of Spacecraft and Rockets*, May-June 1999, pp. 298-99.
4. *Aerospace America*: December 1989, pp. 48-49; December 1990, p. 73.
5. Miller, *X-Planes*, pp. 341-42. Launch crew size: *Acta Astronautica*, February-April 1996, p. 326.
6. NASA SP-4407, Vol. IV, pp. 585-88, 591.
7. *Aviation Week*, 8 July 1996, p. 21.
8. AIAA Paper 95-3777.
9. NASA SP-4407, Vol. IV, pp. 585-86.
10. *Acta Astronautica*, February-April 1996, p. 323.
11. Heppenheimer, *Development*, pp. 365-66; Hanley, *Aerospace America*, July 1991, pp. 34-38.
12. NASA SP-4407, Vol. IV, p. 604; quotes, pp. 605, 617, 624.
13. Ibid., pp. 626-27 (includes extended quote).
14. *Acta Astronautica*, February-April 1996, p. 324.
15. NASA SP-4407, Vol. IV, pp. 628-29.
16. *Aviation Week*, 31 October 1994, pp. 22-23; AIAA Paper 95-3777, p. 1. $941 million: NASA SP-4407, Vol. IV, p. 634.
17. *Aviation Week*, 3 April 1995, pp. 44, 48 (X-34B, p. 48). $60 million: *Aviation Week*, 5 February 1996, p. 86.
18. Miller, *X-Planes*, pp. 352-353. Quote: *Aviation Week*, 6 November 1995, p. 30.
19. Miller, *X-Planes*, p. 353. Concept comparisons: AIAA Paper 95-3777, pp. 3-4; *Aviation Week*, 17 June 1996, p. 31.
20. Miller, *X-Planes*, p. 344. Payloads: *Aviation Week*, 8 July 1996, table, pp. 20-21. 53 percent: Rising, *X-33*, p. 6.
21. General Accounting Office Report NSIAD-99-176, p. 4.
22. *Acta Astronautica*, February-April 1996, pp. 325-26, 329.
23. *Aviation Week*, 26 April 1999, p. 78.
24. *Journal of Spacecraft and Rockets*, March-April 1999, pp. 189, 218; Jenkins, *Space Shuttle*, p. 395.
25. Miller, *X-Planes*, p. 346; AIAA Paper 96-4563.
26. *Aviation Week*, 3 June 1996, p. 92.
27. Air Force Studies Board, *Hypersonic*, p. 5.
28. *Aviation Week*, 8 July 1996, table, pp. 20-21.
29. AIAA Paper 96-4563.
30. *Aviation Week*: 8 July 1996, p. 21; 30 June 1997, p. 28; 10 November 1997, p. 50.
31. Rising, *X-33*, figure 10; author interview, Jerry Rising, 11 January 1999.
32. Miller, *X-Planes*, p. 344; *Aviation Week*, 2 November 1998, pp. 26-27.

33 *Aviation Week*, 25 January 1999, pp. 68-69 (weight, p. 68; quote, p. 68).
34 Sietzen, "Wall Street" (includes quotes).
35 *Aviation Week*, 15 November 1999, pp. 28-30 (quote, p. 28). 2.2 million pounds: *Aviation Week*, 8 July 1996, p. 21.
36 Quote: *Aviation Week*, 15 November 1999, p. 30.
37 Miller, *X-Planes*, pp. 203, 357.
38 Ibid., p. 356; AIAA Paper 98-3516.
39 Miller, *X-Planes*, pp. 354-57.
40 *Aviation Week*, 5 March 2001, p. 24 (includes quote).
41 NIFTA: AIAA Paper 91-5008.
42 *Aviation Week*, 30 March 1992, pp. 18-20.
43 *Aviation Week*, 14/21 December 1992, pp. 70-73.
44 The Incremental Scramjet of the mid-1960s was to achieve transition between Mach 5.4 and 6. See Hallion, *Hypersonic*, pp. VI-xvii to VI-xx. Work at the Applied Physics Laboratory demonstrated transition between Mach 5.9 and 6.2: see AIAA Paper 91-2395.
45 Author interview, Lawrence Huebner, 24 July 2002.
46 NASA/TP 1998-206548, pp. 1, 8-10.
47 AIAA Paper 99-4848.
48 $200,000: *Aviation Week*, 14/21 December 1992, p. 71.
49 AIAA Papers 99-4818, p. 2; 2002-5251, pp. 1-2 (length and range, p. 2).
50 AIAA Paper 99-4818 (quote, p. 1).
51 Ibid.
52 Miller, *X-Planes*, p. 389. Contract awards: AIAA Paper 2001-1910, p. 791.
53 AIAA Paper 99-4818. 13 feet per second: AIAA Paper 2001-1802, p. 257.
54 AIAA Paper 2002-5251, p. 3; Miller, *X-Planes*, pp. 391-92 (2,800 pounds, p. 392).
55 *Journal of Spacecraft and Rockets*, November-December 2001, pp. 846-47.
56 Quote: AIAA Paper 98-1532, p. 4.
57 AIAA Papers 98-1532 (numbers, p. 4), 2001-1809.
58 AIAA Paper 2001-1809, pp. 292-93; Vince Rausch: *Aviation Week*, 9 June 1997, p. 32.
59 *Journal of Spacecraft and Rockets*, November-December 2001, pp. 829-30.
60 *Journal of Spacecraft and Rockets*, November-December 2001, pp. 828-35.
61 *Journal of Spacecraft and Rockets*, November-December 2001, pp. 803-10.
62 *Journal of Spacecraft and Rockets*, November-December 2001, pp. 853-64 (300 runs, table 1).
63 *Journal of Spacecraft and Rockets*, November-December 2001, p. 806.
64 AIAA Paper 2001-1910, p. 793. Mission plan: *Aviation Week*, 11 June 2001, p. 50. Quote: author interview, Lawrence Huebner, 24 July 2002.
65 *Aviation Week*, 11 June 2001, pp. 50-51.
66 *Air & Space*, November 2002, pp. 74-81; AIAA Paper 2003-7029.
67 Quote: *Air & Space*, November 2002, p. 81.

68   Internet postings: Google, "Hyshot Australia." Quote: *Air & Space*, November 2002, p. 79.
69   *Aviation Week*, 28 July 2003, pp. 37-38.
70   *Aviation Week*, 5 April 2004, pp. 28-29.
71   *Aviation Week*, 22 November 2004, pp. 24-26.
72   *Aviation Week*, 5 March 2001, p. 66.
73   *Journal of Spacecraft and Rockets*, November-December 2001, p. 838.
74   AIAA Paper 2002-5127.
75   Dongarra, "Performance."
76   *Science*: 16 January 1981, pp. 268-69; 29 November 2002, p. 1713. Also http://www.top500.org/list/1994/11, http://www.top500.org/list/1996/11
77   AIAA Paper 83-0037, table 1.
78   Author interview, Christian Stemmer, 22 April 2003.
79   Author interview, William Cabot, 23 April 2003.
80   Pope, *Turbulent*, pp. 182-183.
81   Author interview, Christian Stemmer, 23 April 2003.
82   Deardorff: *Journal of Fluid Mechanics*, 13 April 1970, pp. 453-80; *Geophysical Fluid Dynamics*, November 1970, pp. 377-410.
83   *AIAA Journal*, December 1979, pp. 1300-1302.
84   Heppenheimer, "Tractable," pp. 36-37 (includes quote).
85   Jansen: "Preliminary"; "Large-Eddy."
86   Mahesh et al., "Large-Eddy."
87   Author interviews, Parviz Moin and Heinz Pitsch, 22 April 2003.
88   *Scientific American*, May 1968, pp. 59-78.
89   Heppenheimer, "Voyager"; *1989 Yearbook of Science and the Future* (Encyclopedia Britannica), pp. 142-59.
90   *Time*, 29 November 2004, pp. 63-67; *Air & Space*, March 2006, p. 25.
91   Malakoff, *Science*, 9 May 2003, pp. 888-889.
92   Brooks, "Airship"; Botting, *The Giant Airships*.
93   Quotes: Hallion, *Hypersonic*, p. 98-iii; Sänger: *Spaceflight*, May 1973, p. 166.

Bibliography

"Advanced Technology Program: Technical Development Plan for Aerothermodynamic/elastic Structural Systems Environmental Tests (ASSET)." Air Force Systems Command, 9 September 1963.

Air Force Studies Board. "Hypersonic Technology for Military Application." Washington, DC: National Academies Press, 1989.

Allen, Thomas B., and Norman Polmar. *Code-Name Downfall*. New York: Simon & Schuster, 1995.

Anderson, John. *A History of Aerodynamics*. New York: Cambridge University Press, 1997.

Anspacher, William B., Betty H. Gay, Donald E. Marlowe, Paul B. Morgan, and Samuel J. Raff. *The Legacy of the White Oak Laboratory*. Dahlgren, VA: Naval Surface Warfare Center, 2000.

Augenstein, Bruno. "Rand and North American Aviation's Aerophysics Laboratory: An Early Interaction in Missiles and Space." Paper IAA-98-IAA.2.2.06 presented at the 49th Annual International Astronautical Federation, Melbourne, Australia, 1998.

Bagwell, Margaret. "History of the Bomarc Weapon System 1953-1957." Wright-Patterson AFB, OH: Air Materiel Command, Office of Information Services, February 1959.

Baucom, Donald R. *The Origins of SDI, 1944-1983*. Lawrence, KS: University Press of Kansas, 1992.

Becker, John, Norris F. Dow, Maxime A. Faget, Thomas A. Toll, and J.B. Whitten. "Research Airplane Study." NACA-Langley, April 1954.

Becker, Paul R. "Leading-Edge Structural Material System of the Space Shuttle." *American Ceramic Society Bulletin* 60, no. 11 (1981): 1210-1214.

Bond, Aleck C. "Big Joe." Project Mercury Technical History, 27 June 1963.

Botting, Douglas. *The Giant Airships*. Alexandria, VA: Time-Life Books, 1980.

Boyne, Walter J. *The Messerschmitt Me 262: Arrow to the Future*. Washington, DC: Smithsonian Institution Press, 1980.

Broad, William. *Teller's War*. New York: Simon & Schuster, 1992.

Brooks, Peter W. "Why the Airship Failed." *Aeronautical Journal* 79 (1975): 439-449.

Brown, C. E., W. J. O'Sullivan, and C. H. Zimmerman. "A Study of the Problems Related to High Speed, High Altitude Flight." NACA-Langley, 1953.

Chapman, John L. *Atlas: The Story of a Missile*. New York: Harper, 1960.

Clarke, Arthur C. *Profiles of the Future*. New York: Bantam, 1964.

"Columbia Accident Investigation Report." NASA, August 2003.

Cornett, Lloyd H. "History of the Air Force Missile Development Center 1 January–30 June 1964." Holloman AFB, NM. Report AFMDC 64-5791, 1964.

———. "An Overview of ADC Weapons 1946-1972." USAF Aerospace Defense Command, April 1973.

Crickmore, Paul F. *Lockheed SR-71 Blackbird*. London: Osprey Publishing, 1986.

Day, Dwayne A., John M. Logsdon, and Brian Latell. *Eye in the Sky*. Washington, DC: Smithsonian Institution Press, 1998.

Day, Leroy. "NASA Space Shuttle Task Group Report, Volume II: Desired System Characteristics." Washington, DC: Space Shuttle Task Group, NASA, 12 June 1969.

Deardorff, J. W. "A Numerical Study of Three-Dimensional Turbulent Channel Flows at Large Reynolds Numbers." *Journal of Fluid Mechanics* 41 (1970): 453-480.

———. "A Three-Dimensional Numerical Investigation of the Idealized Planetary Boundary Layer." *Geophysical Fluid Dynamics* 1 (1970): 377-410.

"Development of the Bomarc Guided Missile 1950-53." U.S. Air Force. Archives, Maxwell AFB, Montgomery, Alabama.

"Development of the Navaho Guided Missile 1945-1953." U.S. Air Force. Archives, Maxwell AFB, Montgomery, Alabama.

"Development of the SM-64 Navaho Missile 1954-1958." Wright-Patterson AFB, OH: Historical Division, ARDC, January 1961.

Dongarra, Jack J. "Performance of Various Computers Using Standard Linear Equations Software in a Fortran Environment." Argonne, IL: Technical Memorandum 23, Mathematics and Computer Science Division, Argonne National Laboratory, 30 September 1988.

Dornberger, Walter. *V-2*. New York: Bantam Books, 1979.

Drake, Hubert M., and L. Robert Carman. "A Suggestion of Means for Flight Research at Hypersonic Velocities and High Altitudes." In RG 255, National Archives, Philadelphia: NACA-Langley, 26 May 1952.

Emme, Eugene. *The History of Rocket Technology*. Detroit, MI: Wayne State University Press, 1964.

Erb, R. Bryan, and Emily W. Stephens. "Project Mercury: An Analysis of Mercury Heat-Shield Performance During Entry." NASA Project Mercury Working Paper No. 193. Langley Field, VA: NASA Space Task Group, 21 June 1961.

Escher, William J. D. "Cryogenic Hydrogen-Induced Air Liquefaction Technologies for Combined-Cycle Propulsion Applications." European RBCC Workshop, TNO Prins Maurits Laboratory, Delft, Netherlands, 6-9 November 1995.

BIBLIOGRAPHY

"Evaluation Report on X-15 Research Aircraft Design Competition." In RG 255, National Archives, Philadelphia, 5 August 1955.

Faget, Maxime, and Milton Silveira. "Fundamental Design Considerations for an Earth-Surface to Orbit Shuttle." Paper presented at the 21st Annual International Astronautical Congress, Konstanz, West Germany, 4-10 October 1970.

Fahrney, Rear Admiral D. S. "The History of Pilotless Aircraft and Guided Missiles." Washington, DC: Archives, Federation of American Scientists, undated.

Ferri, Antonio. "Possible Directions of Future Research in Air-Breathing Engines." In *Combustion and Propulsion, Proceedings of the Fourth AGARD Colloquium, 1960*. New York: Pergamon Press, 1961.

———. "Statement Regarding the Military Service of Antonio Ferri." Typescript in NASA-Langley Archives.

Ferri, Antonio, and Herbert Fox. "Analysis of Fluid Dynamics of Supersonic Combustion Process Controlled by Mixing." In *Twelfth Symposium on Combustion*. Pittsburgh, PA: Combustion Institute, 1968, 1105-1113.

Fitzgerald, Frances. *Way Out There in the Blue*. New York: Simon & Schuster, 2000.

Fletcher, Edward A. "Early Supersonic Combustion Studies at NACA and NASA." In *Eleventh Symposium on Combustion*. Pittsburgh, PA: Combustion Institute, 1967, 729-737.

Geiger, Clarence J. "Termination of the X-20A Dyna-Soar." In *History of Aeronautical Systems Division*, July-December 1963, Volume III. Air Force Systems Command: Historical Division, Report 64 ASE-39.

Gibson, James N. *The Navaho Missile Project*. Atglen, PA: Schiffer Publishing, 1996.

Greer, Kenneth E. "Corona." In *Corona*, by Kevin C. Ruffner, ed., 3-39. Langley, VA: CIA, Center for the Study of Intelligence, 1995.

Grimwood, James M., and Frances Strowd. "History of the Jupiter Missile System." Redstone Arsenal, AL: U.S. Army Ordnance Missile Command, 27 July 1962.

Gunston, Bill. *Fighters of the Fifties*. Osceola, WI: Specialty Press, 1981.

Hallion, Richard P. "American Rocket Aircraft: Precursors to Manned Flight Beyond the Atmosphere." In *History of Rocketry and Astronautics*. Edited by Kristan R. Lattu. AAS History Series, Volume 8. San Diego: Univelt, 1989, 283-312.

———, ed. *The Hypersonic Revolution*. Bolling AFB, DC: Air Force History and Museums Program, 1998.

Hansen, Harry, ed. *The World Almanac and Book of Facts for 1956*. New York: New York World-Telegram Corporation, 1956.

Harshman, D. L. "Design and Test of a Mach 7-8 Supersonic Combustion Ramjet Engine." Paper presented at the American Institute of Aeronautics and Astronautics Propulsion Specialist Meeting, Washington, DC, 17-21 July 1967.

Heppenheimer, T. A. *Countdown: A History of Space Flight*. New York: John Wiley, 1997.

———. *Development of the Space Shuttle, 1972-1981*. Washington, DC: Smithsonian Institution Press, 2002.

———. *First Flight*. New York: John Wiley, 2003.

———. *Hypersonic Technologies and the National Aerospace Plane*. Arlington, VA: Pasha Publications, 1990.

———. *The Man-Made Sun: The Quest for Fusion Power*. Boston: Little, Brown, 1984.

———. "Some Tractable Mathematics for Some Intractable Physics." *Mosaic*. Vol. 22, No. 1, Spring 1991, pp. 28-39.

———. *Turbulent Skies: The History of Commercial Aviation*. New York: John Wiley, 1995.

Hermann, R. "The Supersonic Wind Tunnel Installations at Peenemunde and Kochel and Their Contributions to the Aerodynamics of Rocket-Powered Vehicles." In *Space: Mankind's Fourth Environment*. Edited by L. G. Napolitano. New York: Pergamon Press, 1982, 435-446.

"History of the Arnold Engineering Development Center." AFSC Historical Publications, U.S. Air Force.

Hunley, J. D. "The Significance of the X-15." NASA-Dryden History Office, 1999.

Hunter, Maxwell W. "The Origins of the Shuttle (According to Hunter)." Sunnyvale, CA: Lockheed, September 1972. Reprinted in part in *Earth/Space News*, November 1976, 5-7.

Huxley, Aldous, *Brave New World*. New York: HarperCollins, 1998.

Jane's All the World's Aircraft. New York: McGraw-Hill. Annual editions; began in 1930.

Jansen, Kenneth. "Preliminary Large-Eddy Simulations of Flow Around a NACA 4412 Airfoil Using Unstructured Grids." Annual Research Briefs. Stanford, CA: Center for Turbulence Research, Stanford University, 1995, 61-72.

———. "Large-Eddy Simulation of Flow Around a NACA 4412 Airfoil Using Unstructured Grids." Annual Research Briefs. Stanford, CA: Center for Turbulence Research, Stanford University, 1996, 225-232.

Jenkins, Dennis. *Space Shuttle*. Stillwater, MN: Voyageur Press, 2001.

Kantrowitz, Arthur. "Shock Tubes for High Temperature Gas Kinetics." Avco-Everett Research Laboratory, Research Report 141, October 1962.

Kaufman, Louis, Barbara Fitzgerald, and Tom Sewell. *Moe Berg: Athlete, Scholar, Spy*. Boston: Little, Brown, 1974.

Korb, L. J., C.A. Morant, R.M. Calland and C.S. Thatcher. "The Shuttle Orbiter Thermal Protection System." *American Ceramic Society Bulletin* 60 (1981): 1188-1193.

Krieth, Frank. *Principles of Heat Transfer*. Scranton, PA: International Textbook Company, 1964.

Kuethe, Arnold M., and Chuen-Yen Chow. *Foundations of Aerodynamics*. New York: John Wiley, 1986.

Lay, Joachim E. *Thermodynamics*. Columbus, OH: C. E. Merrill, 1963.

Lee, John F., Francis W. Sears, and Donald L. Turcotte. *Statistical Thermodynamics*. Reading, MA: Addison-Wesley, 1963.

Ley, Willy. *Rockets, Missiles, and Man in Space*. New York: Signet, 1968.

Lin, C. C.. *Statistical Theories of Turbulence*. Princeton, NJ: Princeton University Press, 1961.

Lukasiewicz, J. *Experimental Methods of Hypersonics*. New York: Marcel Dekker, 1973.

MacKenzie, Donald. *Inventing Accuracy*. Cambridge, MA: MIT Press, 1990.

Mackley, Ernest A. "Historical Aspects of NASA Scramjet Technology Development." NASA-Langley, undated.

Mahesh, Krishnan, et al. "Large-Eddy Simulation of Gas Turbine Combustors." Annual Research Briefs. Stanford, CA: Center for Turbulence Research, Stanford University, 2001, 3-17.

Malia, Martin E. *The Soviet Tragedy*. New York: Free Press, 1994.

Manchester, William. *American Caesar*. New York: Dell, 1978.

Martin, James A. "History of NACA-Proposed High-Mach-Number, High-Altitude Research Airplane." 16 December 1954.

McDougall, Walter A. *The Heavens and the Earth*. New York: Basic Books, 1985.

McPhee, John. *The Curve of Binding Energy*. New York: Ballantine Books, 1975.

Miller, Jay. *The X-Planes, X-1 to X-45*. North Branch, MN: Specialty Press, 2001.

Miller, Ron, and Frederick C. Durant. *Worlds Beyond: The Art of Chesley Bonestell*. Norfolk, VA: Donning, 1983.

Murray, Russ. *Lee Atwood...Dean of Aerospace*, Rockwell International Corporation, 1980.

Naval Research Laboratory. "Upper Atmosphere Research Report No. 1." Report R-2955. 1 October 1946.

Neufeld, Jacob. *The Development of Ballistic Missiles in the United States Air Force, 1945-1960*. Washington, DC: Office of Air Force History, USAF, 1990.

Neufeld, Michael J. *The Rocket and the Reich*. New York: Free Press, 1995.

Ordway, Frederick and Mitchell R. Sharpe. *The Rocket Team*. New York: Thomas Y. Crowell, 1979.

Pace, Scott. "Engineering Design and Political Choice: The Space Shuttle 1969-1972." Masters thesis, MIT, May 1982.

Pedigree of Champions: Boeing Since 1916. Seattle: Boeing, 1985.

Peterson, Lee L. "Evaluation Report on X-7A." Report AFMDC ADJ 57-8184. Holloman AFB, NM, 3 October 1957.

Pfeifer, John L. "The BOMARC Weapon System." Bomarc Service News Article. Seattle, WA: Boeing archives, 4 March 1959.

Pope, Stephen B. . *Turbulent Flows*. New York: Cambridge University Press, 2000.

Preston, Richard. *American Steel*. New York: Prentice Hall, 1991.

"Re-Entry Studies," Volume 1. Redstone Arsenal, AL: Army Ballistic Missile Agency, 25 November 1958.

"Re-Entry Test Vehicle X-17." In "History, Air Force Missile Test Center, 1 July – 31 December 1957." Volume IV, "Supporting Documents," Appendix F. Air Force archives, Maxwell AFB, Montgomery, AL.

"Research-Airplane-Committee Report on Conference on the Progress of the X-15 Project." Langley Field, VA: Langley Aeronautical Laboratory, October 25-26, 1956.

"Research-Airplane-Committee Report on Conference on the Progress of the X-15 Project." Los Angeles: IAS Building, July 28-30, 1958.

Rhodes, Richard. *Dark Sun: The Making of the Hydrogen Bomb*. New York: Simon & Schuster, 1995.

———. *The Making of the Atomic Bomb*. New York: Simon & Schuster, 1986.

Richelson, Jeffrey. *American Espionage and the Soviet Target*. New York: William Morrow, 1987.

———. *America's Secret Eyes in Space*. New York: Harper & Row, 1990.

Riddell, F. R. and J. D. Teare. "The Differences Between Satellite and Ballistic Missile Re-Entry Problems." In *Vistas in Astronautics*, Vol. 2. Edited by Morton Alperin and Hollingsworth F. Gregory. New York: Pergamon Press, 1959, 174-190.

Rising, Jerry J. "The Lockheed Martin/NASA X-33 Program: A Stepping Stone to *VentureStar*." Washington, DC: Lockheed Martin Technology Symposium, 3 November 1998.

Ritchie, William A. "Evaluation Report on X-7A (System 601B)." Report AFMDC DAS 58-8129. Holloman AFB, NM, January 1959. Montgomery, AL: Archives, Maxwell AFB.

Rose, P. H. "Physical Gas Dynamics Research at the Avco Research Laboratory." AGARD Report 145, July 1957.

Ruffner, Kevin C. "Corona: America's First Satellite Program." Washington: Central Intelligence Agency, 1995.

Schlaifer, Robert, and S. D. Heron. *Development of Aircraft Engines and Fuels*. Boston: Harvard University, 1950.

Schnyer, A. Dan and R. G. Voss. "Review of Orbital Transportation Concepts—Low Cost Operations." Washington, DC: NASA-OMSF, 18 December 1968.

Schramm, Wilson. "HRSI and LRSI—The Early Years." *American Ceramic Society Bulletin* 60 (1981): 1194-1195.

Schweikart, Larry. *The Quest for the Orbital Jet*. Volume III of *The Hypersonic Revolution*. Edited by Richard P. Hallion. Bolling AFB, DC: Air Force History and Museums Program, 1998.

Shapiro, Ascher. *The Dynamics and Thermodynamics of Compressible Fluid Flow*. New York: Ronald Press. Two volumes, 1953 and 1954.

Sietzen, Frank. "Wall Street Rejects VentureStar." Washington, DC: News release, 21 May 1999.

Speer, Albert. *Inside the Third Reich*. New York: Macmillan, 1970.

"Standard Missile Characteristics: IM-99A Bomarc." U.S. Air Force, 8 May 1958.

"Standard Missile Characteristics: XSM-64 Navaho." U.S. Air Force, 1 November 1956.

Sunday, Terry L., and John R. London. "The X-20 Space Plane: Past Innovation, Future Vision." In *History of Rocketry and Astronautics*. Edited by John Becklake. AAS History Series, Volume 17. San Diego: Univelt, 1995, 253-284.

Sutton, George P. *Rocket Propulsion Elements*. New York: John Wiley, 1986.

Thompson, Tina D. ed. *TRW Space Log*. Redondo Beach, CA: TRW Space and Electronics Group. Annual editions.

Tsien, Hsue-shen. "Similarity Laws of Hypersonic Flows." *Journal of Mathematics and Physics* 25 (1946): 247-251.

Von Karman, Theodore with Lee Edson. *The Wind and Beyond*. Boston: Little, Brown, 1967.

Walker, Martin. *The Cold War*. New York: Henry Holt, 1994.

Waltrup, Paul J., Griffin Y. Anderson, and Frank S. Stull. "Supersonic Combustion Ramjet (Scramjet) Engine Development in the United States." Paper presented at the Third International Symposium on Air Breathing Engines, Munich, 7-12 March 1976, pp. 42-1 to 42-27.

Wattendorf, Frank L. "German Wind Tunnels." AAF Technical Report 5268, Air Technical Service Command, 20 August 1945.

Wegener, Peter P. *The Peenemunde Wind Tunnels*. New Haven, CT: Yale University Press, 1996.

Facing the Heat Barrier: A History of Hypersonics

Wolfe, Tom. *The Right Stuff.* New York: Bantam Books, 1980.

Yeager, Chuck, and Leo Janos. *Yeager, an Autobiography.* New York: Bantam Books, 1985.

Books, Papers, and Reports by Source

Acta Astronautica

Dale D. Myers. "The Navaho Cruise Missile—A Burst of Technology." 26 (1992): 741-748.

Delma C. Freeman and Theodore A. Talay. "Single-Stage-to-Orbit—Meeting the Challenge." 38 (1996): 323-331.

Aerospace America

Adrian M. Messner. "Solid Rockets." December 1979, pp. 48-49.

___, "Solid Rockets." December 1980, p. 73.

"Hypersonic Flight Revisited." June 1984, p. 1.

Patrick L. Johnston, Allen B. Whitehead, and Gary T. Chapman. "Fitting Aerodynamics and Propulsion Into the Puzzle." September 1987, pp. 32-37.

Douglas L. Dwoyer, Paul Kutler, and Louis A. Povinelli. "Retooling CFD for Hypersonic Aircraft." October 1987, pp. 32-35.

Art Hanley. "Titan IV: Latest in a Family of Giants." July 1991, pp. 34-38.

Alan S. Brown. "NASA Funds Titanium Composite Plant." August 1992, pp. 66-67.

AIAA (American Institute of Aeronautics and Astronautics)

63075. Perry V. Row and Jack Fischel. "Operational Flight-Test Experience with the X-15 Airplane." March 1963.

64-17. Euclid C. Holleman and Elmor J. Adkins. "Contributions of the X-15 Program to Lifting Entry Technology." January 1964.

67-166. B. J. Griffith. "Comparison of Data from the Gemini Flights and AEDC-VKF Wind Tunnels." January 1967.

67-838. Robert L. Perry. "The Ballistic Missile Decisions." October 1967.

68-1142. R. Bryan Erb, D.H. Greenshields, L.T. Chauvin, J.E. Pavlosky and C.L. Statham. "Apollo Thermal-Protection System Development." December 1968.

69-98. Eugene P. Bartlett, Larry W. Anderson, and Donald M. Curry. "An Evaluation of Ablation Mechanisms for the Apollo Heat Shield Material." January 1969.

70-715. A. J. Metzler and T. W. Mertz. "Large Scale Supersonic Combustor Testing at Conditions Simulating Mach 8 Flight." June 1970.

BIBLIOGRAPHY

70-1249. P. P. Antonatos, A. C. Draper, and R. D. Neumann. "Aero Thermodynamic and Configuration Development." October 1970.

71-443. F. M. Anthony, R. R. Fisher, and R. G. Helenbrook. "Selection of Space Shuttle Thermal Protection Systems." April 1971.

71-804. J. F. Yardley. "Fully Reusable Shuttle." July 1971.

71-805. Bastian Hello. "Fully Reusable Shuttle." July 1971.

75-1212. Carl A. Trexler. "Inlet Performance of the Integrated Langley Scramjet Module (Mach 2.3 to 7.6)." September 1975.

78-257. B. S. Baldwin and H. Lomax. "Thin Layer Approximation and Algebraic Model for Separated Turbulent Flows." January 1978.

78-485. M. J. Suppanz and C. J. Schroeder. "Space Shuttle Orbiter Thermal Protection Development and Verification Test Program." April 1978.

79-0219. L. L. Cronvich. "Aerodynamic Development of Fleet Guided Missiles in Navy's Bumblebee Program." January 1979.

79-7045. Robert W. Guy and Ernest A. Macklay. "Initial Wind Tunnel Tests at Mach 4 and 7 of a Hydrogen-Burning, Airframe-Integrated Scramjet." April 1979.

83-0037. P. Kutler. "A Perspective of Theoretical and Applied Computational Fluid Dynamics (Invited)." January 1983.

83-1693. T. J. Coakley. "Turbulence Modeling Methods for the Compressible Navier-Stokes Equations." July 1983.

84-0387. F. S. Billig, M. E. White, and D. M. Van Wie. "Application of CAE and CFD Techniques to a Complete Tactical Missile Design." January 1984.

84-2414. S. A. Tremaine and Jerry B. Arnett. "Transatmospheric Vehicles—A Challenge for the Next Century." November 1984.

85-1509. J. S. Shang and S. J. Scherr. "Navier-Stokes Solution of the Flow Field Around a Complete Aircraft." July 1985.

85-1652. B. Lakshminarayana. "Turbulence Modelling for Complex Flows (Invited Paper)." July 1985.

86-0159. G. Burton Northam and G. Y. Anderson. "Supersonic Combustion Ramjet Research at Langley." January 1986.

86-1680. W. J. D. Escher, R. R. Teeter, and E. E. Rice. "Airbreathing and Rocket Propulsion Synergism: Enabling Measures for Tomorrow's Orbital Transports." June 1986.

88-0230. R. J. Johnston and W. C. Sawyer. "An Historical Perspective on Hypersonic Aerodynamic Research at the Langley Research Center." January 1988.

89-0008. J. Der. "Improved Methods of Characterizing Ejector Pumping Performance." January 1989.

89-5010. T. Ronald. "Materials Challenges for NASP." July 1989.

89-5014. N. Hannum and F. Berkopec. "Fueling the National Aerospace Plane with Slush Hydrogen." July 1989.

90-1480. P. Bradshaw. "Progress in Turbulence Research." June 1990.

90-5206. N. Newman. "NASP Materials and Structures Augmentation Program (NMASAP) Overview." October 1990.

90-5207. J. Sorensen. "Titanium Matrix Composites—NASP Materials and Structures Augmentation Program." October 1990.

90-5231. F. Owen. "Transition and Turbulence Measurements in Hypersonic Flows." October 1990.

90-5232. M. Malik, T. Zang, and D. Bushnell. "Boundary Layer Transition in Supersonic Flows." October 1990.

90-5233. T. Elias and E. Eiswirth. "Stability Studies of Planar Transition in Supersonic Flows." October 1990.

90-5247. T. Gatski, S. Sarkar, C. Speziale, L. Balakrishnan, R. Abid and E. Anderson. "Assessment and Application of Reynolds-Stress Closure Models to High-Speed Compressible Flows." October 1990.

91-1944. Jerry Rosevear. "Liquid Air Cycle Engines." June 1991.

91-2395. G. A. Sullins, D. A. Carpenter, M.W. Thompson, F.T. Kwok and L.A. Mattes. "A Demonstration of Mode Transition in a Scramjet Combustor." June 1991.

91-5008. B. Waldman and F. Harsha. "The First Year of Teaming: A Progress Report." December 1991.

92-3499. L. Q. Maurice, J. L. Leingang, and L. R. Carreiro. "Airbreathing Space Boosters Using In-Flight Oxidizer Collection." July 1992.

92-5023. M. Togawa, T. Aoki, and Y. Kaneko. "On LACE Research." December 1992.

92-5049. M. Malik and P. Balakumar. "Instability and Transition in Three-Dimensional Supersonic Boundary Layers." December 1992.

93-0200. P. G. Huang and T. J. Coakley. "Turbulence Modeling for Complex Hypersonic Flows." January 1993.

93-0309. William H. Dana. "The X-15 Airplane—Lessons Learned." January 1993.

93-2323. Earl H. Andrews and Ernest A. Mackley. "Review of NASA's Hypersonic Research Engine Project." June 1993.

BIBLIOGRAPHY

93-2328. William O. T. Peschke. "Approach to In Situ Analysis of Scramjet Combustor Behavior." June 1993.

93-2329. Frederick S. Billig. "SCRAM—A Supersonic Combustion Ramjet Missile." June 1993.

93-2564. K. E. Dayton. "National AeroSpace Plane Integrated Fuselage/Cryotank Risk Reduction Program." June 1993.

93-5101. E. Kraft and G. Chapman. "A Critical Review of the Integration of Computations, Ground Tests, and Flight Test for the Development of Hypersonic Vehicles." December 1993.

95-3777. A. Elias, D. Hays, and J. Kennedy. "Pioneering Industry/Government Partnerships: X-34." September 1995.

95-6031. R. L. Chase and M. H. Tang. "A History of the NASP Program from the Formation of the Joint Program Office to the Termination of the HySTP Scramjet Performance Demonstration Program." April 1995.

95-6055. R. Voland and K. Rock. "NASP Concept Demonstration Engine and Subscale Parametric Engine Tests." April 1995.

95-6131. T. Ronald. "Status and Applications of Materials Developed for NASP." April 1995.

96-4563. S. Cook. "X-33 Reusable Launch Vehicle Structural Technologies." November 1996.

98-1532. R. T. Voland, K.E. Rock, L.D. Huebner, D.W. Witte, K.E. Fisher and C.R. McClinton. "Hyper-X Engine Design and Ground Test Program." April 1998.

98-2506. R. C. Rogers, D. P. Capriotti, and R. W. Guy. "Experimental Supersonic Combustion Research at NASA Langley." June 1998.

98-3516. R. B. Sullivan and B. Winters. "X-34 Program Overview." July 1998.

99-4818. David E. Reubush. "Hyper-X Stage Separation—Background and Status." November 1999.

99-4848. R. T. Voland, A.H. Auslander, M.K. Smart, A.S. Rousakov, V.L. Semenov and V. Kapchenov. "CIAM/NASA Mach 6.5 Scramjet Flight and Ground Test." November 1999.

99-4896. D. E. Pryor, E. H. Hyde, and W. J. D. Escher. "Development of a 12-Thrust Chamber Kerosene/Oxygen Primary Rocket Subsystem for an Early (1964) Air-Augmented Rocket Ground Test System." November 1999.

2001-1802. David E. Reubush, John G. Martin, Jeffrey S. Robinson, David M. Bose and Brian K. Strovers. "Hyper-X Stage Separation—Simulation Development and Results." April 2001.

2001-1809. Lawrence D. Huebner, Kenneth E. Rock, Edward G. Ruf, David W. Witte and Earl H. Andrews. "Hyper-X Flight Engine Ground Testing for X-43 Flight Risk Reduction." April 2001.

2001-1910. Charles R. McClinton, David R. Rausch, Joel Sitz and Paul Reukauf. "Hyper-X Program Status." April 2001.

2002-5127. M. E. M. Stewart, A. Suresh, M.S. Liou, A.K. Owen and D.G. Messitt. "Multidisciplinary Analysis of a Hypersonic Engine." October 2002.

2002-5251. George F. Orton. "Air-Breathing Hypersonics Research at Boeing Phantom Works." October 2002.

2003-7029. Russell R. Boyce, Sullivan Gerard, and Allan Paull. "The HyShot Scramjet Flight Experiment—Flight Data and CFD Calculations Compared." December 2003.

AIAA Journal

Henry Hidalgo and Leo P. Kadanoff. "Comparison Between Theory and Flight Ablation Data." 1 (1963): 41-45.

Dean R. Chapman. "Computational Aerodynamics Development and Outlook." 17 (1979): 1293-1313.

Air & Space (Smithsonian)

Luba Vangelova. "Outback Scramjet." November 2002, pp. 74-81.

Craig Mellow. "Go Ballistic." March 2006, pp. 20-25.

Air Enthusiast

Richard A. DeMeis. "The Trisonic Titanium Republic." No. 7, July-September 1978, pp. 198-213.

Air Power History

Larry Schweikart. "Hypersonic Hopes: Planning for NASP, 1986-1991." Spring 1994, pp. 36-48.

Roy Houchin. "Hypersonic Technology and Aerospace Doctrine." Fall 1999, pp. 4-17.

T. A. Heppenheimer. "Origins of Hypersonic Propulsion: A Personal History." Fall 2000, pp. 14-27.

American Heritage of Invention and Technology

James I. Killgore. "The Planes That Never Leave the Ground." Winter 1989, pp. 56-63.

James E. Tomayko. "The Airplane as Computer Peripheral." Winter 1992, pp. 19-24.

ARS (American Rocket Society)

Thomas R. Brogan. "Electric Arc Gas Heaters for Re-Entry Simulation and Space Propulsion." ARS paper 724-58, November 1958.

Leo Steg. "Materials for Re-Entry Heat Protection of Satellites." *ARS Journal*, September 1960, pp. 815-822.

Astronautics & Aeronautics (continues as *Aerospace America*)

Melvin W. Root and George M. Fuller. "The Astro Concept." January 1964, pp. 42-51.

John V. Becker. "The X-15 Project." February 1964, pp. 52-61.

Thomas A. Toll and Jack Fischel. "The X-15 Project: Results and New Research." March 1964, pp. 20-28.

Antonio Ferri. "Supersonic Combustion Progress." August 1964, pp. 32-37.

Max Faget, "Space Shuttle: A New Configuration." January 1970, pp. 52-61.

Alfred C. Draper, Melvin L. Buck, and William H. Goesch. "A Delta Shuttle Orbiter." January 1971, pp. 26-35.

John Sloop. "Looking for the Sweet Combination." October 1972, pp. 52-57.

George Strouhal and Douglas J. Tillian. "Testing the Shuttle Heat-Protection Armor." January 1976, pp. 57-65.

Robert A. Jones and Paul W. Huber. "Toward Scramjet Aircraft." February 1978, pp. 38-48.

Astronautics and Aerospace Engineering

Perry V. Row and Jack Fischel. "X-15 Flight-Test Experience." 1 (1963): 25-32.

Aviation Week (includes *Aviation Week & Space Technology*)

"Ford Beats Schedule on J57 Delivery." 3 May 1954: 44.

"How Mercury Capsule Design Evolved." 21 September 1959, pp. 52-59.

"Rolls-Royce RCo. 12, Mark 509." 2 May 1960, p. 145.

Larry Booda. "USAF Plans Radical New Space Plane." 31 October 1960, p. 26.

"Fourth ASSET Glider Gathers Flutter Data." 2 November 1964, p. 25.

William J. Normyle. "Manned Flight Tests to Seek Lifting-Body Technology." 16 May 1966, pp. 64-75.

B. K. Thomas. "USAF Nears Manned Lifting Body Tests." 10 July 1967, pp. 99-101.

"Heading Shift Cited in X-15 Loss." 12 August 1968, pp. 104-117.

"X-15 Crash Spurs Adaptive Control Study." 26 August 1968, pp. 85-105.

Michael L. Yaffee. "Program Changes Boost Grumman Shuttle." 12 July 1971, pp. 36-39.

"Sortie Module May Cut Experiment Cost." 17 January 1972, p. 17.

William S. Hieronymus. "Two Reusable Materials Studied for Orbiter Thermal Protection." 27 March 1972, p. 48.

Benjamin L. Elson. "New Unit to Test Shuttle Thermal Guard." 31 March 1975, pp. 52-53.

"Japan Explores Liquid Air Cycle Engine for Future Rocket Propulsion Needs." 3 April 1989, p. 58.

Stanley W. Kandebo. "Lifting Body Design Is Key to Single-Stage-to-Orbit." 29 October 1990, pp. 36-37.

———. "NASP Team Narrows Its Options As First Design Cycle Nears Completion." 1 April 1991, p. 80.

Michael A. Dornheim. "NASP Fuselage Model Undergoes Tests for Temperatures, Stresses of Flight." 3 February 1992, p. 52.

Stanley W. Kandebo. "Russians Want U.S. to Join Scramjet Tests." 30 March 1992, pp. 18-20.

Michael A. Dornheim. "Mitsubishi Planning to Test Liquid Air Cycle Rocket Engine." 11 May 1992: 54.

Stanley W. Kandebo. "Franco-Russian Tests May Spur New Projects." 14/21 December 1992, pp. 70-73.

———. "Boeing 777 to Incorporate New Alloy Developed for NASP." 3 May 1993, p. 36.

James R. Asker. "New Space Launchers a Tough Test for NASA." 31 October 1994, pp. 22-23.

———. "Racing to Remake Space Economics." 3 April 1995, pp. 44-45.

"X-34 to Be Acid Test for Space Commerce." 3 April 1995, pp. 44-53.

Joseph C. Anselmo and James R. Asker. "NASA Suspends X-34 Booster." 6 November 1995, p. 30.

"Endeavour Tests Metallic TPS." 3 June 1996, p. 92.

Joseph C. Anselmo. "NASA Gives Orbital Second Shot at X-34." 17 June 1996, p. 31.

Michael A. Dornheim. "Follow-on Plan Key to X-33 Win." 8 July 1996, pp. 20-22.

Bruce A. Smith. "Hyper-X Undergoes Design Review." 9 June 1997, p. 32.

James R. Asker. "NASA, Industry Hit Snags on X-33." 30 June 1977, pp. 27-28.

Michael Dornheim. "X-33 Design Gets Go-Ahead." 10 November 1997.

———. "X-33 Flight Slips Half-Year to December 1999." 2 November 1998, pp. 26-27.

———. "Bonding Bugs Delay X-33's First Flight." 25 January 1999, pp. 46-47.

Joseph C. Anselmo. "A Mach 8 Vehicle on the RLV Frontier." 26 April 1999, pp. 78-80.

Michael A. Dornheim. "Engineers Anticipated X-33 Tank Failure." 15 November 1999, pp. 28-30.

Frank Morring. "NASA Kills X-33, X-34, Trims Space Station." 5 March 2001, pp. 24-25.

Edward H. Phillips. "'Mariah' Team Eyes Hypersonic Tunnel." 5 March 2001, p. 66.

Bruce A. Smith. "Elevon Failure Precedes Loss of First X-43A." 11 June 2001, pp. 50-51.

Michael Dornheim. "X-43 to Fly in Fall." 28 July 2003, pp. 36-37.

———. "A Breath of Fast Air." 5 April 2004, pp. 28-29.

———. "But Now What?" 22 November 2004, pp. 24-26.

Bell Aircraft

D143-981-009. "MX-2276: Advanced Strategic Weapon System Progress Report No. 9." 15 February 1955.

D143-945-055. "Brass Bell Reconnaissance Aircraft Weapon System Summary Report." 31 August 1957.

Boeing Airplane Company

D-11532. "Final Summary Report: Development of USAF XIM-99A Weapon System January 1951-March 1959."

CASI (Center for Aerospace Information)

64X-17126. R. J. Bassett and R. M. Knox. "Preliminary Design Analysis of a 10,000 lb Thrust LACE Engine." Marquardt: 5 November 1963.

72A-10764. Richard C. Thuss, Harry G. Thibault, and Arnold Hiltz. "The Utilization of Silica Based Surface Insulation for the Space Shuttle Thermal Protection System." SAMPE National Technical Conference on Space Shuttle Materials, Huntsville, AL, October 1971, pp. 453-464.

75N-75668. J. F. Foote. "Operating Experience and Performance Documentation of a LACE System." Marquardt: 15 November 1963.

77A-35304. David H. Greenshields. "Orbiter Thermal Protection System Development." Canaveral Council of Technical Societies: 14th Space Congress, Cocoa Beach, FL, April 1977, pp. 1-28 to 1-42.

81A-44344. L. J. Korb and H. M. Clancy. "The Shuttle Thermal Protection System—A Material and Structural Overview." SAMPE 26th National Symposium, Los Angeles, April 1981, pp. 232-249.

88N-13231. P. J. Waltrup. "Hypersonic Airbreathing Propulsion: Evolution and Opportunities."

Comptes Rendus

Paul Vieille. "Sur les discontinuites produites par la detente brusque de gaz compreimes." 129 (1988): 1228-1230.

Convair (see also *General Dynamics*)

ZP-M-095. "Project Spaceplane Status Presentation." April 1960.

"Reusable Space Launch Vehicle Systems Study." Montgomery, AL: Archives, Maxwell AFB.

DTIC (Defense Technical Information Center)

AD-098980. Daniel E. Bloxsom. "Production of High-Temperature, Moderate-Pressure Gases by Means of Electrical Spark Discharge." Arnold Engineering Development Center, November 1956.

AD-299774. James C. Sivells. "Aerodynamic Design and Calibration of the VKF 50-inch Hypersonic Wind Tunnels." Arnold Engineering Development Center, March 1963.

AD-318628. F. Herr and R. L. Maxwell. "Hypersonic, Air-Breathing Engine Research Program." Marquardt, 31 August 1960.

AD-320753. "Preliminary Flight Test Report. RVX-1, Flight No. 1. AVCO Vehicle No. 5. Missile No. 133 (Thor-Able Phase II)." General Electric, 29 April 1959.

AD-322199. "Hypersonic, Air-Breathing Engine Research Program." Marquardt, 28 February 1961.

AD-328815. F. A. Costello and J. A. Segletes. "Development and Thermal Performance of the Discoverer Heat Shield." General Electric, 1 March 1962.

AD-335212. J. F. Foote and R. J. Bassett. "Summary of Initial Tests of the Marquardt Model MA117 LACE Engine." Marquardt, 15 February 1963.

AD-336169. J. R. Turner and L. C. Hoagland. "Study of High Temperature Chemical Separation of Air into O_2-N_2 Components." Dynatech Corporation, February 1963.

AD-342828. "Thor-Able Phase II, RVX-1, Flight-3, Re-entry Vehicle No. 172, Missile No. 132, AMR Test No. 15." General Electric, 9 May 1959.

AD-342992. "Spaceplane Technical Status Report." Convair, February 1961.

AD-345178. "Airframe Support of a Boilerplate Air Separator Air Enrichment Program." General Dynamics, October 1963.

AD-346912. "Proceedings of 1962 X-20A (Dyna-Soar) Symposium. Volume III. Structures and Materials." Aeronautical Systems Division, U.S. Air Force, March 1963.

AD-350952. R. J. Bassett and R. M. Knox. "Preliminary Design Analysis of a 10,000-lb Thrust LACE Engine." Marquardt, 5 November 1963.

AD-351039. "Aerospaceplane Propulsion Integration Study. Volume 1." Republic Aviation, 15 February 1963.

AD-351040. "Aerospaceplane Propulsion Integration Study. Volume 3." Republic Aviation, 15 February 1963.

AD-351041. "Aerospaceplane Propulsion Integration Study. Volume 2." Republic Aviation, 15 February 1963.

AD-351207. "Feasibility Study of a High Capacity Distillation Separator for an Air Enrichment System." Union Carbide, February 1964.

Bibliography

AD-351239. "The Dynatech High-Temperature, Chemical Air Separator for ACES." Dynatech Corporation, 15 February 1962.

AD-351581. "Republic Aerospaceplane. Volume 1." Republic Aviation, 28 January 1964.

AD-351582. "Evaluation of P + W Turboramjet Engines in Republic Aerospaceplane. Volume II." Republic Aviation, 28 January 1964.

AD-351583. "Evaluation of G.E. Turboramjet Engines in Republic Aerospaceplane. Volume III." Republic Aviation, 28 January 1964.

AD-351584. "Evaluation of Marquardt SuperLACE Engines in Republic Aerospaceplane. Volume IV." Republic Aviation, 28 January 1964.

AD-353898. L. C. Hoagland and J. R. Turner. "High-Temperature Air Enrichment Program." Dynatech Corporation, 15 September 1964.

AD-357523. "ASSET ASV-3 Flight Test Report." McDonnell Aircraft Corporation, 4 January 1965.

AD-359545. "Phase I Report on Recoverable Orbital Launch System (ROLS)." Holloman AFB, NM, 23 April 1965.

AD-359763. "Experimental and Theoretical Investigation of the Rocket Engine-Nozzle Ejector (RENE) Propulsion System." Martin, April 1965.

AD-362539. T. E. Shaw, W.S. Mertz, N.M. Keough and I.M Musser. "Thermal Flight Test Summary Report for Mark 3 Mod 1 Re-Entry Vehicles." General Electric, 2 August 1960.

AD-366546. "ASSET ASV-4 Flight Test Report." McDonnell Aircraft Corporation, 25 June 1965.

AD-372320. R. L. Doebler and M. G. Hinton. "Current Status and Challenge of Airbreathing Propulsion for Launch Vehicle Systems." Aerospace Corporation, 15 March 1966.

AD-381504. "Air Separator Test Program." Union Carbide, October 1966.

AD-384302. R. J. Bassett and R. M. Knox. "Component Specifications for the MA116 Advanced LACE System." Marquardt, 5 November 1963.

AD-386653. C. Edward Kepler and Arthur J. Karanian. "Hydrogen-Fueled Variable Geometry Scramjet." United Aircraft Research Laboratories, January 1968.

AD-388239. M. L. Brown and R. L. Maxwell. "Scramjet Incremental Flight Test Program." Marquardt, February 1968.

AD-393374. David J. McFarlin and C. Edward Kepler. "Mach 5 Test Results of Hydrogen-Fueled Variable-Geometry Scramjet." United Aircraft Research Laboratories, 24 October 1968.

AD-441740. D. Cona. "Coatings for Refractory Metals." Boeing, 27 December 1963.

AD-449685. Howard J. Middendorf. "Materials and Processes for X-20A (Dyna-Soar)." Air Force Systems Command, June 1964.

AD-461220. A. R. Reti and J. R. Turner. "Study of Oxides for Chemical Separation of Air Into O_2–N_2 Components." Dynatech Corporation, March 1965.

AD-500489. "Air Enrichment Study Program." Union Carbide, February 1969.

AD-609169. William Cowie. "Utilization of Refractory Metals on the X-20A (Dyna-Soar)." Air Force Systems Command, June 1964.

AD-832686. "Flight Test Evaluation Report, Missile 7D." Convair, 3 June 1959.

ADA-274530. "Report of the Defense Science Board Task Force on National Aero-Space Plane (NASP) Program." November 1992.

ADA-303832. Roy Franklin Houchin. "The Rise and Fall of Dyna-Soar: A History of Air Force Hypersonic R&D, 1944-1963." Air Force Institute of Technology, August 1995.

ADB-190937. "NASP Materials and Structures Augmentation Program Final Report. Refractory Composites." General Dynamics, March 1993.

ADB-191402. "1993 National Aero-Space Plane Technology Review. Volume V—Materials." NASP Joint Program Office, July 1993.

ADB-191464. G. James Van Fossen. "Performance Testing and Modelling of a Water Alleviation System." NASA-Lewis, November 1991.

ADB-191627. John Bradley. "Test and Evaluation Report for Carbon/Carbon Wing Box Component." General Dynamics, 21 February 1992.

ADB-191699. "Summary of Results of Carbon-Carbon Composites Technology Maturation Program for NASP." NASA-Langley, May 1992.

ADB-191715. P. C. Davis. "Low-Speed Uncertainty Analysis Report." General Dynamics, 30 September 1992.

ADB-191898. "High Conductivity Composites. Volume 1—Executive Summary, Copper Materials, Beryllium Materials, Coatings, Ceramic Materials and Joining." Rockwell International, March 1993.

ADB-192311. "National Program Office Review." Engine review at West Palm Beach, Florida—6 August 1991. NASP National Program Office, August 1991.

ADB-192559. "NASP Materials and Structures Augmentation Program. Titanium Matrix Composites." McDonnell Douglas, 31 December 1991.

ADB-196085. Lecia L. Mayhew and James L. McAfee. "Full Scale Assembly Summary Status Report." McDonnell Douglas, 8 November 1994.

ADB-197041. R. L. Campbell, T. J. Treinen, and G. H. Ratekin. "Low Speed Propulsion Summary Report." Rocketdyne, 19 May 1993.

ADB-197189. "Composite Configuration Steering Committee Report." NASP Joint Program Office, 14 February 1990.

BIBLIOGRAPHY

ADB-968993. "NASP Security Classification Guide." Department of Defense and NASA, 30 November 1990.

ADC-053100. "Proposed 730 Orbital Airplane." Republic Aviation, November 22, 1960.

General Accounting Office

NSIAD-92-5. Nancy R. Kingsbury. "Aerospace Plane Technology: Research and Development Efforts in Japan and Australia." 4 October 1992.

NSIAD-93-71. Frank C. Conahan. "National Aero-Space Plane: Restructuring Further Research and Development Efforts." 3 December 1992.

NSIAD-99-176. "Space Transportation: Status of the X-33 Reusable Launch Vehicle Program." 11 August 1999.

General Dynamics

GD/C-DCJ 65-004. "Reusable Space Launch Vehicle Study." 18 May 1965.

GDC-DCB-67-031. "Multipurpose Reusable Spacecraft Preliminary Design Effort." 25 September 1967.

GDC-DCB-68-017. "Reusable Launch Vehicle/Spacecraft Concept." November 1968.

GDC-DCB-69-046. "Space Shuttle Final Technical Report." 31 October 1969.

Grumman

B35-43 RP-33. "Space Shuttle System Program Definition. Phase B Extension Final Report." 15 March 1972.

High Technology

T. A. Heppenheimer. "Making Planes from Powder." September 1986, pp. 54-55.

ISABE (International Society for Air Breathing Engines)

97-7004. E. T. Curran. "Scramjet Engines: The First Forty Years."

Journal of the British Interplanetary Society

Joel W. Powell. "Thor-Able and Atlas-Able." (1984): 219-225.

Jet Propulsion

Lester Lees. "Laminar Heat Transfer Over Blunt-Nosed Bodies at Hypersonic Flight Speeds." April 1956, pp. 259-269.

Johns Hopkins APL Technical Digest

William Garten, and Frank A. Dean. "Evolution of the Talos Missile." 3, No. 2 (1982): 117-122.

Harold E. Gilreath. "The Beginning of Hypersonic Ramjet Research at APL." 11, Nos. 3 and 4 (1990): 319-335.

James L Keirsey. "Airbreathing Propulsion for Defense of the Surface Fleet." Vol. 13, No. 1 (1992): 57-67.

Johns Hopkins Magazine

Alan Newman. "Speed Ahead of Its Time." December 1988, pp. 26-31.

Journal of the Aeronautical Sciences.

Theodore von Karman. "Supersonic Aerodynamics—Principles and Applications." Tenth Wright Brothers Lecture. (1947): 373-409.

Charles H. McLellan. "Exploratory Wind-Tunnel Investigation of Wings and Bodies at M = 6.9." (1951): 641-648.

J. A. Fay and F. R. Riddell. "Theory of Stagnation Point Heat Transfer in Dissociated Air." (1958): 73-85.

F. H. Rose and W. I. Stark. "Stagnation Point Heat-Transfer Measurements in Dissociated Air." (1958): 86-97.

Journal of the Aero/Space Sciences

George W. Sutton. "Ablation of Reinforced Plastics in Supersonic Flow." (1960): 377-385.

Mac C. Adams, William E. Powers, and Steven Georgiev. "An Experimental and Theoretical Study of Quartz Ablation at the Stagnation Point." (1960): 535-543.

Journal of Aircraft

A. Ferri, "Review of SCRAMJET Propulsion Technology." 5 (1968): 3-10.

Journal of Applied Physics

John V. Becker. "Results of Recent Hypersonic and Unsteady Flow Research at the Langley Aeronautical Laboratory." 21 (1950): 619-628.

E. L. Resler, Shao-Chi Lin, and Arthur Kantrowitz. "The Production of High Temperature Gases in Shock Tubes." 23 (1952): 1390-1399.

Journal of Guidance and Control

Charles Stark Draper. "Origins of Inertial Navigation." 4 (1981), pp. 449-63.

Journal of the Royal Aeronautical Society

R. L. Schleicher. "Structural Design of the X-15." (1963): 618-636.

Antonio Ferri. "Review of Problems in Application of Supersonic Combustion." Seventh Lanchester

BIBLIOGRAPHY

Memorial Lecture. (1964): 575-597.

Journal of Spacecraft and Rockets

B. J. Griffith and D. E. Boylan. "Postflight Apollo Command Module Aerodynamic Simulation Tests." 5 (1968): 843-848.

F. S. Billig, R. C. Orth, and M. Lasky. "Effects of Thermal Compression on the Performance Estimates of Hypersonic Ramjets." 5 (1968): 1076-1081.

George W. Sutton. "The Initial Development of Ablation Heat Protection, An Historical Perspective." 19 (1982): 3-11.

Steven P. Schneider. "Flight Data for Boundary-Layer Transition at Hypersonic and Supersonic Speeds." 36 (1999): 8-20.

Frank S. Milos and Thomas H. Squire. "Thermostructural Analysis of X-34 Wing Leading-Edge Tile Thermal Protection System." 36 (1999): 189-198.

Kathryn E. Wurster, Christopher J. Riley, and E. Vincent Zoby. "Engineering Aerothermal Analysis for X-34 Thermal Protection System Design." 36 (1999): 216-227.

Walter C. Engelund, Scott D. Holland, Charles E. Cockrell and Robert D. Bittner. "Aerodynamic Database Development for the Hyper-X Airframe-Integrated Scramjet Propulsion Experiments." 38 (2001): 803-810.

Scott D. Holland, William C. Woods, and Walter C. Engelund. "Hyper-X Research Vehicle Experimental Aerodynamics Test Program Overview." 38 (2001): 828-835.

Charles E Cockrell, Walter C. Engelund, Robert D. Bittner, Tom N. Jentink, Arthur D. Dilley and Abdelkader Frendi. "Integrated Aeropropulsive Computational Fluid Dynamics Methodology for the Hyper-X Flight Experiment." 38 (2001): 836-843.

Lawrence D. Huebner, Kenneth E. Rock, Edward G. Ruf, David W. Witte and Earl H. Andrews. "Hyper-X Flight Engine Ground Testing for Flight Risk Reduction." 38 (2001): 844-852.

Scott A. Berry, Aaron H. Auslender, Arthur D. Dilley and John F. Calleja. "Hypersonic Boundary-Layer Trip Development for Hyper-X." 38 (2001): 853-864.

Langley Working Papers

LWP-54. James L. Raper. "Preliminary Results of a Flight Test of the Apollo Heat Shield Material at 28,000 Feet per Second." NASA-Langley, 27 October 1964.

Lockheed

LAC 571375. "Aerospace Plane." 20 April 1962.

LMSC-A959837. "Final Report: Integrated Launch and Re-Entry Vehicle." 22 December 1969.

Lockheed Horizons

Clarence "Kelly" Johnson. "Development of the Lockheed SR-71 Blackbird." Issue 9, Winter 1981/82, pp. 2-7, 13-18.

Wilson B. Schramm, Ronald P. Banas, and Y. Douglas Izu. "Space Shuttle Tile—The Early Lockheed Years." Issue 13, 1983, pp. 2-15.

Martin Marietta

M-63-1. C. W. Spieth and W. T. Teegarden, "Astrorocket Progress Report." December 1962.

ER 14465. "SV-5D PRIME Final Flight Test Summary." September 1967.

Marquardt Corporation

"Scramjet Flight Test Program." Brochure, September 1965.

McDonnell Douglas

MDC E0056. "A Two-Stage Fixed Wing Space Transportation System." 15 December 1969.

MDC E0308. "Space Shuttle System Phase B Study Final Report." 30 June 1971.

MDC E076-1. "External LH_2 Tank Study Final Report." 30 June 1971.

"Interim Report to OMSF: Phase B System Study Extension." 1 September 1971.

NACA (National Advisory Committee for Aeronautics)

1381. H. Julian Allen and A. J. Eggers. "A Study of the Motion and Aerodynamic Heating of Ballistic Missiles Entering the Earth's Atmosphere at High Supersonic Speeds." 28 April 1953.

RM L51D17. Charles H. McLellan, Mitchel H. Bertram, and John A. Moore. "An Investigation of Four Wings of Square Plan Form at a Mach Number of 6.86 in the Langley 11-inch Hypersonic Tunnel." 29 June 1951.

RM L51J09. Ralph D. Cooper and Raymond A. Robinson. "An Investigation of the Aerodynamic Characteristics of a Series of Cone-Cylinder Configurations at a Mach Number of 6.86." 17 December 1951.

RM E51K26. Irving Pinkel, John S. Serafini, and John L. Gregg, "Pressure Distribution and Aerodynamic Coefficients Associated with Heat Addition to Supersonic Air Stream Adjacent to Two-Dimensional Supersonic Wing." February 14, 1952.

RM L54F21. Charles H. McLellan. "A Method for Increasing the Effectiveness of Stabilizing Surfaces at High Supersonic Mach Numbers." 3 August 1954.

RM A55E26. Alvin Seiff and H. Julian Allen, "Some Aspects of the Design of Hypersonic Boost-Glide Aircraft." August 15, 1955.

RM A55L05. A. J. Eggers and Clarence Syvertson, "Aircraft Configurations Developing High Lift-Drag Ratios at High Supersonic Speeds." March 5, 1956.

BIBLIOGRAPHY

RM L58E07a. Maxime A. Faget, Benjamine J. Garland, and James J. Buglia. "Preliminary Studies of Manned Satellites. Wingless Configuration: Nonlifting." 11 August 1958.

TN 2171. Charles H. McLellan, Thomas W. Williams, and Mitchel H. Bertram. "Investigation of a Two-Step Nozzle in the Langley 11-inch Hypersonic Tunnel." September 1950.

TN 2206. Irving Pinkel and John S. Serafini. "Graphical Method for Obtaining Flow Field in Two-Dimensional Supersonic Stream to Which Heat Is Added." 1 November 1950.

TN 2223. Charles H. McLellan, Thomas W. Williams, and Ivan E. Beckwith,. "Investigation of the Flow Through a Single-Stage Two-Dimensional Nozzle in the Langley 11-inch Hypersonic Tunnel." December 1950.

TN 2773. Mitchel H. Bertram. "An Approximate Method for Determining the Displacement Effects and Viscous Drag of Laminar Boundary Layers in Two-Dimensional Hypersonic Flow." September 1952.

TN 3302. Charles H. McLellan and Thomas W. Williams. "Liquefaction of Air in the Langley 11-inch Hypersonic Tunnel." October 1954.

TN 4046. Alfred J. Eggers, H. Julian Allen, and Stanford E. Neice. "A Comparative Analysis of the Performance of Long-Range Hypervelocity Vehicles." October 1957.

TN 4386. Richard J. Weber and John S. MacKay. "An Analysis of Ramjet Engines Using Supersonic Combustion." September 1958.

NASA (National Aeronautics and Space Administration)

CP-3105. "Proceedings of the X-15 First Flight 30th Anniversary Celebration." 1991.

RP-1028. Joseph Adams Shortal. "A New Dimension. Wallops Island Flight Test Range: The First Fifteen Years." December 1978.

RP-1132. "Aeronautical Facilities Catalogue Volume 1. Wind Tunnels." January 1985.

SP-60. Wendell H. Stillwell. "X-15 Research Results." 1965.

SP-292. "Vehicle Technology for Civil Aviation. The Seventies and Beyond." 1971.

SP-440. Donald D. Baals and William R. Corliss. "Wind Tunnels of NASA." 1981.

SP-2000-4029. Richard W. Orloff. "Apollo By the Numbers: A Statistical Reference." 2000.

SP-4201. Loyd S. Swenson, James M. Grimwood, and Charles C. Alexander. *This New Ocean: A History of Project Mercury.* 1966.

SP-4206. Roger E. Bilstein. *Stages to Saturn: A Technological History of the Apollo/Saturn Launch Vehicles.* 1980.

SP-4220. R. Dale Reed and Darlene Lister. *Wingless Flight: The Lifting Body Story.* 1997.

SP-4221. T. A. Heppenheimer. *The Space Shuttle Decision*. 1999.

SP-4222. J. D. Hunley, ed. *Toward Mach 2: The Douglas D-558 Program*. 1999.

SP-2004-4236. Dennis R. Jenkins, *The X-15*. 2004.

SP-4302. Edwin P. Hartman. *Adventures in Research: A History of Ames Research Center, 1940-1976*. 1970.

SP-4303. Richard P. Hallion. *On the Frontier: Flight Research at Dryden, 1946-1981*. 1981.

SP-4305. James R. Hansen. *Engineer in Charge: A History of the Langley Aeronautical Laboratory, 1917-1958*. 1987.

SP-4308. James R. Hansen. *Spaceflight Revolution: NASA Langley Research Center from Sputnik to Apollo*. 1995.

SP-4404. John L. Sloop. *Liquid Hydrogen as a Propulsion Fuel, 1945-1959*. 1978.

SP-4407. John M. Logsdon, ed. *Exploring the Unknown: Selected Documents in the History of the U.S. Civil Space Program*. Volume 1, *Organizing for Exploration*, 1995; Volume IV, *Accessing Space*, 1999.

SP-2000-4408. Asif A. Siddiqi, *Challenge to Apollo: The Soviet Union and the Space Race, 1945-1974*. 2000.

SP-2000-4518. Dennis R. Jenkins, *Hypersonics Before the Shuttle: A Concise History of the X-15 Research Airplane*. June 2000.

TM X-490. Robert L. O'Neal and Leonard Rabb. "Heat-Shield Performance During Atmosphere Entry of Project Mercury Research and Development Vehicle." 12 October 1960.

TM X-1053. Richard C. Dingeldein. "Flight Measurements of Re-entry Heating at Hyperbolic Velocity (Project Fire)." 14 October 1964.

TM X-1120. Elden S. Cornette. "Forebody Temperatures and Total Heating Rates Measured During Project Fire 1 Reentry at 38,000 Feet per Second." 20 April 1965.

TM X-1130. William T. Schaefer. "Characteristics of Major Active Wind Tunnels at the Langley Research Center." July 1965.

TM X-1182. James Raper. "Results of a Flight Test of the Apollo Heat-Shield Material at 28,000 Feet per Second." 2 September 1965.

TM X-1222. Dona L. Cauchon. "Project Fire 1 Radiative Heating Experiment." 8 December 1965.

TM X-1305. Elden S. Cornette. "Forebody Temperatures and Calorimeter Heating Rates Measured During Project Fire II Reentry at 11.35 Kilometers per Second." 7 July 1966.

TM X-1407. Richard M. Raper. "Heat-Transfer and Pressure Measurements Obtained During Launch and Reentry of the First Four Gemini-Titan Missions and Some Comparisons with Wind-Tunnel Data." 1 March 1967.

Bibliography

TM X-2273. "NASA Space Shuttle Technology Conference." Volume 2—Structures and Materials. 1 April 1971.

TM X-2572. "Hypersonic Research Engine Project Technological Status 1971." September 1972.

TM X-2895. John R. Henry and Griffin Y. Anderson. "Design Considerations for the Airframe-Integrated Scramjet." December 1973.

TM X-56008. Lawrence W. Taylor and Elmor J. Adkins. "Adaptive Flight Control Systems—Pro and Con." 15 May 1964.

TN D-1157. Lawrence W. Taylor and George B. Merrick. "X-15 Airplane Stability Augmentation System." March 1962.

TN D-1159. Robert G. Hoey and Richard E. Day. "Mission Planning and Operational Procedures for the X-15 Airplane." March 1962.

TN D-1278. Joseph Weil. "Review of the X-15 Program." 6 April 1962.

TN D-1402. Robert A. Tremant. "Operational Experiences and Characteristics of X-15 Flight Control System." December 1962.

TN D-2996. William I. Scallion and John L. Lewis. "Flight Parameters and Vehicle Performance for Project Fire Flight 1, Launched April 14, 1964." 3 June, 1965.

TN D-4713. Randolph A. Graves and William G. Witte. "Flight-Test Analysis of Apollo Heat-Shield Material Using the Pacemaker Vehicle System." 21 May 1968.

TN D-6208. "Experience with the X-15 Adaptive Flight Control System." March 1971.

TN D-7564. James E. Pavlosky and Leslie G. St. Leger. "Apollo Experience Report—Thermal Protection Subsystem." January 1974.

TP-1998-206548. Alexander S. Roudakov, Vyacheslav L. Semenov, and John W. Hicks. "Recent Flight Test Results of the Joint CIAM-NASA Mach 6.5 Scramjet Flight Test Program." April 1998.

Nature

R. N. Hollyer, A.C. Hunting, Otto Laporte and E.B. Turner. "Luminosity Generated by Shock Waves." 171 (1953): 395-396.

North American Aviation

AL-1347. "Development of a Strategic Missile and Associated Projects." October 1951.

North American Rockwell

SD 69-573-1. "Study of Integral Launch and Reentry Vehicle System." December 1969.

SD 71-114-1. "Space Shuttle Phase B Final Report." 25 June 1971.

SV 71-28. "Fully Reusable Shuttle." 19 July 1971.

SV 71-50. "Space Shuttle Phase B Extension, 4th Month Review." 3 November 1971.

Pratt & Whitney

"Dependable Engines…Since 1925." Product summary, July 1990.

Proceedings of the Royal Society

William Payman and Wilfred Charles Furness Shepherd, "Explosion Waves and Shock Waves VI. The Disturbances Produced by Bursting Diaphragms with Compressed Air." Series A, 186 (1946): 293-321.

Quarterly Review (Air Research and Development Command)

Lt. Col. D. D. Carlson. "Research at AEDC." Summer 1959, pp. 80-88.

Leon S. Jablecki. "Development of High-Priority Aerial Weapons Advanced in AEDC's New Mach 8 Tunnel." Fall 1959, pp. 56-61.

Rand Corporation

R-217. J. E. Lipp, R. M. Salter, and R. S. Wehner. "The Utility of a Satellite Vehicle for Reconnaissance." April 1951.

SM-11827. "Preliminary Design of an Experimental World-Circling Spaceship." 2 May 1946. Reprinted in part in NASA SP-4407, Vol. I, pp. 236-244.

Raumfahrtforschung

John V. Becker. "The X-15 Program in Retrospect." (1969): 45-53.

Republic Aviation News

"Kartveli Envisions a Mach 25 Vehicle As 'Ultimate Airplane.'" 9 September 1960, pp. 1, 5.

Review of Scientific Instruments

Walter Bleakney, D. K. Weimer, and C. H. Fletcher. "The Shock Tube: A Facility for Investigations in Fluid Dynamics." 20 (1949): 807-815.

Rocketdyne

RI/RD87-142. "Space Shuttle Main Engine." 1 May 1987.

"Thirty Years of Rocketdyne." Company publication, 1985.

Science

Gina Bari Kolata. "Who Will Build the Next Supercomputer?" 211 (1981): 268-269.

R. Jeffrey Smith. "Estrangement on the Launch Pad." 224 (1984): 1407-1409.

Antony Jameson. "Computational Aerodynamics for Aircraft Design." 245 (1989): 361-371.

Richard E. Young, Martha A. Smith, and Charles K. Sobeck. "Galileo Probe: In Situ Observations of Jupiter's Atmosphere." 272 (1996): 837-838.

"World's Fastest Supercomputers." 298 (2002): 1713.

David Malakoff. "Mach 12 by 2012?" 300 (2003): 888-889.

Scientific American

Ellis Levin, Donald D. Viele, and Lowell B. Eldrenkamp. "The Lunar Orbiter Missions to the Moon." May 1968, pp. 58-78.

Spaceflight

Irene Sanger-Bredt. "The Silver Bird Story." (1973): 166-181.

Curtis Peebles. "The Origin of the U.S. Space Shuttle—1." (1979): 435-442.

———. "Project Bomi." (1980): 270-272.

Space World

"Sophisticated Insulation Barrier to Protect Crew of Space Shuttle." June-July 1979, pp. 18, 23-24.

Time

"Journey Into Space." 8 December 1952, pp. 62-73.

"The Rough & the Smooth." 18 November 1957, pp. 19-20.

"'Like a Bullet.'" 8 December 1958, p. 15.

"The Great Capsule Hunt." 27 April 1959, pp. 16-17.

"Educated Satellites." 27 April 1959, p. 65.

"Back from Space." 13 June 1960, pp. 70-76.

"A Place in Space." 27 October 1961, pp. 89-94.

"A Decade of Deadly Birds." 22 May 1964, p. 25.

"Moonward Bound." 17 November 1967, pp. 84-85.

"The Sky's the Limit." 29 November 2004, pp. 63-67.

U.S. Patents

2,735,263. J. O. Charshafian. "Combination Rocket and Ram-Jet Power Plant." Filed 6 December 1947.

2,883,829. Alfred Africano. "Rocket Engine Convertible to a Ramjet Engine." Filed 24 January 1955.

2,922,286. Randolph Rae. "Production of Liquid Oxygen." Filed 13 August 1954.

3,040,519. Randolph Rae. "Jet Propulsion Unit with Cooling Means for Incoming Air." Filed 13 August 1954.

3,040,520. Randolph Rae. "Jet Power Unit for an Aircraft." Filed 22 November 1954.

3,172,253. Helmut Schelp and Frederick Hughes. "Combination Turbo and Ramjet Propulsion Apparatus." Filed 2 January 1959.

3,525,474. Hans von Ohain, Roscoe M. Mills, and Charles A. Scolatti. "Jet Pump or Thrust Augmentor." Filed 9 December 1968.

3,690,102. Anthony A. duPont. "Ejector Ram Jet Engine." Filed 29 October 1970.

Vought Missiles and Space Company

"Technical Overview: Oxidation Resistant Carbon-Carbon for the Space Shuttle." Undated; circa 1971.

NASA History Series

Reference Works, NASA SP-4000:

Grimwood, James M. *Project Mercury: A Chronology.* NASA SP-4001, 1963.

Grimwood, James M., and C. Barton Hacker, with Peter J. Vorzimmer. *Project Gemini Technology and Operations: A Chronology.* NASA SP-4002, 1969.

Link, Mae Mills. *Space Medicine in Project Mercury.* NASA SP-4003, 1965.

Astronautics and Aeronautics, 1963: Chronology of Science, Technology, and Policy. NASA SP-4004, 1964.

Astronautics and Aeronautics, 1964: Chronology of Science, Technology, and Policy. NASA SP-4005, 1965.

Astronautics and Aeronautics, 1965: Chronology of Science, Technology, and Policy. NASA SP-4006, 1966.

Astronautics and Aeronautics, 1966: Chronology of Science, Technology, and Policy. NASA SP-4007, 1967.

Astronautics and Aeronautics, 1967: Chronology of Science, Technology, and Policy. NASA SP-4008, 1968.

Ertel, Ivan D., and Mary Louise Morse. *The Apollo Spacecraft: A Chronology, Volume I, Through November 7, 1962.* NASA SP-4009, 1969.

Morse, Mary Louise, and Jean Kernahan Bays. *The Apollo Spacecraft: A Chronology, Volume II, November 8, 1962–September 30, 1964.* NASA SP-4009, 1973.

Brooks, Courtney G., and Ivan D. Ertel. *The Apollo Spacecraft: A Chronology, Volume III, October 1, 1964–January 20, 1966.* NASA SP-4009, 1973.

Ertel, Ivan D., and Roland W. Newkirk, with Courtney G. Brooks. *The Apollo Spacecraft: A Chronology, Volume IV, January 21, 1966–July 13, 1974.* NASA SP-4009, 1978.

Astronautics and Aeronautics, 1968: Chronology of Science, Technology, and Policy. NASA SP-4010, 1969.

Newkirk, Roland W., and Ivan D. Ertel, with Courtney G. Brooks. *Skylab: A Chronology.* NASA SP-4011, 1977.

Van Nimmen, Jane, and Leonard C. Bruno, with Robert L. Rosholt. *NASA Historical Data Book, Volume I: NASA Resources, 1958–1968*. NASA SP-4012, 1976, rep. ed. 1988.

Ezell, Linda Neuman. *NASA Historical Data Book, Volume II: Programs and Projects, 1958–1968*. NASA SP-4012, 1988.

Ezell, Linda Neuman. *NASA Historical Data Book, Volume III: Programs and Projects, 1969–1978*. NASA SP-4012, 1988.

Gawdiak, Ihor Y., with Helen Fedor, compilers. *NASA Historical Data Book, Volume IV: NASA Resources, 1969–1978*. NASA SP-4012, 1994.

Rumerman, Judy A., compiler. *NASA Historical Data Book, 1979–1988: Volume V, NASA Launch Systems, Space Transportation, Human Spaceflight, and Space Science*. NASA SP-4012, 1999.

Rumerman, Judy A., compiler. *NASA Historical Data Book, Volume VI: NASA Space Applications, Aeronautics and Space Research and Technology, Tracking and Data Acquisition/Space Operations, Commercial Programs, and Resources, 1979–1988*. NASA SP-2000-4012, 2000.

Astronautics and Aeronautics, 1969: Chronology of Science, Technology, and Policy. NASA SP-4014, 1970.

Astronautics and Aeronautics, 1970: Chronology of Science, Technology, and Policy. NASA SP-4015, 1972.

Astronautics and Aeronautics, 1971: Chronology of Science, Technology, and Policy. NASA SP-4016, 1972.

Astronautics and Aeronautics, 1972: Chronology of Science, Technology, and Policy. NASA SP-4017, 1974.

Astronautics and Aeronautics, 1973: Chronology of Science, Technology, and Policy. NASA SP-4018, 1975.

Astronautics and Aeronautics, 1974: Chronology of Science, Technology, and Policy. NASA SP-4019, 1977.

Astronautics and Aeronautics, 1975: Chronology of Science, Technology, and Policy. NASA SP-4020, 1979.

Astronautics and Aeronautics, 1976: Chronology of Science, Technology, and Policy. NASA SP-4021, 1984.

Astronautics and Aeronautics, 1977: Chronology of Science, Technology, and Policy.

NASA SP-4022, 1986.

Astronautics and Aeronautics, 1978: Chronology of Science, Technology, and Policy. NASA SP-4023, 1986.

Astronautics and Aeronautics, 1979–1984: Chronology of Science, Technology, and Policy. NASA SP-4024, 1988.

Astronautics and Aeronautics, 1985: Chronology of Science, Technology, and Policy. NASA SP-4025, 1990.

Noordung, Hermann. *The Problem of Space Travel: The Rocket Motor.* Edited by Ernst Stuhlinger and J. D. Hunley, with Jennifer Garland. NASA SP-4026, 1995.

Astronautics and Aeronautics, 1986–1990: A Chronology. NASA SP-4027, 1997.

Astronautics and Aeronautics, 1990–1995: A Chronology. NASA SP-2000-4028, 2000.

Management Histories, NASA SP-4100:

Rosholt, Robert L. *An Administrative History of NASA, 1958–1963.* NASA SP-4101, 1966.

Levine, Arnold S. *Managing NASA in the Apollo Era.* NASA SP-4102, 1982.

Roland, Alex. *Model Research: The National Advisory Committee for Aeronautics, 1915–1958.* NASA SP-4103, 1985.

Fries, Sylvia D. *NASA Engineers and the Age of Apollo.* NASA SP-4104, 1992.

Glennan, T. Keith. *The Birth of NASA: The Diary of T. Keith Glennan.* J. D. Hunley, editor. NASA SP-4105, 1993.

Seamans, Robert C., Jr. *Aiming at Targets: The Autobiography of Robert C. Seamans, Jr.* NASA SP-4106, 1996.

Garber, Stephen J., editor. *Looking Backward, Looking Forward: Forty Years of U.S. Human Spaceflight Symposium.* NASA SP-2002-4107, 2002.

Mallick, Donald L. with Peter W. Merlin. *The Smell of Kerosene: A Test Pilot's Odyssey.* NASA SP-4108, 2003.

Iliff, Kenneth W. and Curtis L. Peebles. *From Runway to Orbit: Reflections of a*

NASA Engineer. NASA SP-2004-4109, 2004.

Chertok, Boris. *Rockets and People, Volume 1.* NASA SP-2005-4110, 2005.

Laufer, Alexander, Todd Post, and Edward Hoffman. *Shared Voyage: Learning and Unlearning from Remarkable Projects.* NASA SP-2005-4111, 2005.

Dawson, Virginia P. and Mark D. Bowles. *Realizing the Dream of Flight: Biographical Essays in Honor of the Centennial of Flight, 1903-2003.* NASA SP-2005-4112, 2005.

Project Histories, NASA SP-4200:

Swenson, Loyd S., Jr., James M. Grimwood, and Charles C. Alexander. *This New Ocean: A History of Project Mercury.* NASA SP-4201, 1966; rep. ed. 1998.

Green, Constance McLaughlin, and Milton Lomask. *Vanguard: A History.* NASA SP-4202, 1970; rep. ed. Smithsonian Institution Press, 1971.

Hacker, Barton C., and James M. Grimwood. *On the Shoulders of Titans: A History of Project Gemini.* NASA SP-4203, 1977.

Benson, Charles D., and William Barnaby Faherty. *Moonport: A History of Apollo Launch Facilities and Operations.* NASA SP-4204, 1978.

Brooks, Courtney G., James M. Grimwood, and Loyd S. Swenson, Jr. *Chariots for Apollo: A History of Manned Lunar Spacecraft.* NASA SP-4205, 1979.

Bilstein, Roger E. *Stages to Saturn: A Technological History of the Apollo/Saturn Launch Vehicles.* NASA SP-4206, 1980, rep. ed. 1997.

SP-4207 not published.

Compton, W. David, and Charles D. Benson. *Living and Working in Space: A History of Skylab.* NASA SP-4208, 1983.

Ezell, Edward Clinton, and Linda Neuman Ezell. *The Partnership: A History of the Apollo-Soyuz Test Project.* NASA SP-4209, 1978.

Hall, R. Cargill. *Lunar Impact: A History of Project Ranger.* NASA SP-4210, 1977.

Newell, Homer E. *Beyond the Atmosphere: Early Years of Space Science.* NASA SP-4211, 1980.

THE NASA HISTORY SERIES

Ezell, Edward Clinton, and Linda Neuman Ezell. *On Mars: Exploration of the Red Planet, 1958–1978.* NASA SP-4212, 1984.

Pitts, John A. *The Human Factor: Biomedicine in the Manned Space Program to 1980.* NASA SP-4213, 1985.

Compton, W. David. *Where No Man Has Gone Before: A History of Apollo Lunar Exploration Missions.* NASA SP-4214, 1989.

Naugle, John E. *First Among Equals: The Selection of NASA Space Science Experiments.* NASA SP-4215, 1991.

Wallace, Lane E. *Airborne Trailblazer: Two Decades with NASA Langley's Boeing 737 Flying Laboratory.* NASA SP-4216, 1994.

Butrica, Andrew J., editor. *Beyond the Ionosphere: Fifty Years of Satellite Communication.* NASA SP-4217, 1997.

Butrica, Andrew J. *To See the Unseen: A History of Planetary Radar Astronomy.* NASA SP-4218, 1996.

Mack, Pamela E., editor. *From Engineering Science to Big Science: The NACA and NASA Collier Trophy Research Project Winners.* NASA SP-4219, 1998.

Reed, R. Dale, with Darlene Lister. *Wingless Flight: The Lifting Body Story.* NASA SP-4220, 1997.

Heppenheimer, T. A. *The Space Shuttle Decision: NASA's Search for a Reusable Space Vehicle.* NASA SP-4221, 1999.

Hunley, J. D., editor. *Toward Mach 2: The Douglas D-558 Program.* NASA SP-4222, 1999.

Swanson, Glen E., editor. *"Before this Decade Is Out . . .": Personal Reflections on the Apollo Program.* NASA SP-4223, 1999.

Tomayko, James E. *Computers Take Flight: A History of NASA's Pioneering Digital Fly-by-Wire Project.* NASA SP-2000-4224, 2000.

Morgan, Clay. *Shuttle-Mir: The U.S. and Russia Share History's Highest Stage.* NASA SP-2001-4225, 2001.

Leary, William M. *"We Freeze to Please": A History of NASA's Icing Research Tunnel and the Quest for Flight Safety.* NASA SP-2002-4226, 2002.

Mudgway, Douglas J. *Uplink-Downlink: A History of the Deep Space Network 1957–1997.* NASA SP-2001-4227, 2001.

Dawson, Virginia P. and Mark D. Bowles. *Taming Liquid Hydrogen: The Centaur Upper Stage Rocket, 1958-2002.* NASA SP-2004-4230, 2004.

Meltzer, Michael. *Mission to Jupiter: A History of the Galileo Project.* NASA SP-2007-4231.

Center Histories, NASA SP-4300:

Rosenthal, Alfred. *Venture into Space: Early Years of Goddard Space Flight Center.* NASA SP-4301, 1985.

Hartman, Edwin P. *Adventures in Research: A History of Ames Research Center, 1940-1965.* NASA SP-4302, 1970.

Hallion, Richard P. *On the Frontier: Flight Research at Dryden, 1946-1981.* NASA SP-4303, 1984.

Muenger, Elizabeth A. *Searching the Horizon: A History of Ames Research Center, 1940-1976.* NASA SP-4304, 1985.

Hansen, James R. *Engineer in Charge: A History of the Langley Aeronautical Laboratory, 1917-1958.* NASA SP-4305, 1987.

Dawson, Virginia P. *Engines and Innovation: Lewis Laboratory and American Propulsion Technology.* NASA SP-4306, 1991.

Dethloff, Henry C. *"Suddenly Tomorrow Came . . .": A History of the Johnson Space Center.* NASA SP-4307, 1993.

Hansen, James R. *Spaceflight Revolution: NASA Langley Research Center from Sputnik to Apollo.* NASA SP-4308, 1995.

Wallace, Lane E. *Flights of Discovery: 50 Years at the NASA Dryden Flight Research Center.* NASA SP-4309, 1996.

Herring, Mack R. *Way Station to Space: A History of the John C. Stennis Space Center.* NASA SP-4310, 1997.

Wallace, Harold D., Jr. *Wallops Station and the Creation of the American Space Program.* NASA SP-4311, 1997.

Wallace, Lane E. *Dreams, Hopes, Realities: NASA's Goddard Space Flight Center, The First Forty Years.* NASA SP-4312, 1999.

Dunar, Andrew J., and Stephen P. Waring. *Power to Explore: A History of the Mar-

The NASA History Series

shall Space Flight Center. NASA SP-4313, 1999.

Bugos, Glenn E. *Atmosphere of Freedom: Sixty Years at the NASA Ames Research Center.* NASA SP-2000-4314, 2000.

Schultz, James. *Crafting Flight: Aircraft Pioneers and the Contributions of the Men and Women of NASA Langley Research Center.* NASA SP-2003-4316, 2003.

General Histories, NASA SP-4400:

Corliss, William R. *NASA Sounding Rockets, 1958–1968: A Historical Summary.* NASA SP-4401, 1971.

Wells, Helen T., Susan H. Whiteley, and Carrie Karegeannes. *Origins of NASA Names.* NASA SP-4402, 1976.

Anderson, Frank W., Jr. *Orders of Magnitude: A History of NACA and NASA, 1915–1980.* NASA SP-4403, 1981.

Sloop, John L. *Liquid Hydrogen as a Propulsion Fuel, 1945–1959.* NASA SP-4404, 1978.

Roland, Alex. *A Spacefaring People: Perspectives on Early Spaceflight.* NASA SP-4405, 1985.

Bilstein, Roger E. *Orders of Magnitude: A History of the NACA and NASA, 1915–1990.* NASA SP-4406, 1989.

Logsdon, John M., editor, with Linda J. Lear, Jannelle Warren-Findley, Ray A. Williamson, and Dwayne A. Day. *Exploring the Unknown: Selected Documents in the History of the U.S. Civil Space Program, Volume I, Organizing for Exploration.* NASA SP-4407, 1995.

Logsdon, John M., editor, with Dwayne A. Day and Roger D. Launius. *Exploring the Unknown: Selected Documents in the History of the U.S. Civil Space Program, Volume II, Relations with Other Organizations.* NASA SP-4407, 1996.

Logsdon, John M., editor, with Roger D. Launius, David H. Onkst, and Stephen J. Garber. *Exploring the Unknown: Selected Documents in the History of the U.S. Civil Space Program, Volume III, Using Space.* NASA SP-4407, 1998.

Logsdon, John M., general editor, with Ray A. Williamson, Roger D. Launius, Russell J. Acker, Stephen J. Garber, and Jonathan L. Friedman. *Exploring the Unknown: Selected Documents in the History of the U.S. Civil Space Program, Volume IV,*

Accessing Space. NASA SP-4407, 1999.

Logsdon, John M., general editor, with Amy Paige Snyder, Roger D. Launius, Stephen J. Garber, and Regan Anne Newport. *Exploring the Unknown: Selected Documents in the History of the U.S. Civil Space Program, Volume V, Exploring the Cosmos*. NASA SP-2001-4407, 2001.

Siddiqi, Asif A. *Challenge to Apollo: The Soviet Union and the Space Race, 1945–1974*. NASA SP-2000-4408, 2000.

Hansen, James R., editor. *The Wind and Beyond: Journey into the History of Aerodynamics in America, Volume 1, The Ascent of the Airplane*. NASA SP-2003-4409, 2003.

Monographs in Aerospace History, NASA SP-4500:

Launius, Roger D. and Aaron K. Gillette, compilers, *Toward a History of the Space Shuttle: An Annotated Bibliography*. Monograph in Aerospace History, No. 1, 1992.

Launius, Roger D., and J. D. Hunley, compilers, *An Annotated Bibliography of the Apollo Program*. Monograph in Aerospace History, No. 2, 1994.

Launius, Roger D. Apollo: A Retrospective Analysis. Monograph in Aerospace History, No. 3, 1994.

Hansen, James R. *Enchanted Rendezvous: John C. Houbolt and the Genesis of the Lunar-Orbit Rendezvous Concept*. Monograph in Aerospace History, No. 4, 1995.

Gorn, Michael H. Hugh L. *Dryden's Career in Aviation and Space*. Monograph in Aerospace History, No. 5, 1996.

Powers, Sheryll Goecke. *Women in Flight Research at NASA Dryden Flight Research Center, from 1946 to 1995*. Monograph in Aerospace History, No. 6, 1997.

Portree, David S. F. and Robert C. Trevino. *Walking to Olympus: An EVA Chronology*. Monograph in Aerospace History, No. 7, 1997.

Logsdon, John M., moderator. *Legislative Origins of the National Aeronautics and Space Act of 1958: Proceedings of an Oral History Workshop*. Monograph in Aerospace History, No. 8, 1998.

Rumerman, Judy A., compiler, *U.S. Human Spaceflight, A Record of Achievement 1961–1998*. Monograph in Aerospace History, No. 9, 1998.

THE NASA HISTORY SERIES

Portree, David S. F. *NASA's Origins and the Dawn of the Space Age.* Monograph in Aerospace History, No. 10, 1998.

Logsdon, John M. *Together in Orbit: The Origins of International Cooperation in the Space Station.* Monograph in Aerospace History, No. 11, 1998.

Phillips, W. Hewitt. *Journey in Aeronautical Research: A Career at NASA Langley Research Center.* Monograph in Aerospace History, No. 12, 1998.

Braslow, Albert L. *A History of Suction-Type Laminar-Flow Control with Emphasis on Flight Research.* Monograph in Aerospace History, No. 13, 1999.

Logsdon, John M., moderator. *Managing the Moon Program: Lessons Learned From Apollo.* Monograph in Aerospace History, No. 14, 1999.

Perminov, V. G. *The Difficult Road to Mars: A Brief History of Mars Exploration in the Soviet Union.* Monograph in Aerospace History, No. 15, 1999.

Tucker, Tom. *Touchdown: The Development of Propulsion Controlled Aircraft at NASA Dryden.* Monograph in Aerospace History, No. 16, 1999.

Maisel, Martin D., Demo J. Giulianetti, and Daniel C. Dugan. *The History of the XV-15 Tilt Rotor Research Aircraft: From Concept to Flight.* NASA SP-2000-4517, 2000.

Jenkins, Dennis R. *Hypersonics Before the Shuttle: A Concise History of the X-15 Research Airplane.* NASA SP-2000-4518, 2000.

Chambers, Joseph R. *Partners in Freedom: Contributions of the Langley Research Center to U.S. Military Aircraft in the 1990s.* NASA SP-2000-4519, 2000.

Waltman, Gene L. *Black Magic and Gremlins: Analog Flight Simulations at NASA's Flight Research Center.* NASA SP-2000-4520, 2000.

Portree, David S. F. *Humans to Mars: Fifty Years of Mission Planning, 1950–2000.* NASA SP-2001-4521, 2001.

Thompson, Milton O., with J. D. Hunley. *Flight Research: Problems Encountered and What They Should Teach Us.* NASA SP-2000-4522, 2000.

Tucker, Tom. *The Eclipse Project.* NASA SP-2000-4523, 2000.

Siddiqi, Asif A. *Deep Space Chronicle: A Chronology of Deep Space and Planetary Probes, 1958–2000.* NASA SP-2002-4524, 2002.

Merlin, Peter W. *Mach 3+: NASA/USAF YF-12 Flight Research, 1969–1979.* NASA SP-2001-4525, 2001.

Anderson, Seth B. *Memoirs of an Aeronautical Engineer—Flight Tests at Ames Research Center: 1940–1970.* NASA SP-2002-4526, 2002.

Renstrom, Arthur G. *Wilbur and Orville Wright: A Bibliography Commemorating the One-Hundredth Anniversary of the First Powered Flight on December 17, 1903.* NASA SP-2002-4527, 2002.

No monograph 28.

Chambers, Joseph R. *Concept to Reality: Contributions of the NASA Langley Research Center to U.S. Civil Aircraft of the 1990s.* SP-2003-4529, 2003.

Peebles, Curtis, editor. *The Spoken Word: Recollections of Dryden History, The Early Years.* SP-2003-4530, 2003.

Jenkins, Dennis R., Tony Landis, and Jay Miller. *American X-Vehicles: An Inventory—X-1 to X-50.* SP-2003-4531, 2003.

Renstrom, Arthur G. *Wilbur and Orville Wright: A Chronology Commemorating the One-Hundredth Anniversary of the First Powered Flight on December 17, 1903.* NASA SP-2003-4532, 2002.

Bowles, Mark D. and Robert S. Arrighi. *NASA's Nuclear Frontier: The Plum Brook Research Reactor.* SP-2004-4533, 2003.

Matranga, Gene J. and C. Wayne Ottinger, Calvin R. Jarvis with D. Christian Gelzer. *Unconventional, Contrary, and Ugly: The Lunar Landing Research Vehicle.* NASA SP-2006-4535.

McCurdy, Howard E. *Low Cost Innovation in Spaceflight: The History of the Near Earth Asteroid Rendezvous (NEAR) Mission.* NASA SP-2005-4536, 2005.

Seamans, Robert C. Jr. *Project Apollo: The Tough Decisions.* NASA SP-2005-4537, 2005.

Lambright, W. Henry. *NASA and the Environment: The Case of Ozone Depletion.* NASA SP-2005-4538, 2005.

Chambers, Joseph R. *Innovation in Flight: Research of the NASA Langley Research Center on Revolutionary Advanced Concepts for Aeronautics.* NASA SP-2005-4539, 2005.

Phillips, W. Hewitt. *Journey Into Space Research: Continuation of a Career at NASA Langley Research Center.* NASA SP-2005-4540, 2005.

Index

A

ablation, 39-40, 50, 152, 153, 165, 173, 174, 257.
    See also carbon-carbon; hot structure.
ACES (Air Collection and Enrichment System), 111, 116, 122-25, 127-28
Adams, Major Mike, 85-86, 204.
Aerospaceplane, xi-xii, 91, 112-28, 178-79, 198, 200
    Aerospaceplane (Lockheed), 117-19
    Aerospaceplane (Republic), 119
    Astrorocket (Martin), 126-27, 179
    POBATO, 120
    Point Design, 1964 (Convair), 122
    press coverage, 115
    PROFAC (Northrop), 114
    ROLS (Holloman AFB), 122-23
    Space Plane (Convair), 112-14, 115, 117, 125
aircraft, speed of, ix, 65, 197
Allen, H. Julian, 29, 137-38
Allen-Eggers blunt-body principle, 29-30, 34, 55, 59-60, 6l, 133, 144,
    151, 168, 173, 175
Apollo program, 133, 154-55, 156-57, 159, 173, 282
    Project Fire, 157-58
    Saturn V, 159, 284
Applied Physics Laboratory (Johns Hopkins), 95, 100-02, 112, 199, 202
arc tunnel, 40-42
Armstrong, Neil 76, 150
Arnold Engineering Development Center, 65-67, 70, 120,
    125, 156, 273, 277-78
ASSET, 165-68, 182, 191
Atlas ICBM, 25-29, 36, 44, 48-49, 72, 96, 135, 152, 153-54, 157, 261
Atwood, Lee, quoted, 96
Avery, William, 101, 102
A-4b, 5-6
A-9, 4-5, 24

B

Becker, John, xi, 2, 14-15, 20, 71, 82, 203, 206
    and boost-glide studies, 136, 138-40, 173
    and 11-inch wind tunnel, 2, 14-20, 30, 61, 70, 152, 155, 156
    and X-15 feasibility study, 58-61, 70, 72

Billig, Frederick, 102, 107, 199, 202, 216, 231
Bogdonoff, Seymour, 116, 220
Bomarc, 94-95, 197
Bomi, 57, 70, 133-35
boundary layers, 223, 224, 238. See also turbulence.
Branson, Richard, 283
Bush, Vannevar, quoted, 24
B-50 bomber, 92, 93, 200
B-52 bomber, 59, 83, 197, 258, 272, 274, 275, 277
B-58 bomber, 116, 144, 183, 198

C
carbon-carbon, 165, 191-93, 222, 249-52
combined-cycle engines, 108, 215. See also LACE; ejectors.
computational fluid mechanics, 229, 230-32, 257, 273, 278-82.
   See also boundary layers; Direct Numerical Simulation,
   Large Eddy Simulation, Navier-Stokes equations; turbulence.
Concorde, ix, 284
Cooper, Robert, 214, 215, 216-17, 218, 222
Corona program, 152-53, 282
Czysz, Paul, quoted, 239, 244

D
DARPA, 214-16, 218
DC-X, 258-59, 263
Direct Numerical Simulation, 280, 281
dirigibles, 284
Dornberger, Walter, 5, 57, 70, 133
Douglas Skyrocket, 56, 71, 79, 80
Draper, Alfred, 182-85
Dryden, Hugh, 68, 69, 203
Dugger, Gordon, 102, 199
DuPont, Anthony, 203-04, 215, 216, 218, 219, 222, 229
Dyna-Soar, xii, 12, 126, 127, 133, 137-43, 150, 159-60, 173, 174, 182
   early concepts, 133-37, 138-41, 148-49, 151
   technology, 143-50, 191. See also Rene 41, Q-felt.
Dynatech, 124-25, 127-28

E
Eastham, James, quoted, 106
Eggers, Alfred, 29, 116, 137-38
Eisenhower, President Dwight, 46
ejectors, xii, 108-10, 120, 124, 215, 217, 229, 239, 242-44

F

Faget, Maxime, 59, 155, 173
   Project Mercury, 151-52, 153
   space shuttle, 179-85, 186, 190
Fay-Riddell theory of heating, 35-36, 151
Ferri, Antonio, xi, 91, 103-07, 116, 198, 199, 201-02, 207
Flax, Alexander, 116, 143
flow visualization, 17-18
flutter, aerodynamic, 145-46, 165, 168, 245
F-86 (also XP-86), 56, 65
F-100A, 65
F-104, ix, 64, 65, 197, 198
F-105 (also XF-105), 56, 65, 70, 114, 198

G

Galileo (mission to Jupiter), 257, 283
Gardner, Trevor, 27, 28, 69
Garrett AiResearch, 120, 203-04
Gemini program, 152, 155-56, 173
General Applied Science Laboratories (GASL), 103, 198, 201-02, 208, 210, 217, 242-43, 270

H

Hall, Colonel Edward, 115
Hallion, Richard, quoted, x-xi, 167-68, 204, 283, 284
Holloman AFB, 93, 122
hot structures, 142-48, 165-66, 174, 181, 185-87, 189, 222, 264
   columbium, 186
Huebner, Lawrence, 274, 275
Hunter, Maxwell, 57, 175, 187
hydrogen bomb, xii, 23, 27-29
hypersonics, xii, 29
   and Air Force, 31, 58, 66-67, 68-69, 200-01
   defined, ix
   difficulty of, 128, 237
   and Naxi Germany, 1-2, 3
   origin of term, x
   significance, 282-84
ICBM, 20, 23-24. See also Atlas; Minuteman.
Inconel X, 59, 71, 72, 74, 150, 266

J

Jenkins, Dennis, quoted, 71, 73
Johnson, Clarence "Kelly", 68-69, 197
Jupiter missile, 37, 46, 48
Jupiter-C missile, 44

K

Kantrowitz, Arthur, 23, 31-33, 40, 41, 50
Kartveli, Alexander, 56, 70, 91, 104, 114
Kennedy, President John F., quoted, 86
Keyworth, George, 216-17
Kincheloe, Iven, 67, 75
Knight, William "Pete", 79, 81, 204-05

L

Large Eddy Simulation (LES), xii, 257, 280, 281-82
Lees theory of heating, 35, 36, 38, 151
lifting body, 165, 168-73, 179-82.
    See also PRIME.
Linde (division of Union carbide), 125, 127-28
Liquid Air Cycle Engine (LACE), 110-11, 114, 116, 119, 120-21, 123-24, 127-28, 240-41
    deicing, 239-40
    DuPont version, 215-16, 217, 269
    See also ACES, Dynatech, Linde.
Lombard, Albert, quoted, 57-58

M

Manned Orbiting Laboratory, 141, 143
Marquardt Aviation, 93, 94, 98, 101, 120-21, 123-24, 199, 201, 202, 203-04, 216, 242, 243
McLellan, Charles, 19-20, 55, 61, 173
McNamara, Robert, 70, 126, 133, 141-43
Mercury program, 80, 151-52, 153-54, 158, 166, 173
Messerschmitt Me-262, 12
Minuteman ICBM, 141, 200
Missiles, and aviation, 197
National Aerospace Plane (NASP), xii, 198, 218, 219
    Copper Canyon, 216-17, 219, 229-30
    engine design, 238
    funding, 218, 222, 224
    program reviews, 222-24, 252

See also ejectors; LACE; X-30.
Navaho missile, 24-25, 56, 57, 65, 66, 72, 95-96, 101, 151, 197
Navier-Stokes equations, 230-31, 233, 234-35, 278, 282
Nicholls, James, 102, 104
nose cone, 29-30, 36-38, 41
    heat sink, 36, 37, 38-39, 44, 153, 189-90, 210
    Mark 3, 48-49
    RVX-1, 47, 49
    RVX-2, 48, 49
    See also ablation.
Nucci, Louis, 104, 201-02
nuclear freeze, 212-13

O

Orbital Sciences Corp., 258, 261-62, 270
Ordnance Aerophysics Laboratory, 13, 98, 201, 204

P

Paull, Allan, 274-75
Pegasus rocket, 258, 259, 270, 272, 273-74, 275, 277
Post-shuttle launch vehicles, 258-61. See also
    Transatmospheric vehicles; X-30; X-33; X-34.
Prandtl, Ludwig, 2
PRIME, 165, 170, 172-73, 174

Q

Q-felt, 146, 147, 177, 247

R

ramjets, 56, 66, 91-99, 114-15, 147, 197-98, 200, 202, 214-15
Rand Corp., 26, 27, 57, 122, 214
Rausch, Vince, 214, 272
Reagan, President Ronald, xi, 198, 212-13, 218, 224
re-entry, x, 23, 30-31, 34-36.
    of a satellite, 152
    See also hot structures; nose cones.
Rene 41, 143-44, 146, 150, 174
RL10 rocket engine, 123-24, 258
Roshko, Anatol, 236-37
Rutan, Burt, 283

S

Sanger, Eugen, xi, 2, 8-12, 137, 284
scramjets, xi, xii, 98-99, 103-07, 111, 114, 117, 199-200, 215-16, 220, 221, 222, 231
  Australian, 274-75
  engine-airframe integration, 119, 206-11
  at General Electric, 199, 200, 203-04, 205
  Hypersonic Research Engine, 202-06, 210, 215
  inlets, 106-07, 200, 207, 208, 210, 231, 273
  origins, 100-02
  at Pratt & Whitney, 199, 201, 205, 243-44
  produces net thrust, 211
  and ramjets
  and rockets, 99, 253, 284
  Russian, 267-69
  significance, 284
  struts for fuel injection, 207-08, 210
  thermal compression, 105-06, 107, 207
  thrust and drag, 232, 272, 276, 277
  transition to supersonic combustion, 107-08, 268, 277, 286
  See also Ferri, Antonio; X-43A.
Shock tubes and tunnels, 23, 32-34, 231-32, 277
Single-stage-to-orbit, 121-22, 128, 259-61, 263
Skantze, General Lawrence, 198, 213-14, 217, 218
Soviet activities, 121, 151
SpaceShipOne, 283
Space Shuttle, xii, 121, 123, 127, 160, 165, 175-76, 193, 260, 284
  Columbia, 165, 191, 192-93
  configuration, 179-87, 190
  first stage, 187, 189-90
  early concepts, 175-76, 178-79
  Space Shuttle Main engine, 213-14, 219
  tests thermal protection
  thermal protection "tiles", 165, 174-78, 181, 186-88, 191
space station, 193, 218, 284
specific impulse, 99, 123, 219, 220, 221
SR-71, ix, 43, 94, 106-07, 109, 110, 183, 186, 197, 198, 242
Stack, John, 13, 15
Storms, Harrison, 116, 151
Strategic Defense Initiative (SDI), xii, 198, 212-13, 216, 218, 224, 258, 266
  See also X-ray laser.
supercomputers, 230, 278-80
Sutton, George, 23, 36, 38-40, 47-48

INDEX

T
Talos, 95, 101
Tether, Anthony, 215, 216
Thermal protection, 174
    See also ablation, hot structure nose cone re-entry space shuttle "tiles."
Thomas, Arthur, 107, 112, 201, 202, 216, 218
Thompson, Floyd (NACA-Langley), 58
Thor missile, 38, 45, 166
Thor-Able missile, 47-48, 166-67
Thor-Delta launch vehicle, 166-67
Titan rockets, 44, 140, 147, 182, 200, 260
titanium, xii, 65, 72, 74, 134, 186, 187, 222
    Beta-21S, 246-47, 252
    powder metallurgy, 245-46
Transatmospheric Vehicles (TAVs), 213-14, 215, 217, 218
turbojets, 62-64, 66, 91-92, 116, 197-98, 199
turboramjet, 66, 108
turbulence, 223, 229, 232-37, 280-81
Twining, General Nathan, 28, 153

U
U. S. Air Force, 56, 61-62, 70, 112, 114-15
    Scientific Advisory Board, 31, 33, 58, 116-17, 121, 125-26, 145
    and space shuttle, 182-85, 198, 214
Von Braun, Wernher, 36-37
Von Karman, Theodore, 13, 27, 50
V-2, ix-x, 2, 3-6, 14, 24, 122, 133

W
Watkins, Admiral James, 212-13, 217
Weber, Richard and John MacKay, 101, 104
Weinberg, Steven, 50
Weinberger, Caspar, 217-18
Williams, Robert, 215-18, 229-30
wind tunnels, 2-3, 6-8, 13, 20, 66-67, 156, 201, 208, 231-32, 238, 271-73, 277-78
wings, 182-85, 190
Woods, Robert, 57, 70
Worth, Weldon, 114, 199
Wright Aeronautical Corp., 93, 95, 97, 98

X

X-ray laser, 212
X-l, 56, 57, 70, 80
X-lA, 57, 61, 93, 173
X-lB, 75-76
X-2, 6, 56, 57, 58, 65, 67, 70, 75, 80-81, 93
X-3, 29, 55, 71
X-7, ix, 65, 92-94
X-15, x, xi, xii, 6, 20, 43, 55-56, 61, 65, 67-68, 116, 137, 138, 150, 151, 173, 189-90, 200-01, 266, 283
    aerodynamics and heating, 83-84
    auxiliary power units, 85
    engines, 80-81
    flight controls, 67, 74-78
    flight simulators, 78-79
    and Hypersonic Research Engine, 202-03, 204-05
    Langley design concept, 58-60
    lubricanta, 74, 85
    piloted, 80, 85
    pressure suits, 79
    and Q-ball, 67-68, 74
    selection of contractor, 70-73
    skin thickness, 56, 60
    stability of, 55, 61
    X-15A-2, 81-82
    X-15B, 151
    See also Becker, John; Inconel X.
X-17, 23, 42-44, 48
X-30, 198, 212, 218, 241, 244, 284
    materials, 248-53
    NIFTA, 247-48, 266-67
    test program, 249
    weight and performance, 219-22, 264
    See also National Aerospace Plane.
X-33, xii, 257, 259, 261, 262-66
X-34, xii, 257, 259, 261-62, 263, 266
X-43A, x, xii, 269-74, 275-77, 278, 284
XB-70 bomber, 116, 122, 183, 197
XF-91 fighter, 70
XF-103 fighter, 56, 65, 66, 96-97, 197

Y

Yeager, Chuck, 61, 85, 93, 173